LANDSLIDES

LANDSLIDES

Editor

Ortwin Penzig

scitus academics

Landslides

Edited by **Ortwin Penzig**

ISBN: 9781681178837
Library of Congress Control Number: 2017948910

SCITUS Academics LLC,
616, Corporate Way, Suite 2, 4766,
Valley Cottage, NY 10989

www.scitusacademics.com

Printed in United States of America.

Preface

Landslides are among the many natural disasters causing massive destructions and loss of lives across the globe. Landslides have always existed on our planet. While landslides are a naturally occurring environmental hazard they have recently increased in frequency in certain areas due to human activity. Although there are many different causes of landslides, they all have two things in common. How these factors interrelate is important in understanding what causes landslides along with an understanding of the impact humans have on these factors by altering natural processes. Landslides are among the many natural disasters causing massive destructions and loss of lives across the globe. According to a survey study by the International Landslide Centre at Durham University, UK, 2,620 fatal landslides occurred between 2004 and 2010. These landslides resulted in the death of over 32,322 people. It is, thus, paramount to know the causes and warning signs of a potential landslide to minimize losses. Seismic activities have always been a main cause of landslides throughout the world. Any time plate tectonics move the soil that covers those moves with it. When earthquakes occur on areas with steep slopes, many times the soil slips causing landslides. Furthermore, ashen debris flows caused by earthquakes can also trigger mass movement of soil. One of the most serious and costliest post earthquake problems is the stabilization of earthquake-induced landslide and debris masses. Shaking from the January 12, 2010, Haiti earthquake caused devastating structural damage and triggered many landslides that blocked roads, dammed rivers and streams, and threatened infrastructure in many parts of Haiti.

This Volume 'Landslides' covers landslide dynamics, mechanisms and processes; volcanic, urban, marine and reservoir landslides; related tsunamis and seiches; hazard assessment and mapping; remedial or preventive measures; early warning and evacuation and a global landslide database. This book also investigates landslide caused disturbances to ecosystem-vegetation burial and overturning of soil horizons which can accelerate vegetation loss, or even lead to complete deforestation of the landscape. It is also presented of the historical development and recent advances in understanding and prediction of landslides/debris flows, which are major global natural hazards that have caused great loss of life and damage to property and infrastructure. It is important for engineers and geologists to evaluate slope stability and any landslide threat during development assessments so that effective and timely remedial measures can be implemented.

Table of Contents

CHAPTER 1

Landslides Caused by Climate Change and Groundwater Movement in Permafrost Mountain

Wei Shan*, Ying Guo, Zhaoguang Hu, Chunjiao Wang and Chengcheng Zhang

[1] *Institute of Cold Regions Science and Engineering, Northeast Forestry University, Harbin, China*

ABSTRACT

Climate change induced warming results in permafrost degradation. Melting permafrost subsequently leads to an increased incidence of landslides. The study area was within the northwest section of the Lesser Khingan Range in northern China along the Bei'an-Heihe Highway. We analyzed the impact of climate change on landslide movement in the permafrost zone via a combination of geological survey and meteorological data. The average annual temperature of the study area has increased by 3.2°C in last 60 years, and permafrost degradation is severe. Loose soil on the hillside surface provides appropriate conditions for the infiltration of atmospheric precipitation and snowmelt, and seepage from thawing permafrost. As it infiltrates downwards, water is blocked by the underlying permafrost or dense soil, and infiltrates along this barrier layer toward lower positions, forming a potential sliding zone. The combination of high density resistivity (HDR) methods based on soil resistivity values, ground-penetrating radar (GPR) methods based on characteristics of radar wave reflection, respectively, and geological drilling can be utilized to determine the regional stratigraphic distribution. This will allow the exact location of the landslide sliding surface to be precisely determined. Field test results indicate that radar reflectivity characteristics and the resistivity values of the soil in the landslide mass is significantly different from surrounding soil. There are sudden decreases in the apparent resistivity values at the sliding surface location. In addition, the radar exhibits strong reflection at the sliding surface position, with a sudden increase in the amplitude of the radar wave. Drilling results indicate that the soil has high water content at the location of the sliding surface of the landslide mass in the study area, which is entirely consistent with the GPR and HDR results. Thus, abnormal radar wave reflection and abrupt changes in apparent resistivity values can be used in practice to identify the location of landslide sliding surfaces in this region. We produce a detailed analysis of a representative landslide within the study area. Displacement monitoring locations were positioned at the trailing edge of the landslide mass and on the landslide mass surface. We then used this data to determine the relationships of landslide movement with both ground temperature and the trailing edge pore water pressure. The results suggest seasonal variation in the landslide movement process and characteristics of an annual cyclical trend. Landslide movement can be described by intermittence and low angles. The slip rate and the timing of slide occurrence exhibit relationships with the trailing edge pore water pressure of the landslide mass. The seepage of thaw water into the landslide mass will impact the trailing edge pore water pressure of the landslide mass. This phenomenon is identified as the primary cause of landslide movement.

Keywords: climate change, groundwater movement, permafrost, landslides, mountains

1. INTRODUCTION

The area north of 47° N in northeast China exhibits widespread permafrost. This represents the sole high-latitude permafrost region in China. This region also represents the second largest permafrost region in China [1–3]. In recent years, the southern boundary of this high-latitude permafrost region has gradually moved to the north as a result of climate change. Permafrost proximate to the southern boundary exhibits accelerated degradation and a distribution that is discontinuous and island-like [4–6].

Climate change refers to the temporal alteration of the baseline state of climate. The mean global surface temperature has been increasing throughout the last century. The rate of warming during 1957–2007 was almost double the rate during 1957–2007 [7]. Temperatures in a majority of the permafrost regions throughout the world have risen since the early

1980s. The recorded temperature rise has reached 3°C in some regions in northern Alaska, and the rise has reached 2°C in the northern regions of Europe and Russia. During the period of 1975–2005, there were significant reductions in the range and thickness of the permafrost layer in the northern area of Russia [8]. Climate change and its impact are an important research issue receiving extensive global attention.

A number of studies have demonstrated that climate change in China has presented the same tendency as the global trend [9]. The temperature increase rate in China for the last 54 years was approximately 0.25°C/10a. This value is far exceeds the hemispheric and global average rates of warming. Northeast China is among the national areas exhibiting the most notable warming and permafrost degradation [10].

Accumulated snow and vegetation have extremely significant impacts on the thawing process of the underlying seasonal frozen ground and permafrost [11]. Temperature shifts caused by snow, vegetation, water, topography, atmospheric inversion, and other local factors can be substantial in the Great and Lesser Khingan Ranges of northeast China and the Outer Baikal region of Russia. This results in the "Khingan–Baikal type" permafrost that differs from high-altitude and polar permafrost distributions [12, 13]. The form and process of high-latitude permafrost degradation in northeast China are also affected by the occurrence conditions. Therefore, the form and process of degradation in this region differ from their analogs in high-altitude and polar [14].

Landslide events tied to extreme weather and climate change have increased in recent years [15], and have gradually received attention from national governments and relevant international academic organizations [16, 17]. In particular, landslides in cold areas are becoming a hot issue in landslide research [18, 19].

Landslides are geological phenomena that naturally occur in mountainous areas. The triggering mechanism and evolution of landslides are closely tied to environmental factors and geological conditions. Landslides are influenced by geological forces, lithologic structure, and other crustal internal factors as well as by land cover, topography, human activities, precipitation, and environmental conditions. The temporal and spatial distribution of landslides exhibits characteristics of continuation, uncertainty, intermittency, and irreversibility. Landslides are the result of environmental and geological changes and their occurrence can dramatically alter the geology of an area [20]. Scientists have applied various methodologies to analyze the relationships between climate change in cold areas and landslide evolution and mechanisms [21]. These studies have discussed the landslides induced by permafrost and glacier degradation and their impact on the geology, topography, biodiversity, and water resources. Climate has also been investigated as a main factor influencing the landslide evolution in cold regions [22–28]. However, the preponderance of these studies were conducted at large-scales due to a lack of observational data [29]. To date, there have been no studies on the movement characteristics, mechanisms, and patterns of landslides caused by the combined effects of extreme weather events and permafrost thawing.

In the present study, we designate the area surrounding the intersection of the Bei'an to Heihe Highway with the northwest region of the Lesser Khingan Range as the study area. We conducted field measurements, a geological survey, and an engineering survey. These data sets were plotted on a geological cross section and a topographic map of the landslide area. We investigated the impact of climate change on geological environment, permafrost thaw, and landslide mechanisms in the study area via meteorological data from Sunwu County (30 km from the study area) provided by the China Meteorological Data Sharing Service System (http://cdc.cma.gov.cn/home). In order to conduct a comprehensive analysis of the impact of climate change on the landslide movement process and the pore water pressure, we utilized point displacement on the landslide mass, trailing edge ground temperature, and pore water pressure data sets collected from the landslide area of the Landslide K178 + 530. In addition, in this paper we apply a combination of geophysical techniques (HDR and GPR) and traditional methods (drilling and mapping) to Landslide K178 + 530 that occurred in the Lesser Khingan Range of northeast China. Landslides occur frequently in this region due to atmospheric precipitation and permafrost degradation. We aim to determine the thickness and internal structure of the study landslide and thereby ascertain the applicability of GPR and HDR for these regional types of landslide.

2. BACKGROUND

The study area ranges between 127° 17′ 31″–127° 21′ 24″ east longitude and 49° 30′ 57″–49° 41′ 50″ north latitude (**Figure 1**). In this area, the northwest section of the Lesser Khingan Range intersects the Bei'an-Heihe Highway at the border between Sunwu County and the Aihui District. The area is located on the southern fringe of China's high-latitude permafrost region and has typical periglacial landforms. The island-like permafrost in this region is the result of residual paleo-glacial deposition and is currently in the degradation stage. The geological conditions are extremely unstable.

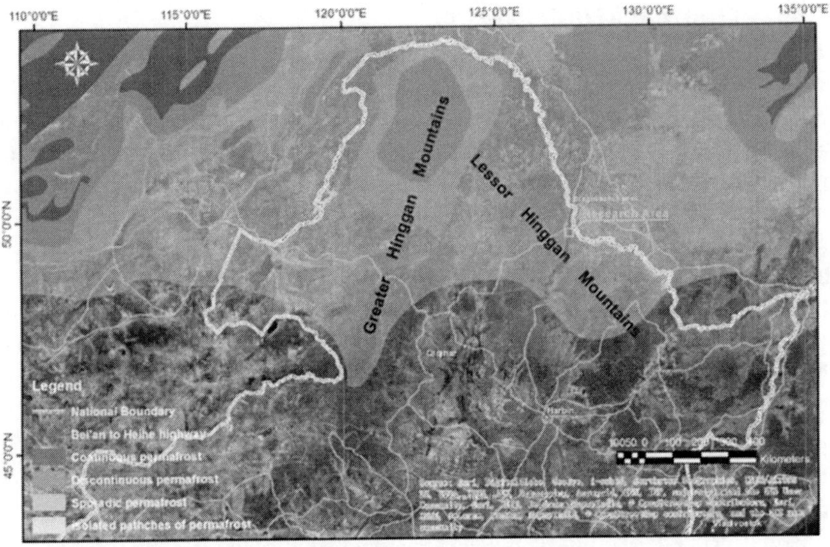

FIGURE 1. High-latitude permafrost distribution in Northeast China (data from Zurich University, Switzerland), and the location of study area.

2.1. Climate Conditions

The study area is positioned in the transition area between the cold-temperate zone and the northern area of the middle temperate zone. The area is subject to monsoons as well as alternating influences from high and low pressures originating from inland and the ocean. Generally, the study area experiences dry, cold, and long winters as well as hot, humid, and short summers. It is classified as a continental monsoon climate zone.

In the study area, the average annual temperature is within the range of −2 to 1°C, with an extreme minimum of −48.1°C and an extreme maximum of 38.6°C. The average annual precipitation is within the range of 530–552 mm, with a maximum of 800 mm. Precipitation is mostly concentrated from July to September during the summer. The precipitation that falls during this period accounts for 61–67% of the total yearly precipitation. The first snow fall is typically in mid-October, and the final snow fall is often in late March or early April of the following year. The annual average wind speed is 2.7–4.0 m/s. The minimum annual evaporation is 850 mm and the maximum annual evaporation exceeds 1000 mm. The average annual total radiation is 1148 kcal/cm^2, with a minimum of 1039 kcal/cm^2 and a maximum of 1229 kcal/cm^2. The average annual sunshine is 2500 h, with a minimum of 2200 h and a maximum of 2800 h.

2.2. Permafrost Distribution

The climate in Northeast China is subject to the Siberia-Mongolia high-pressure system. This region has a widely distributed inversion layer. This layer has important impacts on the regional distribution and development process of permafrost. Meanwhile, the majority of forest area in China is distributed in the northeast region. In this area, permafrost coexists and is influenced by wetlands, grasslands, and forests. Forests, mosses, shrubs, and other ground covers reduce absorption of solar radiation, thereby decreasing ground and air temperatures. The valleys, low-lying river terraces, shady slopes, and wetlands provide suitable conditions for slow permafrost degradation as well as the development of island-like permafrost.

In the study area, the seasonal frozen ground reaches its maximum depth at the end of May. The recorded maximum depth of seasonal frozen ground is in the range of 2.26–2.67 m. In the mountains, the maximum depth of seasonal frozen ground can exceed this range. The thawing period of seasonal frost is April to September. All the seasonal frost in dry areas thaws by early July. However, seasonal frosts do not completely thaw in swampy areas with thick peat and humus zones until the end of October. Island-like permafrost is mainly distributed in valley areas and on shaded slopes.

2.3. Climate Change In The Study Area

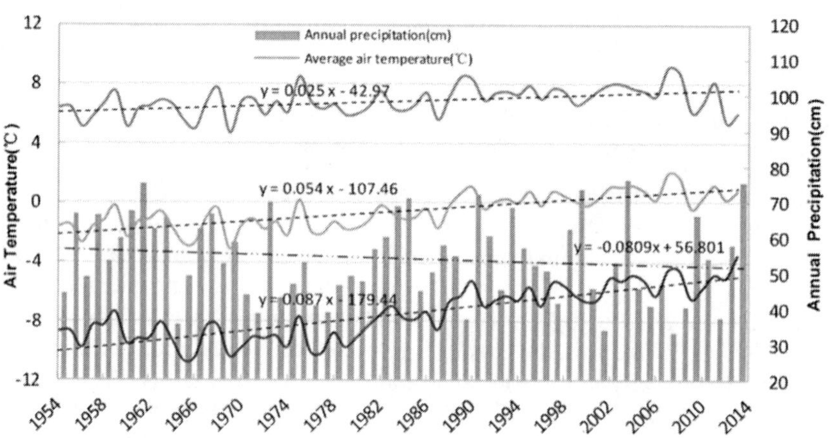

FIGURE 2. Annual average maximum temperatures, average temperatures, average minimum temperatures and average precipitation in Sunwu County (1954–2013).

Due to global climate change, the study area is among the regions in northeast China that have experienced the largest temperature increases during the past 50 years. Sunwu County meteorological data indicate that during the past 60 years, change in temperature was the most significant among all climate indicators. Using meteorological data from Sunwu County, **Figure 2** shows the yearly average temperatures, average maximum temperatures, average minimum temperatures, and average precipitation from 1954 to 2013. Linear regression indicates that during the 60 year period from 1954 to 2013 the annual average temperature increased by 3.2°C; the average annual minimum temperature increased by 5.2°C (69.04%); the average annual maximum temperature rose by 1.5°C (21.99%); and the average annual precipitation decreased by 4.85 cm (8.93%). The average annual minimum temperature increase was 3.45 times the increase in the average annual maximum temperature.

The permafrost distribution map generated from the Enhanced Thematic Mapper (ETM+) data collected in 2009 via the Landsat7 satellite (**Figure 3**) indicates that the landslide locations match the permafrost distribution to a high degree [30]. We therefore conjectured that these landslides were the result of thawing permafrost.

FIGURE 3. Permafrost distribution map in K176 + 500–K180 + 000 section of Bei'an-Heihe Highway.

2.4. Climate Change And Landslides

Climate change in the study area has led to permafrost degradation, and engineering projects have exacerbated the permafrost degradation process [31]. In 1999, a planning survey conducted for the secondary road from Bei'an to Heihe indicated that along the entire road length there were 17 permafrost road segments [32, 33]. However, a survey conducted right before the construction of the Bei'an-Heihe Highway in 2009 showed that there were only six permafrost road segments remaining; the other 11 segments had been completely degraded.

The survey conducted in 2009 showed that within 10 m of the left of the roadbed in the K177 + 400 to K179 + 200 section there were four landslides with a total volume that exceeded 20,000 m^3 (**Figure 6**). In addition, there were numerous landslides within 3 km of this road section (**Figure 4**).

The phase change of water during the thawing of permafrost has a significant impact on the soil's mechanical properties [34]. Permafrost degradation causes a multitude of geological engineering issues for road construction. In 1999, the Bei'an-Heihe secondary road construction began. A landslide was caused by thawing of permafrost in the roadbed during August 2000 in the K176 + 900 to 178 + 200 section. This necessitated the abandonment of this section due to instability over the entire roadbed. The road had to be redirected along the left side of the ridge (**Figure 5A**).

FIGURE 4. Full view of K175 + 500–K180 + 200 section of Bei'an-Heihe Highway (blue line is the boundary of the landslides, the orange area is corn field).

FIGURE 5. Satellite photos of K176 + 500–K179 + 900 section of Bei'an-Heihe Highway (Google earth). (A) Landslide in K178 + 530 section (in 2000); (B) satellite photos of K178 + 530 (2004.6); and (C) satellite photos of K178 + 530 (2010.9).

2.5. Topography And Geological Structure

The study area encompasses a hilly landscape with rolling terrain. The slope is typically within the range of 10–20°. The upper sections of the slopes are usually within the range of 25–30° but can reach up to 40°. The area can also be divided into hills and valley bottomland. The altitude range over the entire area is 210–330 m. A 2010 survey also showed that, in the section between K177 + 400 and K179 + 200, there were four landslides within 10 m of the left of the roadbed with a total surface area of over 2000 m^2 (**Figure 6**). The four landslides are located at the A, B, C and D position, respectively. C illustrates Landslide K178 + 530. **Figure 6** shows the geologic and geomorphic map of the landside road area of the study area plotted from the field survey conducted in June 2010.

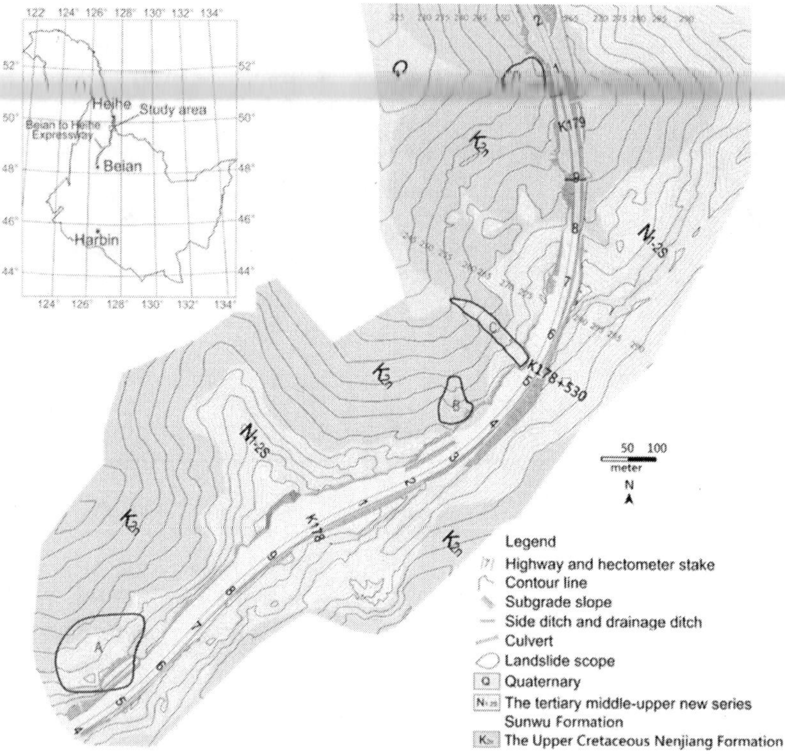

FIGURE 6. Geological map of K177 + 400–K179 + 200 section of Bei'an-Heihe Highway.

The geology of the study area can be characterized as belonging to the Khingan-Haixi fold belt. Beginning at the surface and moving down, the stratigraphy is composed of powdery sandstone, silty mudstone, Tertiary pebbly sandstone, and Cretaceous mudstone. The Lesser Khingan Range experienced block uplift during a period from the late Tertiary to the early Quaternary. Due to leveling and long-term erosion, loose sediment has gradually thinned at the summit and on the slopes of hills. The current residual layer is typically only 1–2 m thick. The loose deposits mainly accumulate with a thickness of approximately 10 m in the valley and basin areas between mountains. The soil is mainly composed of mild clay, clayey silt, and gravelly sand. The ground is covered with a relatively thick layer of turf and grass peat. The surface vegetation is composed of woodland and grassland, and there are inverted trees in the woodlands.

Landslide K178 + 530 is positioned at the widened embankment on the left side of the Bei'an-Heihe Highway (**Figure 7**). The surface soil and the roadbed soil slide together along the valley, and the maximum slippage is 200 m distance from the road. This landslide mass is 20–30 m wide and covers an area of approximately 6000 m^2. The distance between the leading and trailing edges is 200 m. The height of the trailing edge is 285 m, and the height of the leading edge is 254 m. The landslide trailing edge has an arc-shaped dislocation that is positioned within the widened embankment.

FIGURE 7. Landslide scene photos: (a) Panoramic view of landslide (in November 2014); (b) the trailing edge of landslide (in June 2010); and (c) the leading edge of landslide (in October 2013).

For prospecting purposes, four boreholes were drilled into the K178 + 530 section. The borehole depths were within the range of 14–26 m, and the distribution of boreholes is presented in **Figure 8**. The drilling revealed the soil profile as follows, from bottom to top: Cretaceous mudstones, siltstones, Tertiary pebbly sandstones, and Quaternary loose Embankment: yellow, primarily composed of loosely mixed Tertiary pebbly sandstones, sandy mudstones, and Cretaceous mudstones; the soil is plastic when saturated with water and loose when dry. Clay: yellow, plastic and with high rigidity and strength when dry. The downstream region of the landslide mass has a depth distribution within the range of 0–6.7 m. The upstream region of the landslide mass has a depth distribution within the range of 1.5–3.8 m. There are numerous sandwiched grit layers, and one of these layers is approximately 1–10 cm thick. These layers greatly enhance the soil water seepage capacity.

Tertiary pebbly sandstones: distributed in the embankment within the range of 2.0–3.4 m depth. The upstream region of the landslide mass is within the range of 3.8–4.5 m depth. This area is mainly composed of weathered feldspar stone and mineral sands that are well-graded with high permeability. Fully weathered siltstones: yellow, distributed in the upstream region of the landslide mass at a depth of 4.5–9.7 m, sandy, of bedding structure and poor water seepage capacity.

Fully weathered mudstone: yellow or gray-green and pelite. They exhibit a layered structure and poor water seepage capacity. They are easily softened by water. Strongly weathered mudstones: dark gray, pelite, of layered structure, weakly cemented rock. Moderately weathered mudstones: brown, black and gray, pelite, layered structure.

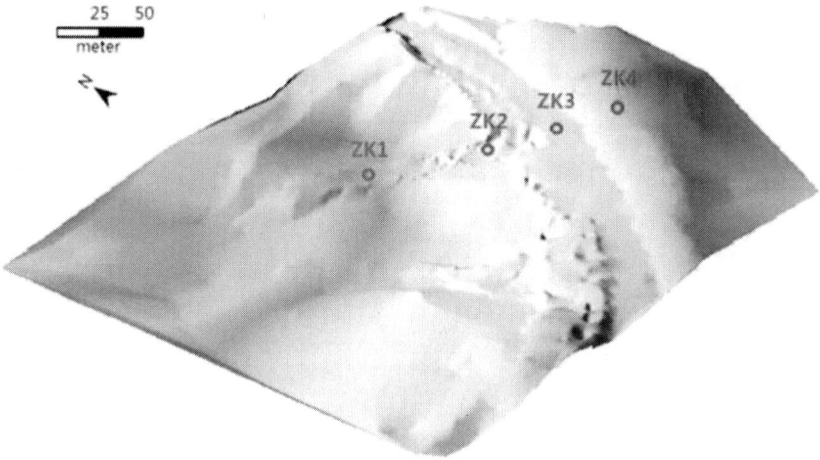

FIGURE 8. Drilling borehole layout on Landslide K178 + 530 and the geomorphological map of the landslide area (using SURFER software to draw the geomorphological map based on GPS terrain data in June 2010).

3. GEOPHYSICAL RESEARCH METHODS

In this paper we present combination geophysical techniques of HDR and GPR on the landslide K178 + 530 on the Bei'an-Heihe Highway in Lesser Khingan Range of northeast China. The survey lines were established, as shown in **Figure 9**.

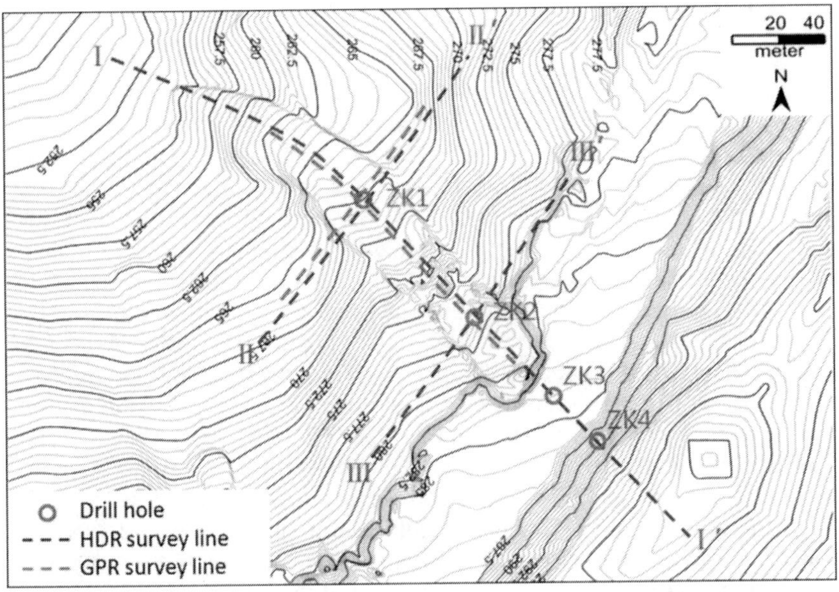

FIGURE 9. The layout of the Landslide K178 + 53 geophysical survey lines. (1) HDR survey lines. A total of three HDR survey lines, i.e., I–I′, II–II′ and III–III′, were established on the Landslide K178 + 530 road section. Line I–I′ passed through the center of the landslide mass and was oriented along the sliding direction. The survey line start point was 40 m from the leading edge of the landslide mass. The survey line passed the drilling boreholes ZK1, ZK2, ZK3 and ZK4 in sequence. The survey line has a total length of 357 m (300 m horizontal distance). Line II–II′ was oriented to be perpendicular to the sliding direction and began 110 m away from the trailing edge. Point ZK1 was the midpoint of this line with a total length of 177 m. Along this line, 1–60 numbered electrodes were arranged in order from left to right. Line III–III′ was also oriented perpendicular to the sliding direction but was had a start position 50 m away from the trailing edge. Point ZK2 served as the midpoint of this transect with a total length of this survey 177 m. A similar set up of electrodes was also deployed in this transect. The data collection date is September 3, 2012. (2) GPR survey lines: we set up two survey lines with the locations that corresponded to the HDR survey lines. However, the start and end points were different. The HDR survey lines were longer than the GPR survey lines. The two survey lines (I–I′ and II–II′) had lengths of 150 m and 118 m, respectively. The data collection date is October 1, 2013.

3.1. Hdr Method

The WGMD-9 Super HDR system produced by the Chongqing Benteng Digital Control Technical Institute (Chongqing, China) was utilized in this study. Within this system, the WDA-1 super digital direct current electric device is used for control host and measurement. Centralized two-dimensional HDR measurements can be generated via the optional WDZJ-4 multi-channel electrode converter, electrodes, and centralized high-density cables. The RES2DINV software was used to perform the inversion of the apparent resistivity data sets. This software package generates a two-dimensional model of the subsurface via apparent resistivity pseudo-sections [35]. A Wenner configuration was employed to acquire the relevant data. The method utilizes the inversion of pseudo-section data via smoothness-constrained least squares [35–37]. The subsurface in this algorithm is subdivided into rectangular blocks with constant resistivity. The resistivity of each block is then assessed via an iterative scheme that minimizes the difference between calculated and observed pseudo-sections using an iterative scheme. The smoothness-constraint leads to results with smooth resistivity changes. The derived pseudo-sections can be generated either via finite-element or finite-difference methods [38, 39]. In this study, we employed the finite element scheme due to the changes in the study area topography.

We utilized the smoothness-constraint least squares method in the inversion model. Fundamentally, this method is used to constantly adjust the model resistivity through model correction to reduce the difference between the calculated apparent resistivity and the measured resistivity, and to describe the degree of fit between the two using the mean square error. The smoothness-constraint least squares method, which has been widely applied, has a number of advantages, such as adaptability to different types of data and models, relatively small noise influence on the inversion data, rapid inversion, high sensitivity to deep units, and a small number of iterations. In tests utilizing the HDR method, the maximum exploration depth of the survey lines was determined to be 30 m and the spacing between unit electrodes was 3.0 m.

We established three HDR survey lines on the Landslide K178 + 530 road section, i.e., I–I′, II–II′ and III–III′ (**Figure 9**). The date of data collection is September 3, 2012.

3.2. Gpr Method

The GPR instrument used was the RIS-K2 FastWave Ground Penetrating Radar produced by IDS Corporation (Italy). We utilized an unshielded dual radar antenna that operated at low-frequency 40 MHz. The time window of detection was set at 600 ns, the sampling rate was set at 1024, and a 0.05 m data acquisition track pitch was employed. Two GPR survey lines were set up (**Figure 9**) with orientations that coincided with the HDR survey lines, but termination points of the GPR survey lines were different. The two survey lines (I–I′and II–II′) had lengths of 150 m and 118 m, respectively. The date of data collection is October 1, 2013. Raw GPR data were processed via the REFLEXW software produced by Sandmeier Scientific Software (Karlsruhe, Germany). The coordinates of each trace were calculated at equal distances. The surface signal reflection was set to time zero. Noise and low frequency parts of the spectrum were filtered by applying bandpass and dewow methods. In the following step of processing, we eliminated temporally consistent signals via a background removal technique. Topographical correction was then performed. The picks were exported along with the two-way travel time attribute. The propagation velocity of the wave in this case appears to be approximately 0.10 m/ns.

4. RESULTS, ANALYSIS, AND DISCUSSION OF HDR

4.1. Survey Line I–I′

The apparent electrical resistivity profile data for line I–I′ as collected on September 3, 2012 is presented in **Figure 10**. The figure clearly shows distinct layering in the soil resistivity values of the landslide mass. The RES2DINV software is able to extract the resistivity curve value for any transect point. This allows further analysis of the relationship between soil resistivity changes and depth. **Figure 11** presents the curves of soil resistivity vs. depth at boreholes ZK1 and ZK2.

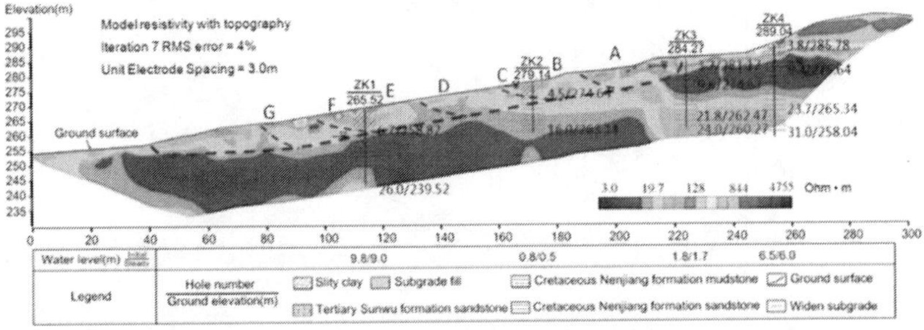

FIGURE 10. Comparison of drilling results and electrical resistivity profile of the survey line I–I′. Geological drilling date is in June 2010, HDR detection date is on September 3, 2012. The dashed thick black line indicates the position of the major sliding surface, and the dashed thin black lines illustrate the secondary sliding surfaces.

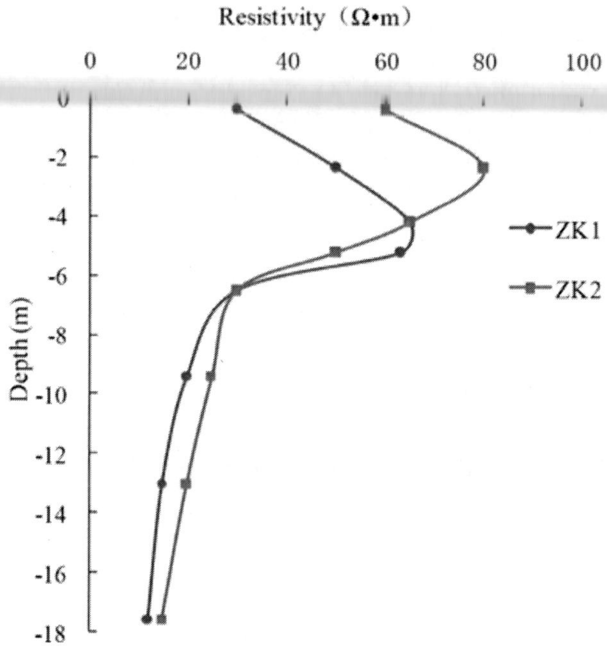

FIGURE 11. The electrical resistivity curves at positions ZK1 and ZK2.

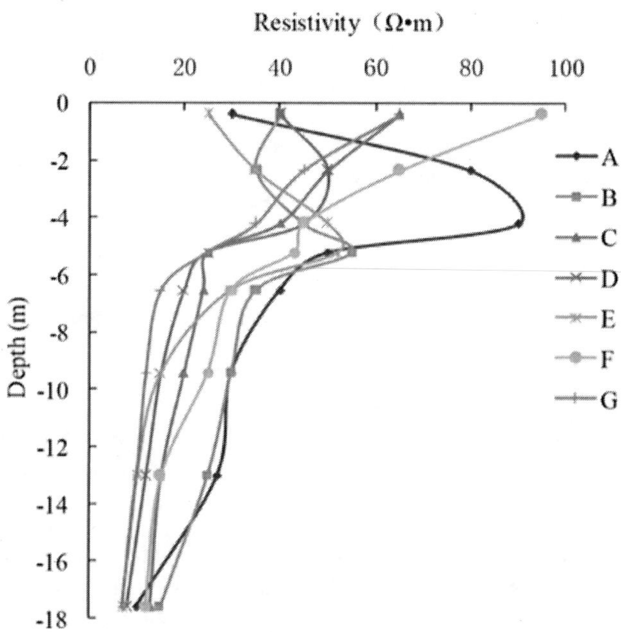

FIGURE 12. The electrical resistivity curves at different points on the survey line I–I′.

The ZK1 borehole is located 115 m from the beginning point of the HDR survey line. At a depth of 0–2.1 m, the soil is rather loose silty clay and is composed of approximately 15% grass roots and other organic matter. The resistivity values at this depth are in the range of 25–45 Ohm m. At the 2.1–6.7 m depth, the soil is primarily silty clay with localized weathered sand layers. The resistivity values for this depth are within the range of 45–65 Ohm m. Within the range of 6.7–8.0 m depth, the soil is composed of yellow mudstone. It is difficult for the water to infiltrate downward as the permeability coefficient is small. This forms a layer where water easily gathers. There are relatively low resistivity values in this layer, i.e., 20–30 Ohm m. At a depth of 8.0–26 m, the soil is gray mudstone, close to or below the water table. The resistivity value is relatively low, i.e., 10–25 Ohm m. As shown in the curve, silty clay contacts mudstone at a depth of 6.7 m, and resistivity exhibits apparent layering, the resistivity value decreased suddenly.

The ZK2 borehole is located 175 m from the beginning point of the survey line. At depths in the range of 0–4.5 m, the soil is somewhat loose. The resistivity values in this layer are within the range of 45–80 Ohm m. The surface embankment soil is dominated by silty clay (depth 0–3.8 m) and gravelly sand (depth 3.8–4.5 m) with resistivity values within the range of 60–80 Ohm m and 45–60 Ohm m, respectively. At the 4.5–9.7 m depth range, the soil is siltstone and primarily composed of relatively small particles. This layer exhibits poor permeability, forming a watertight layer where water can easily gather. The resistivity values are within the range of 25–35 Ohm m. At the 9.7–14.6 m depth range, the soil is composed of sandstone, and the resistivity values are within the range of 15–25 Ohm m. As the gravelly sand contacts the siltstone at a depth of 4.5 m, the resistivity value decreases suddenly.

The RES2DINV software was used to generate the soil resistivity curves along survey line I–I′ at points A, B, C, D, E, F and G (**Figure 10**). The horizontal distances between the starting point and the points A, B, C, D, E, F and G of line I–I′ are 80 m, 100 m, 120 m, 140 m, 160 m, 180 m and 200 m, respectively. The obtained soil resistivity curves are shown in **Figure 12**. At position A, the soil resistivity value decreased abruptly at a depth of 4.5 m. In other words, the soil resistivity values exhibited abrupt stratification at this depth. It can thus be determined that the sliding surface is located at a depth of 4.5 m. Similarly, the depths of the sliding surface at points B, C, D, E, F and G on line I–I′ were determined to be 5 m, 4.7 m, 5.5 m, 6.5 m, 6 m, and 5.5 m, respectively.

The above resistivity curves and HDR profiles illustrate that the soil resistivity values are clearly different and exhibit an abrupt stratification above and below the sliding surface of the landslide mass. The locations of the major sliding surfaces along the line I–I′ were deduced according to this typical sliding surface characteristic. These are illustrated with the thick dashed black line in **Figure 10**. The sliding power originated from the trailing edge of the landslide and the sliding type was propelled sliding. The minimum slip rate occurred at the leading edge. The maximum slip rate was at the trailing edge, followed by the intermediate portion of the landslide [40]. Secondary sliding in the landslide mass occurred as a result. Combining the results of the drilling exploration as well as the changes in the soil resistivity values of different positions in the landslide mass and, the secondary sliding surface was obtained, as shown via the thin dashed black line in **Figure 10**.

4.2. Survey Line II–II′

The apparent electrical resistivity profile measured along survey line II–II′ is presented in **Figure 13**. **Figure 14(a)** shows the soil resistivity curves of points C, D ZK1, E and F of line II–II′. The distances between points C, D ZK1, E and F and the beginning point of the survey line are 80 m, 85 m, 90 m, 95 m and 100 m, respectively. Based on **Figures 13** and **14(a)**, we can know the changes in the soil resistivity values, there are apparent resistivity layering at the depths of the sliding surfaces, the resistivity value decreased suddenly. According to this characteristic of the sliding surface, the positions of the sliding surfaces along line II–II′ were deduced, as shown by the black dotted line in **Figure 13**. Based on the changes in the soil resistivity values, it can be inferred that the depths of the sliding surfaces at positions C, D, ZK1, E and F on line II–II′ were 4 m, 6 m, 6.5 m, 5.5 m and 3.5 m, respectively. **Figure 14(b)** illustrates the soil resistivity curves at the A, B, G and H locations (all outside the landslide mass) along the II–II′ survey line. The distances between the start point of the survey line and the points A, B, G and H were 40 m, 60 m, 110 m and 130 m, respectively. As can be seen in **Figure 14(b)**, the soil resistivity values of the stable soil body outside the landslide only showed stratification in the surface loose layer. As depth increased, the resistivity basically exhibited a monotonic decline, and there was no abrupt stratification.

This landslide represents a recurring landslide that has once again slipped [40]. The dashed black line in **Figure 13** illustrates the current sliding surface. By combing the geological survey at the site, and the resistivity change characteristics, we can infer the location of the sliding surface for the paleo-landslide, as illustrated by the dashed yellow line in **Figure 13**.

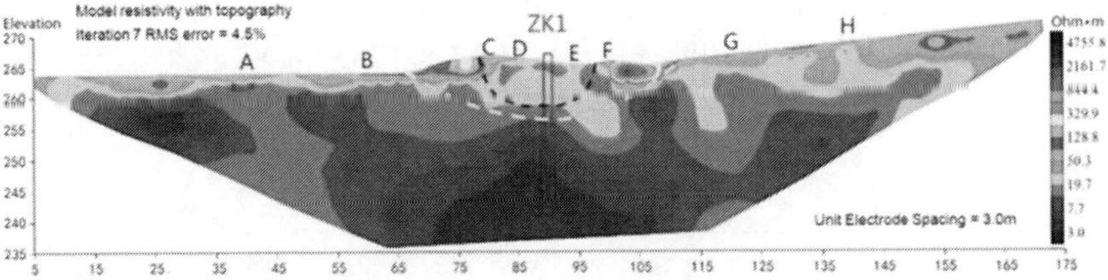

FIGURE 13. Survey line II–II′ electrical resistivity profile. HDR collection date of September 3, 2012. The dashed black line represents the current sliding surface. The dashed yellow line illustrates the sliding surface position of the paleo-landslide.

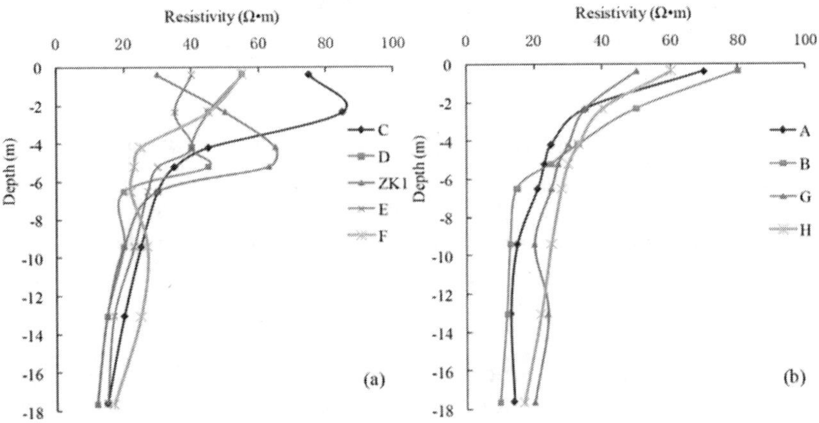

FIGURE 14. The electrical resistivity curves at different points on the survey line II–II′ (points C, D, ZK1, E and F are all on the landslide mass, points A, B, G and H are all outside the landslide mass).

5. RESULTS, ANALYSIS, AND DISCUSSION OF GPR

5.1. Survey Line I–I′

The profile determined by the GPR survey line I–I′ is illustrated in **Figure 15** (due to the constraints under field conditions, this GPR survey line can only be used to measure this long section). The layer picking option (phase follower) of the REFLEXW software can identify a continuous reflector (the thick dashed red line in **Figure 15**). The radar-wave reflection intensity differed significantly from the reflection intensity of the surrounding medium. The strong reflected-wave signal exhibits characteristics of a distinctive horizon. The signal presents a low-frequency high-amplitude sync-phase axis. We conclude that this represents the sliding surface. The sliding surfaces exhibit good continuity and essentially reflect the landslide mass development depth range [41, 42].

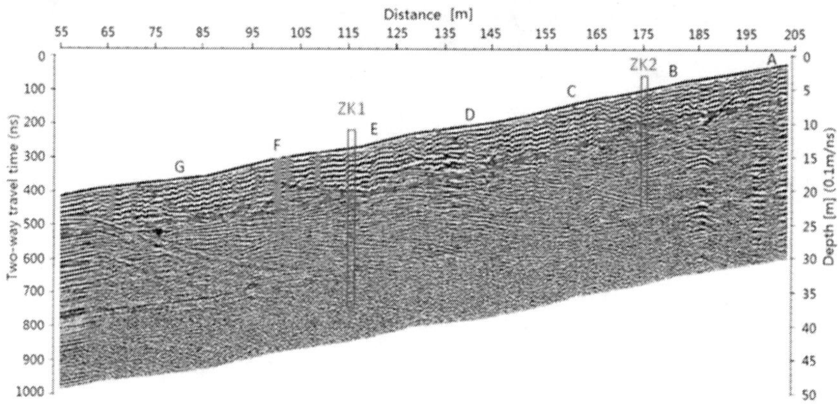

FIGURE 15. The GPR profile of the survey line I–I′. The radar antenna was a low-frequency 40-MHz unshielded dual antenna, the measuring date is October 1, 2013. The thick red dashed line show the position of the major sliding surface, the thin red dashed lines show the secondary sliding surfaces. (see text for discussion).

Using REFLEXW radar data processing software, the radar waves amplitude values of all the data acquisition track points in the profile at different depths can be extracted. In order to better understand the changes in the reflected radar wave intensity, we plotted the radar-wave amplitude curves for positions A, B, ZK2, C, D, E, ZK1, F and G (as depicted in **Figure 15**) on the survey line I–I′ (**Figure 16**). Greater reflected radar wave intensity is indicated by a higher radar-wave amplitude value [41,43]. As shown in **Figure 16**, most curves showed relatively large amplitudes in the depth range of 0–2.5 m as the surface soil body is quite loose at this depth. At position A, the radar-wave amplitude increased substantially and abruptly at a depth of 4.5 m. We can conclude that the soil moisture content was relatively high at this position based on the characteristic differences between different types of soil bodies with regard to their radar-wave reflection [41,43]. We can therefore deduce the location of the sliding surface of the landslide mass. Likewise, an abrupt increase in radar-wave amplitudes occurred at depths of 4.7 m, 4.7 m, 5.5 m, 5.5 m, 6.7 m, 6.5 m, 5.6 m and 5 m on line I–I′ at positions B, ZK2, C, D, E, ZK1, F and G, respectively, and the position of the sliding surface can be inferred. This is very close to the position of the sliding surface as denoted by thick the red line in **Figure 16**.

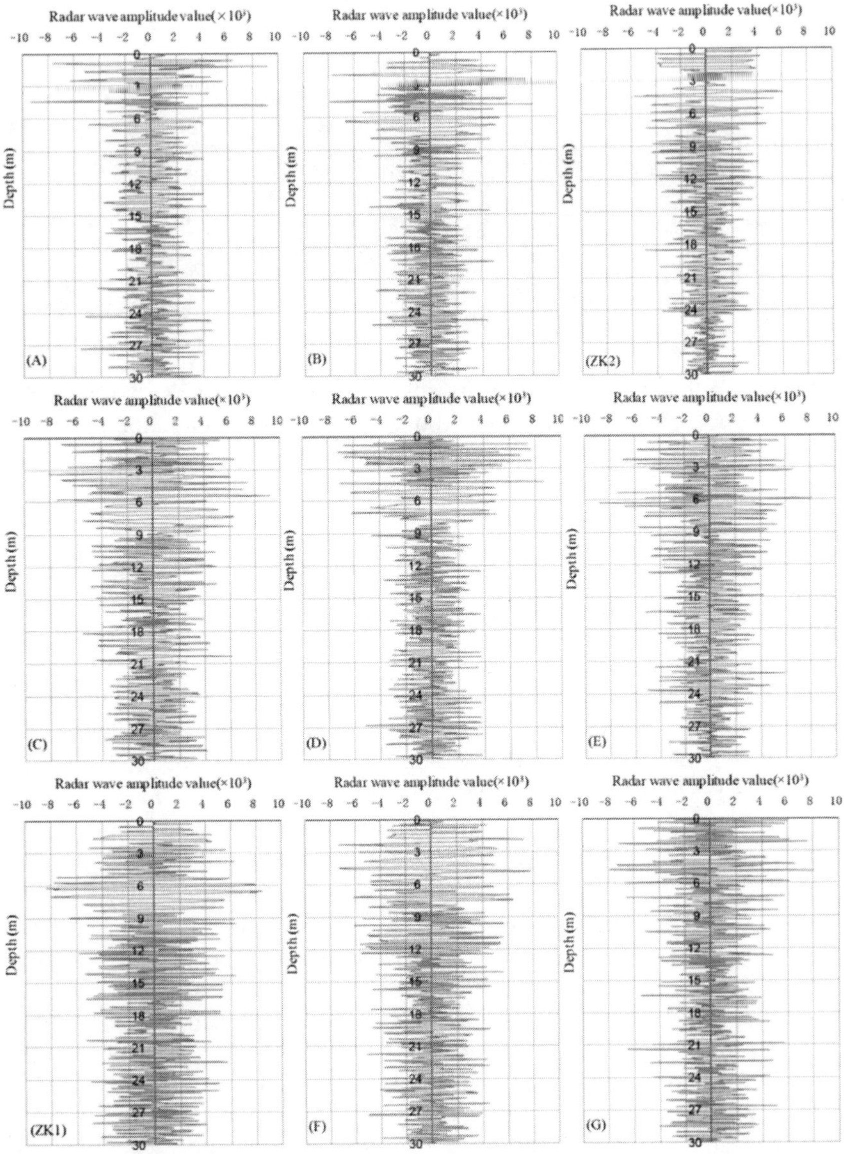

FIGURE 16. The radar-wave amplitude curves on the survey line I–I′. Positions A, B, ZK2, C, D, E, ZK1, F and G are all on the survey line I–I′ (as shown in Figure 15).

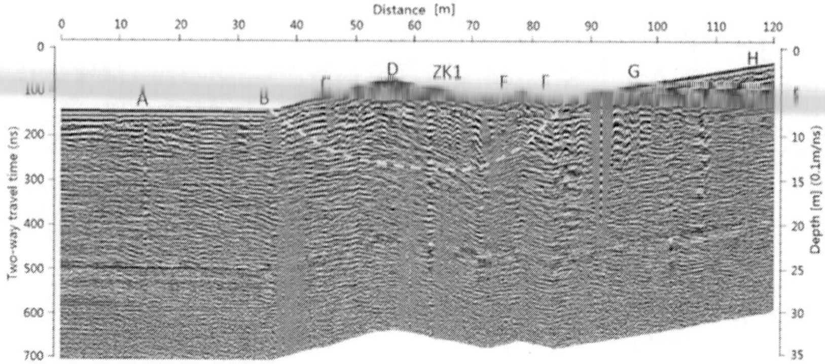

FIGURE 17. The GPR profile of the survey line II–II′. GPR detection date is on October 1, 2013. The red dashed line shows the current sliding surface. The yellow dashed line shows the position of the sliding surface for the paleo-landslide.

5.2. Survey Line II–II′

The GPR profile recorded for survey line II–II′ is presented in **Figure 17**. The dotted red line is the low-frequency high-amplitude sync-phase axis, and we conclude that this represents the profile's sliding surface. This axis illustrates the landslide mass development depth range.

Using REFLEXW radar data processing software, we can obtain the radar-wave amplitude curves at positions A, B, C, D, ZK1, E, F, G and H (as shown in **Figure 17**) on the survey line II–II′, as shown in **Figure 18**. As can be seen from **Figure 18**, because the surface soil body is rather loose in this area, most curves showed relatively large amplitudes in the depth range of 0–2.5 m. D, ZK1 and E are all on the landslide mass, in positions D, ZK1 and E, a sudden and substantial increase in the radar-wave amplitude occurred at depths of 3.5 m, 6.5 m and 3.2 m, respectively, exhibiting abrupt changes. According to the differences in characteristic radar-wave reflection for different classes of soil bodies [41, 43], we inferred a relatively high soil moisture content at this location. This also implies the location of the landslide sliding surface. The deduced position of the sliding surface is about the same as that denoted by the red dashed line in **Figure 17**. Positions A, B, G and H were all located outside of the landslide mass. Except in the surface layer, i.e., in the depth range of 0–2.5 m, there were abrupt changes in the radar-wave amplitudes; at deeper depths, no abrupt changes were observed in the radar-wave amplitude curves.

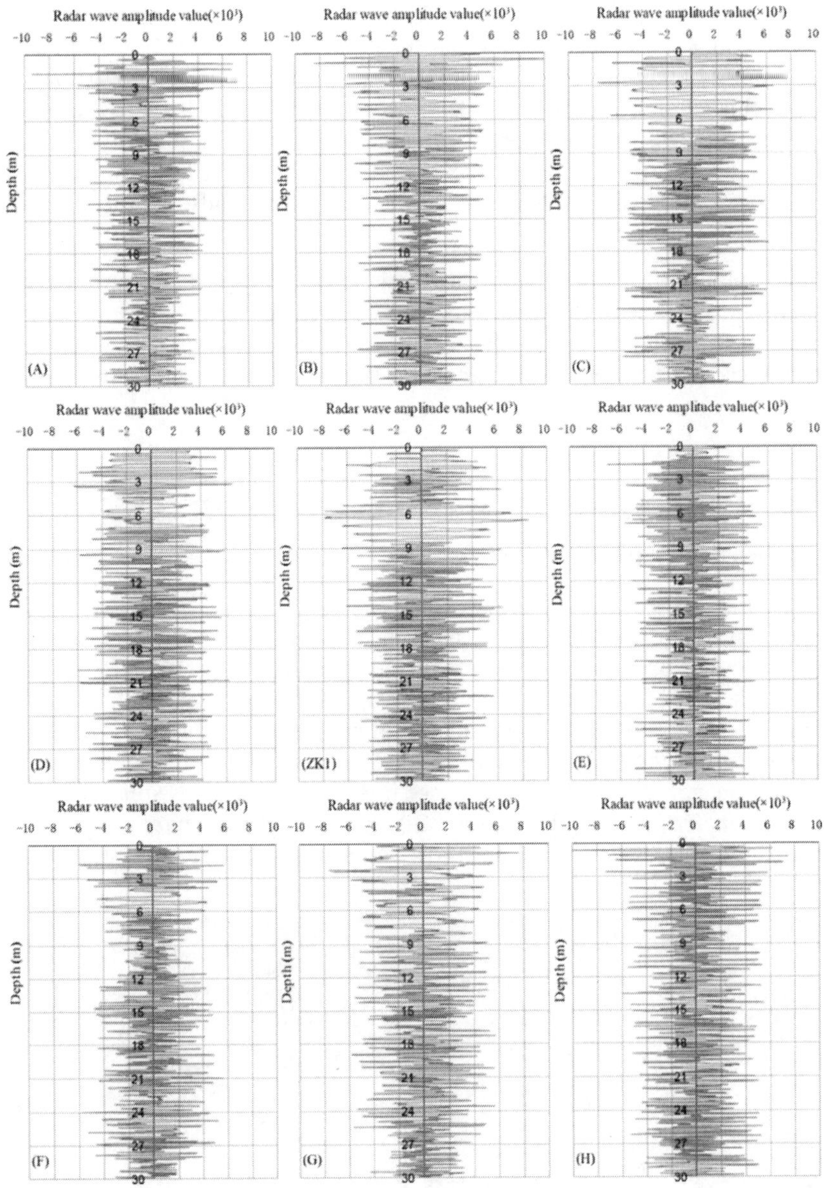

FIGURE 18. The radar-wave amplitude curves on the survey line II–II′. Positions A, B, C, D, ZK1, E, F, G and H are all on the survey line II–II′ (as shown in **Figure 17**).

As shown in **Figure 18**, in positions C, D, ZK1, E and F, the amplitude values showed substantial increases at depths of 6 m, 9.5 m, 9 m, 7 m and 3.5 m, respectively, exhibiting abrupt changes. These locations can be utilized to infer the sliding surface location of the paleo-landslide (illustrated in **Figure 17** with yellow dashed lines). Meanwhile, the magnitude of the sudden alteration in the radar-wave amplitudes at the sliding surface location of the paleo-landslide was less than the magnitude at the current sliding surface location.

The results of the GPR indicate relatively high soil moisture content at the sliding surface of the landslide mass. The drilling data also indicate very high moisture content of the sliding surface in the study area. These drilling results are in complete agreement with the results obtained from the HDR profile and the GPR profile.

6. UNDERLYING MECHANISM OF LANDSLIDE DEVELOPMENT

A geological survey of the study area in May 2010 revealed permafrost in the shaded slopes on the two sides of the landslide [40]. Permafrost locations were identified by drilling into the profile of the survey line III–III′ (**Figure 19**). The HDR method was also used on June 2, 2010 for prospecting along survey line III–III′. Permafrost layer range in the profile of the survey line III–III′ can be inferred based on soil resistivity characteristics (in [41]). **Figure 20** presents the apparent electrical resistivity profile of survey line III–III′ collected in June 2, 2010 (as shown by the black dotted line).

FIGURE 19. The photo of high temperature permafrost. Sampling location is in the survey line III–III′, at 40 m position on the *X* axis, the drilling time is in May 2010.

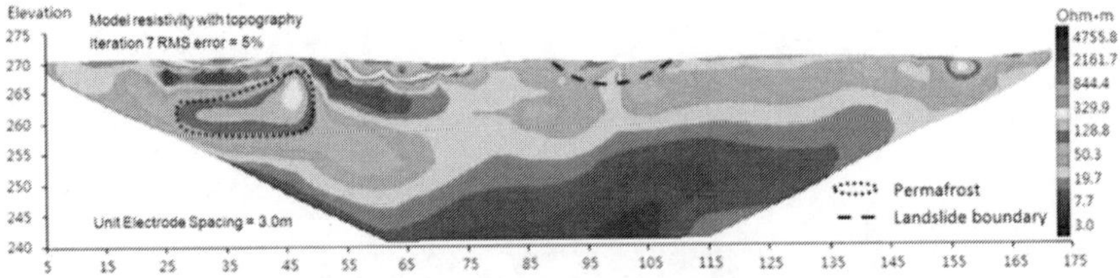

FIGURE 20. The electrical resistivity profile of the survey line III–III′ (2010-6-2).

Due to permafrost melting and concentrated summer rainfall, the landslide mass started to slip near the end of July in 2010 [40]. Borehole drilling and site survey data establish Landslide K178 + 530 as a shallow creeping consequent landslide. Water infiltration generated from concentrated summer precipitation in conjunction with water seepage produced from permafrost melting increase the local moisture content within the soil of hillsides. In the northwest section of the Lesser Khingan Range in China, this is the main cause of landslide formation. During the rainy season and the spring melting season, instability can easily occur. The relatively high number of bulging cracks on the landslide mass facilitates the infiltration and accumulation of water. The highly permeable surface soil, sand and gravel layer, and the silty clay with a weathered sand interlayer produce an expedient route for water infiltration. The mudstone and siltstone layers beneath exhibit low permeability and produce watertight layers. Water generated from melting snow, precipitation, and permafrost melting is hindered by the impermeable layer during its downward infiltration, and the local moisture content sharply increases. Water therefore infiltrates along the interface of the permeable layer and the impermeable layer. This generates a slip zone. Combining the geophysical and drilling data, the position of the sliding surface can be determined, as shown by the red line in **Figure 21**.

7. LANDSLIDE MOVEMENT MONITORING

Comprehensive examination and dynamic monitoring of the slip movement were performed for a representative landslide that could threaten the roadbed. This was done on the basis of geological survey and topographic measurements of the study area with the aim of investigating the impact of permafrost thaw on landslide movement.

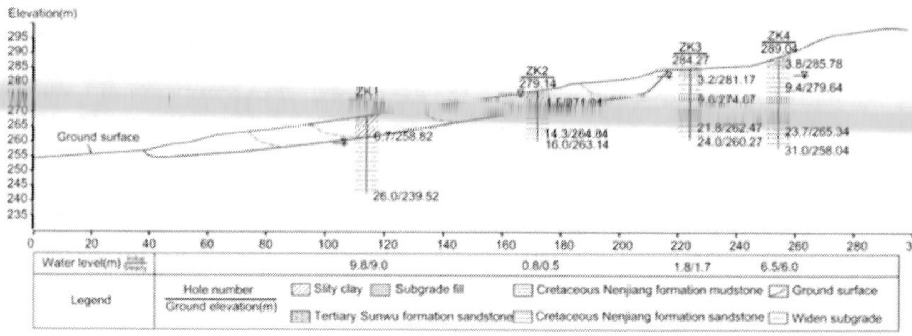

FIGURE 21. The stratigraphic distribution of section K178 + 530 (the thick red dashed line shows the position of the major sliding surface, the thin red dashed lines show the secondary sliding surfaces).

Landslide K178 + 530 on the Bei'an-Heihe Highway (**Figure 6C**) is among the closest landslides to the roadbed and exhibits the most rapid slip rate. Due to the unavailability of monitoring and survey data prior to 2009, we could not determine the date of the initial slip. A satellite photo taken on June 15, 2004 indicates a straight-line distance of 101.26 m between the trailing edge and the leading edge of the landslide mass (**Figure 5B**). A satellite photo taken on September 12, 2010 indicates that the straight-line distance between the trailing and leading edges had increased to 145.05 m (**Figure 5C**). During this six year period, there was 43.79 m of movement in the leading edge of the landslide, whereas there was basically no movement in the position of the trailing edge.

Figure 22 presents the locations of the geophysical measurement lines, the drilling points, and monitoring points for ground temperature, displacement, and pore water pressure. The points ZK1, ZK2, ZK3, ZK4, E, and F in the figure denote the positions of the drilling, and lines I and III denote the high-density resistivity measurement line and the ground-penetrating radar (GPR) measurement line, respectively. During the drilling in holes E and F, permafrost was found at a depth 2.2 m below the ground surface, and the permafrost thicknesses at points E and F were 3.7 m and 2.4m, respectively. The area of permafrost distribution in **Figure 22** is inferred according to **Figure 3**.

Through drilling, GPR, and high-density resistivity prospecting; the profiles of Landslide K178 + 530 in cross section H along the sliding direction and in cross section G perpendicular to the slide direction were obtained (**Figures 23** and **24**).

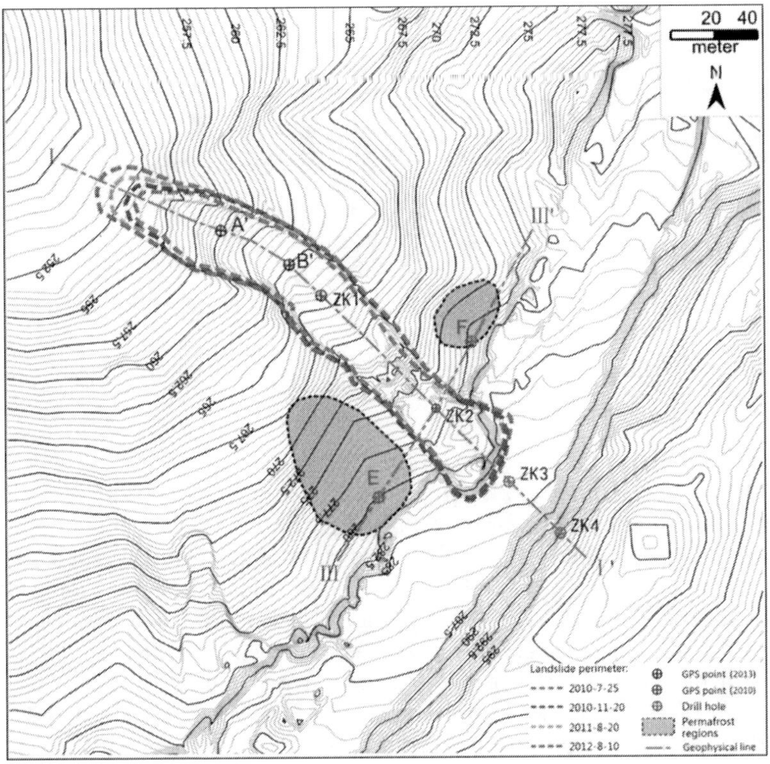

FIGURE 22. Distribution of the drillings, the geophysical measurement lines and points for monitoring displacement, ground temperature, and pore water pressure in Landslide K178 + 530.

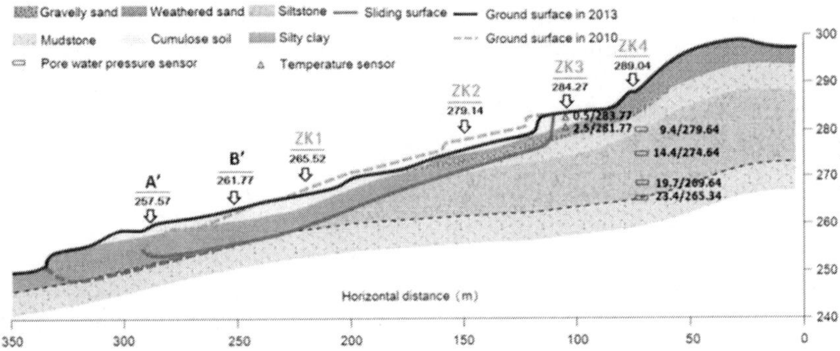

FIGURE 23. The profiles of Landslide K178 + 530 along the sliding direction.

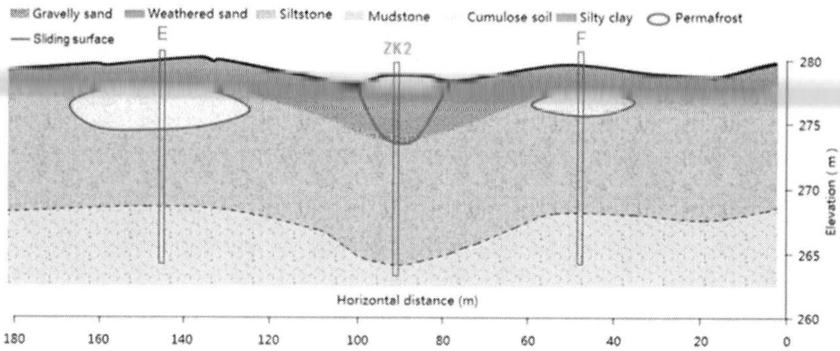

FIGURE 24. The profiles of Landslide K178 + 530 perpendicular to the slide direction.

In the boreholes ZK1 and ZK2 (**Figure 22**), we buried plastic displacement measurement piles that were 400 cm long with an above-ground height of 30 cm, The top of the pile was utilized to quantify the landslide mass slope displacement. ZK3 was used as the monitoring point for the ground temperature. Thermistor soil temperature sensors were set up at 0.5 m intervals from 0.5 m to 2.5 m below the ground surface. ZK4 served as the monitoring point of pore water pressure; in the borehole, steel-wire soil pore water pressure sensors were set up at 3.8 m, 9.4 m, 19.4 m, and 23.7 m depth. The relevant parameters were regularly monitored starting in July 2010.

We utilized a Real Time Kinematic-Global Positioning System (RTK-GPS) to monitor the movement of data collection points on the landslide surface. Point ZK1 horizontally slid by 69.03 m from July 2010 to August 2014, and its elevation decreased by 7.95 m. Point ZK2 horizontally slid by 113.97 m, and its elevation decreased by 17.37 m. From these values, we conclude that the angle between the horizontal plane and the slope dropped from 8.07° to 6.02°. Points A′ and B′ in **Figures 22** and **23** illustrate the locations of points ZK1 and ZK2 after sliding, respectively. These positions reflect data collected in August 2014.

Figure 25 illustrates the daily average atmospheric precipitation and ground temperature at point ZK3 on Landslide K178 + 530 at depths of 0.5 m and 2.0 m. The ground temperature data represents monitoring data collected in the study area. The daily average atmospheric precipitation and temperature represent the meteorological data collected via the Sunwu County weather station.

To facilitate the analysis of data, we defined the soil freeze period as the period from the time when the thermistor sensor at 0.5 m below ground surface detected temperature below 0°C in the autumn to the time when the thermistor sensor at 2.0 m below ground surface detected temperature changing from below to above zero degree Celsius in the spring. This period is displayed in light blue in **Figure 25**. We calculated the cumulative atmospheric precipitation for soil non-freeze and freeze periods, and these are illustrated in **Figure 23** via the dark blue bars. During the frozen soil period, snowmelt water and atmospheric precipitation cannot typically infiltrate to deep soil layers due to the seasonal frozen soil layer and **Figure 26** illustrates the pore water pressure recorded 19.4 m below the ground surface at point ZK4. At the time of drilling, water was first observed for point ZK4 at a depth of 3.8 m. However, the stable water level recorded during the study period was within the range of 15.8–16.6 m. Unsaturated soil was recorded at 3.8 m and 9.4 m below the ground surface during the study period. Measurements from the pore water pressure sensor placed at a 23.7 m depth were unstable near the end of the study period. Data collected from the sensor placed at 19.4 m depth were therefore utilized to reflect pore water pressure at point ZK4.

Figure 27 shows the displacement rates at point ZK1 and point ZK2 as well as the pore water pressure at point ZK4 over time. Displacement values collected via the RTK-GPS real-time dynamic deformation monitoring system were divided by the time interval to obtain displacement rates.

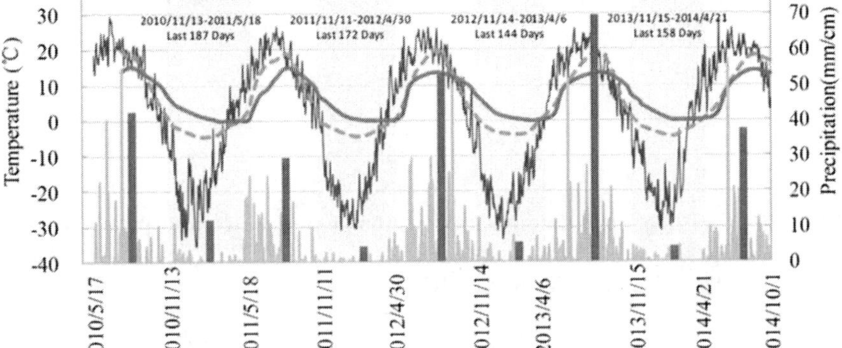

FIGURE 25. Daily average air temperature/precipitation/soil temperature in the depth of 50 cm, 200 cm in K178 + 530 section of Bei'an-Heihe Highway.

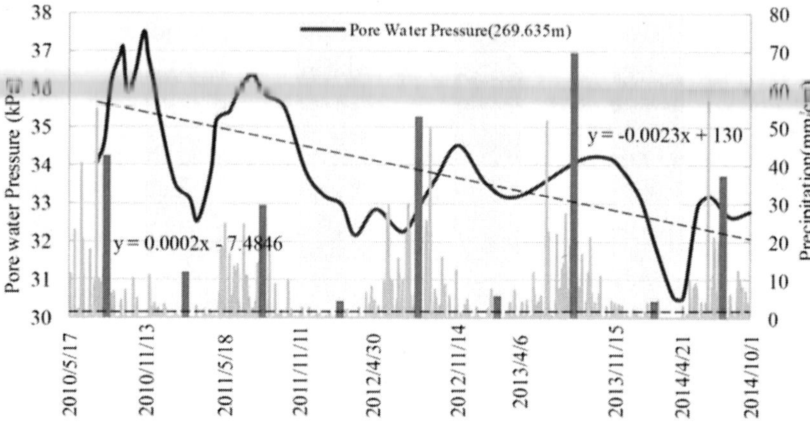

FIGURE 26. Daily precipitation and pore water pressure in the depth of 19.4 cm of ZK4 point in K178 + 530 section of Bei'an-Heihe Highway.

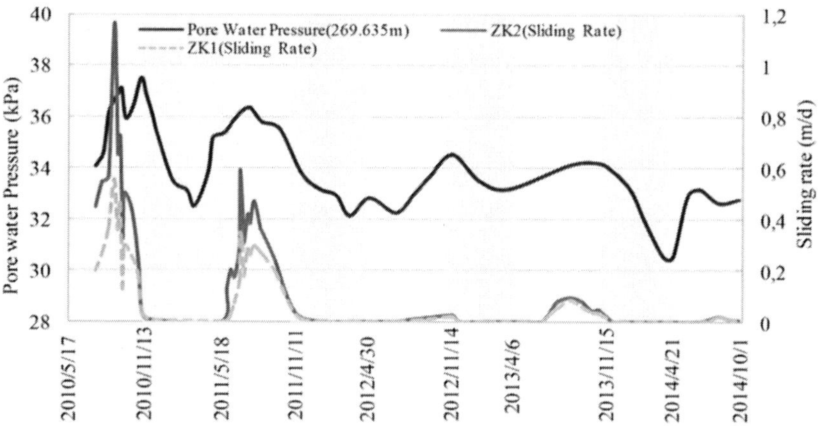

FIGURE 27. Horizontal displacement rates at point ZK1 and point ZK2 on the landslide mass and pore water pressure at point ZK4 (19.4 m depth) in K178 + 530 section of Bei'an-Heihe Highway.

8. RESULTS, ANALYSIS, AND DISCUSSION OF DATA

The permafrost distribution determined by the present study is consistent with previous investigations [44, 45]. In the current study, we analyzed meteorological data from Sunwu County (collected 30 km to the south from the study area) to examine the climate change. The causes of landslide in the study area were evaluated preliminarily by means of permafrost distribution, geological survey of the study area, and topography. The roadbed instability that occurred during August 2000 in the section K176 + 900 to 178 + 200 of the Bei'an-Heihe secondary road was used as a case study.

The meteorological data from Sunwu County indicate that the variation in atmospheric precipitation was 8.93% during the 60 year period from 1954 to 2013. In the study area, this variation had a negligible long-term impact on the thermal state of the soil. As depicted in **Figure 2**, the increase in the average annual maximum temperature was substantially lower than the increase in the average annual minimum temperature. This indicates that the primary cause of the temperature increase was due to a reduction in the soil heat demand from the atmosphere and a rise in the soil's ability to radiate heat rather than an increase in the amount of direct radiation [46]. The latent heat of permafrost thaw caused the difference in the amplitude of temperature rise between the average annual minimum temperature and the average annual maximum temperature. The fact that these amplitudes differed by 47.05% over a 60 year period indicates rapid degeneration of the permafrost in the study.

Permafrost is not found in the section K176 + 900 to 178 + 200 (where overall roadbed instability occurred) of the previous Bei'an-Heihe secondary road (**Figure 3**). This is because the roadbed instability occurred in August 2000, while the ground temperature data used for **Figure 3** represent e ETM+ data collected by Landsat7 in 2009. The permafrost in this road section had entirely degraded entirely after nine years. In **Figure 3**, there are permafrost areas along the road on both sides of point ZK2, i.e., the right side of the K178 + 530 road section. However, no permafrost was found in borehole ZK2. Additionally, no aberrant data was detected below point ZK2 within 15 m along line III via the GPR and high-density resistivity methods. This suggests the absence of permafrost. This could be due to severe degradation of permafrost at this location was at a stage of severe degradation. In other words, the phase transition of soil moisture had already occurred with the associated transition to warm permafrost or melting permafrost. It could also be associated with the resolution of the ETM+ data from the Landsat7 satellite.

Figure 25 shows the relatively concentrated time period of the initial soil freezing in the study area. This typically fell in within 11 November to 15 November. However, the thawing period required for seasonal frost exhibited large annual variations at the same measurement point. This could be related to the level of winter snowfall and autumn precipitation in the previous year. It could also be related to the soil water thermal condition proximate to the maximum seasonal frost depth. However, these potential relationships require further investigation.

As shown in **Figure 26**, the pore water pressure at point ZK4 exhibited annual periodic changes. The pore water pressure began to increase in the spring of each year, reached the maximum value in the summer, and then subsequently decreased. At an interannual time scale, there was an overall downward trend in pore water pressure. This trend exhibited no correlation with atmospheric precipitation. Linear analysis indicates that the atmospheric precipitation in the study area rose by 0.318 mm during the study period, however, the pore water pressure at point ZK4 declined by 3.675 kPa. This resulted in a 36.75 cm decrease in its groundwater level. The groundwater reduction level was 1155 times the rise in atmospheric precipitation for the same period.

The slip rates measured at points ZK1 and ZK2 agree well with the pore water pressure results from point ZK4. During the study period, the landslide mass began to slip with thawing of seasonal frost and snowmelt each spring due to the gradual increase in the pore water pressure of the soil. The pore water pressure reached annual the peak during the concentrated rainfall of the summer, and the slip rate of the landslide obtained its maximum value at the same time. The soil pore water pressure gradually decreased with the landslide movement, and the sliding of the slope gradually stalled. The landslide movement exhibited notable seasonal activity and annual periodicity.

Figures 23 and **27** illustrate consistent trends in the slip rates of points ZK1 and ZK2 on the landslide mass. The displacement of point ZK2 was 1.65 times the movement at point ZK1 during the monitoring period. We conjecture that the displacement at point ZK1 might be due to force coming from the rear area of the landslide mass. For the period of 2010–2014 the annual maxima of pore water pressures at point ZK4 were 37.52 kPa, 36.37 kPa, 34.53 kPa, 34.14 kPa, and 33.19 kPa, respectively. The maxima decline each year, and the decline in amplitude was relatively large in 2012. The maximum slip rates of point ZK2 from 2010 to 2014 were 115.7 cm/d, 59 cm/d, 2.55 cm/d, 9.63 cm/d, and 2.2 cm/d, respectively. Rapid sliding primarily occurred in the summer of 2011 and in 2010. There were decreases in the slip rate each year, and the largest reduction occurred in 2012. However, there was a slight rise in the slip rate in the summer of 2013. The underlying cause of these phenomena may be related to thawing permafrost which supplemented water to the landslide mass at points E and F (**Figure 22**), which are exterior to the lateral edge of the slope at point ZK2. Additionally, these phenomena could be related to the level of summer precipitation. As shown in **Figure 26**, the summer atmospheric precipitation in 2013 reached the largest value for the entire monitoring period (672 mm). This level was 14.33 mm higher than the second largest summer precipitation (which occurred in 2012) and it was 1.46 times the average summer atmospheric precipitation for the entire monitoring period. We therefore concluded that the slip rate of Landslide K178 + 530 during the monitoring period was mainly impacted by pore water pressure at the trailing edge of the landslide mass. At the same time, the slip rate was also impacted by the level of summer precipitation. Furthermore, the thawing of permafrost at the lateral edge of the slope was related to the pore water pressure at the trailing edge of the landslide mass. As permafrost thawed each year, its volume was reduced. This trend gradually decreased the pore water pressure at the trailing edge of the slope. Consequently, there were gradual decreases in the slip rate of the landslide mass each year.

9. CONCLUSION

In this study, we performed a comprehensive analysis via data on the study area climate as well as ground temperature, displacement, and pore water pressure data that were collected at transect points along Landslide K178 + 530 on the Bei'an-Heihe Highway. We arrived at the following main conclusions based on this analysis:

The northwest section of China's Lesser Khingan Range is positioned along the southern boundary of the high-latitude permafrost region. This area has been subject to temperature increases due to global climate change. The increase in yearly average minimum temperature is 3.45 times the increase in the average yearly maximum temperature. Permafrost degradation in the region is quite significant. There will be additional temperature increases in this region due to permafrost degradation. These patterns will also result in an acceleration of the northward migration of the southern boundary of the permafrost region.

Permafrost thaw results in increases in the number of landslides in the northwest section of China's Lesser Khingan Range. This subsequently results in regional topographical changes. Anthropogenic activity can accelerate the thawing of permafrost.

Infiltration of concentrated summer precipitation and water seepage from thawing permafrost both increase the moisture content in the soil of local hillsides. In the northwest section of the Lesser Khingan Range in China, this is the main casual factor for landslides.

Landslide movement in the northwest section of China's Lesser Khingan Range begins in the summer season with concentrated precipitation. The movement gradually stops when the soil freezes in the autumn. The landslide movement exhibits clear seasonal and annual periodicity.

The thawing process of permafrost on and near the landslide mass control the landslide slip rate and movement process in the northwest area of the Lesser Khingan Range. As the permafrost thaws each year and the volume is correspondingly reduced, there is an associated decrease in its ability to supplement water to the landslide mass. This results in a decrease in the landslide slip rate each year until it reaches zero. The landslides then tend towards stability.

The soil resistivity values above and below the sliding surface of the landslide mass are clearly different and exhibit an abrupt stratification. There is apparent resistivity layering at the position of the sliding surface in the landslide mass, the resistivity value decreased suddenly. In the profile of the GPR, the sliding surface is expressed as a low-frequency high-amplitude sync-phase axis, and there is an associated rapid rise in the radar wave amplitude. In practice, the abnormal abrupt changes in the GPR and HDR results can be used in this region to identify the sliding surface location of shallow landslides.

The results from the three methods we used to investigate landslides in the study area, i.e., HDR, GPR, and drilling, generally agree with regard to the sliding surface position. This indicates that the GPR and HDR methods are economical, rapid, and reliable methods for prospecting in landslide areas. These methods can be applied to shallow landslides in high-latitude permafrost regions for accurate and rapid identification of the sliding surface position. These techniques can also aid in the generation of accurate data for engineering projects to ensure that appropriate measures are adopted.

10. ACKNOWLEDGEMENTS

We thank the Science and Technology Project of the Chinese Ministry of Transport (2011318223630) and the International Landslide Research Program (IPL-167) for funding support. We are also grateful to an anonymous referee who helped improve the manuscript.

REFERENCES

1. Guo D X, Wang S L, Lu G, Dai J B, and Li E Y. Regionalization of permafrost in the Da and XiaoXing'anling Mountains in northeastern China. Journal of Glaciology and Geocryology. 1981; 3: 1–9.
2. Zhou Y W, Wang Y X, Gao X W, and Yue H S. Ground temperature, permafrost distribution and climate warming in northeastern China. Journal of Glaclology and Gecryology. 1996; 18: 139–147.

3. Sun G Y, Yu S P, and Wang H X. Causes south borderline and subareas of permafrost in Da Hinggan Mountains and Xiao Hinggan Mountains. Scientia Geographica Sinica. 2007; 27: 68–74.
4. Jin H J, Li S X, Wang S L, and Zhao L. Impacts of climatic change on permafrost and cold regions environments in China. Acta Geographica Sinica. 2000; 55: 161–173.
5. Wei Z, Jin H J, and Zhang J M. Prediction of permafrost changes in Northeastern China under a changing climate. Science China: Earth Sciences. 2010; 41: 74–84.
6. He R X, Jin H J, Lv L Z, Yu S P, Chang X L, Yang S Z, Wang S L, and Sun G Y. Recent changes of permafrost and cold regions environments in the northern part of Northeastern China. Journal of Glaciology and Geocryology. 2009; 31: 525–531.
7. Solomon S, Qin D, and Manning M. Summary for policymakers. Climate Change 2007: The physical science basis (Contribution of Working Group I to the Fourth Assessment Report of the Intergovernmental Panel on Climate Change). Intergovernmental Panel on Climate Change. 2007 (www.ipcc.ch/).
8. IPCC. Summary for policymakers. Working group I contribution to the IPCC fifth assessment report climate change 2013: the physical science basis. Cambridge, UK. Cambridge University Press; 2013.
9. Ding Y H, Ren G Y, and Shi G. National assessment report of climate change (I): climate change in China and its future trend. Advances in Climate Change Research. 2006; 2: 3–8.
10. Shi P J, Sun S, and Wang M. Climate change regionalization in China (1961-2010. Science China: Earth Sciences. 2014;44: 2294–2306.
11. Chang X L, Jin H J, Yu S P, Sun H B, He R X, Luo D L, Sun G Y, and Lu L Z. Influence of vegetation on frozen ground temperatures the forested area in the Da Xing'anling Mountains, Northeastern China. Acta Ecologica Sinica. 2011; 31: 5138–5147.
12. Zhou Y W, and Guo D X. Principal characteristics of permafrost in China. Journal of Glaciology and Geocryology. 1982; 4: 1–19.
13. Chang X L, Jin H J, and He R X. Review of permafrost monitoring in the northern Da Hinggan Mountains, Northeast China. Journal of Glaciology and Geocryology. 2013; 35: 93–100.
14. Jin H J, Wang S L, and Lv L Z. Features of permafrost degradation in Hinggan Mountains, northeastern China. Scientia Geographica Sinica. 2009; 2: 223–228.
15. Arndt D S, Blunden J, and Willett K. State of the Climate in 2011 Special Supplement to the Bulletin of the American Meteorological Society. Bulletin of the American Meteorological Society. 2012; 93(7).
16. Eu-Fp7. A summary for policymakers. PD Investigación. ACQWA: Assessing Climate impacts on the Quantity and quality of Water. A large integrating project under EU Framework Programme 7 (FP7). 2008.
17. Sassa K, Yin Y, and Canuti P. The Third World Landslide Forum, Beijing, 2014. Landslides. 2015; 12(1): 1–16 (http://iplhq.org/category/home/)
18. Sassa K. ICL strategic plan 2012–2021—To create a safer geo-environment. Landslides. 2012; 9(2): 155–164 (http://iplhq.org/category/home/).
19. Guo Y, Canuti P, Strom A, Hideaki M, and Shan W. The First Meeting of ICL Landslides in Cold Regions Network, Harbin, 2012. Landslides. 2013; 10: 99–102.
20. Shan W, Guo Y, Zhang C C, Hu Z G, Jiang H, and Wang C J. Climate-change impacts on embankments and slope stability in permafrost regions of Bei'an-Heihe Highway. Landslide Science for a Safer Geoenvironment. 2014; 1: 155–160.
21. Shan W, Guo Y, Wang F, Marui H, and Strom A. Landslides in cold regions in the context of climate change (Springer: Environmental Science and Engineering). ISBN 978-3-319-00866-0. ISBN 978-3-319-00867-7 (eBook). 2014;105–109.
22. Grab S W, and Linde J H. Mapping exposure to snow in a developing African context: implications for human and livestock vulnerability in Lesotho. Natural Hazards. 2014; 71: 1537–1560.
23. Ballantyne C K, Sandeman G F, Stone J O, and Wilson P. Rock-slope failure following Late Pleistocene deglaciation on tectonically stable mountainous terrain. Quaternary Science Reviews. 2014; 86: 144–157.

24. Nussbaumer S, Schaub Y, Huggel C, and Nat AW. Risk estimation for future glacier lake outburst floods based on local land-use changes. Hazards and Earth System Sciences. 2014; 14: 1611–1624.
25. Haeberli W. Mountain permafrost – research frontiers and a special long-term challenge. Cold Regions Science and Technology. 2013; 96: 71–76.
26. Kliem P, Buylaert J P, Hahn A, Mayrd C, Murray A S, Ohlendorf C, Wastegård S, Zolitschka B, and Team T P S. Magnitude geomorphologic response and climate links of lake level oscillations at Laguna Potrok Aike, Patagonian steppe. Quaternary Science Reviews. 2013; 71: 131–146.
27. Fischer L, Hugge C, Kaab A, and Haeberli. Slope failures and erosion rates on a glacierized high-mountain face under climatic changes Earth Surf. Landforms. 2013; 38: 836–846.
28. Starnberger R, Drescher S R, Reitner J M, Rodnight H, Reimer P J, and Spötl C. Late Pleistocene climate change and landscape dynamics in the Eastern Alps: the inner-alpine Unterangerberg record (Austria) Quaternary Science Reviews. 2013; 68, 17–42.
29. Stoffel M, Tiranti D, and Huggel C. Climate change impacts on mass movements – case studies from the European Alps. Science of the Total Environment 2014; 22014: 1255–1266.
30. Wang C J, Shan W, Guo Y, Hu Z G, and Jiang H. Permafrost Distribution Study Based on Landsat ETM+ Imagery of the Northwest Section of the Lesser Khingan Range. Landslide Science for a Safer Geoenvironment. 2014a; 3: 529–534.
31. He R X, Jin H J, Chang X L, Lv L Z, Yu S P, Yang S Z, Wang S L, and Sun G Y. Degradation of permafrost in the northern part of Northeastern China: Present state and causal analysis. Journal of Glaciology and Geocryology. 2009; 31: 829–834.
32. Zhang Y, Wu Q B, and Liu J P. Distribution characteristics of the permafrost in the section from Heihe to Bei' an in the Xiao Hinggan Mountains. Journal of Glaciology and Geocryology. 2001;23:312–317.
33. Wang B, Sheng Y, and Liu J P. Distribution and degradation of permafrost in Xiao Hinggan Mountains along the Heihe-Dalian Highway. Journal of Glaciology and Geocryology. 2001; 23: 302–306.
34. Wang S H, Qi J L, Yin Z Y, Zhang J M, and Ma W. A simple rheological element based creep model for frozen soils. Cold Regions Science and Technology Cold Regions Science and Technology. 2014; 106: 47–54.
35. Loke M H, and Barker R D. Rapid least-squares inversion of apparent resistivity pseudosections using a quasi-Newton method. Geophys Prospect. 1996; 44: 131–152.
36. Tripp A C, Hohmann G W, and Swift C M. Two-dimensional resistivity inversion. Geophysics. 1984; 49: 1708–1717.
37. Sasaki Y. Resolution of resistivity tomography inferred from numerical simulation. Geophys Prospect. 1992; 40: 453–463.
38. Coggon J H. Electromagnetic and electrical modelling by the finite element method. Geophysics. 1971; 36: 132–155.
39. Dey A, and Morrison H F. Resistivity modelling for arbitrarily shaped two-dimensional structures. Geophys Prospect. 1979; 27: 106–136
40. Shan W, Hu ZG, Guo Y, Zhang C C, Wang C J, Jiang H, Liu Y, and Xiao J T. The impact of climate change on landslides in southeastern of high-latitude permafrost regions of China. Frontiers in Earth Science. 2015; 3(7): 1–11.
41. Telford W M, Geldart L P, and Sheriff R E. Applied geophysics. Geological Journal. 1992; 27(1): 97.
42. Daniels D J. The Institution of Electrical Engineers, London, United Kingdom, editors. Ground Penetrating Radar. 2nd ed. London: IEE Radar; 2004.
43. Benedetto A, Benedetto F, and Tosti F. GPR applications for geotechnical stability of transportation infrastructures. Nondestructive Testing and Evaluation. 2013; 27(3): 253–262.

44. Wang C J, Shan W, Guo Y, Hu Z G, and Jiang H. Permafrost distribution study based on Landsat ETM+ Imagery of the northwest section of the lesser Khingan Range. Landslide Science for a Safer Geoenvironment. 2014; 3: 529–534.

45. Wang C J, Shan W, Guo Y, Hu Z G, and Jiang H. Permafrost distribution research based on remote sensing technology in northwest section of Lesser Khingan Range in China. Engineering Geology for Society and Territory. 2015; 1: 285–290.

46. Chen J, Sheng Y, and Cheng G D. Discussion on protection measures of permafrost under the action of engineering from the point of earth surface energy balance equation in Qinghai Tibetan Plateau. Journal of Glaciology and Geocryology. 2006; 28: 223–228.

CHAPTER 2

Shallow Landslides as Drivers for Slope Ecosystem Evolution and Biophysical Diversity

Alejandro Gonzalez-Ollauri [1], Slobodan B. Mickovski [1]

[1].School of Engineering and Built EnvironmentGlasgow Caledonian UniversityGlasgowUK

ABSTRACT

Shallow landslides may be seen as local disturbances that foster the evolution of slope landscapes as part of their self-regulating capacity. Gaining insight into how slope ecosystems function and evolve could make eco-engineering interventions on slopes more successful. The objective of the present study is to detect traits of shallow landslide-triggered ecosystem evolution, self-regulation and biophysical diversity in a small-scale landslide-prone slope in Northeast Scotland. A protocol was defined to explore the emergence of landslide-driven slope habitats. This protocol studied plant diversity, species richness and plant biomass differences and their interactions with certain soil and topographic attributes at three slope strata during two consecutive growing seasons following an assemblage of shallow landslide events. Plant species and soil properties with potential as indicators of the different landslide-driven slope habitats and landscape evolution were also considered. Shallow landslides contributed to biophysical diversity and created distinct slope habitats within the landscape. Habitat differences in terms of species richness and composition were a direct consequence of the slope self-regulation. Certain plant species were found to be valid indicators of landslide-driven biophysical diversity. Soil total nitrogen and resistance to penetration were related to slope habitat and landscape evolution. As expected, plant establishment relied upon light and nitrogen trade-offs, which in turn were influenced by landscape topography. The insights derived from this study will be useful in slope restoration, particularly in harmonising effective actions with the functioning of landslide-prone ecosystems. Further research directions to clarify the observed variability and interactions are highlighted.

Keywords: Shallow landslide Ecosystem Self-organisation Plant diversity Biophysical diversity Slope restoration

INTRODUCTION

Landslides are normally seen as catastrophic geomorphological processes that lead to dramatic losses of soil, human property and life globally. However, from an ecological perspective, they are natural disturbance episodes of varying frequency and intensity that contribute to the natural evolution of sloped ecosystems (Walker and Shield 2013). The ecology of landslides has been relatively well studied (see Walker and Shield 2013 for review), but it is still an emergent discipline. During a landslide, nutrient-rich soil materials usually move downwards due to the action of gravity (Guariguata 1990; Walker et al. 1996). This leads to the differentiation of clear slope zones on the basis of the accumulation of soil materials (Velázquez and Gómez-Sal 2008; Elias and Dias 2009; Walker et al. 2009; Neto et al. 2017). Additionally, the biophysical diversity of the landslide-prone landscape increases (Geertsema and Pojar 2007) due to marked changes that can occur in soil properties and vegetation cover (Shiels et al. 2008; Elias and Dias 2009) and the emergence of novel topographical shapes (Cendrero and Dramis 1996).

Shallow landslides increase the openness (Odum 1969) of the landscape by displacing the topsoil and established vegetation downwards on the slope at the time of failure (Walker et al. 1996). This is evident in the bare scars and patches produced after landslides. These landscape gaps present unique ecological features (e.g. areas with low levels of light competition between plant individuals; Walker et al. 1996; Myster and Walker 1997) for colonisation by plants present within the surrounding landscape (Velázquez and Gómez-Sal 2008). Thus, a given slope will tend to self-organise after a landslide (Walker et al. 1996) and the most visible evidence of this can be found in a slope's re-colonisation by vegetation (Velázquez and Gómez-Sal 2008) and its subsequent succession (Myster and Walker 1997; Walker et al. 2010). These processes tend to be mimicked in human-driven slope restoration actions (e.g. Norris et al. 2008), but the process of slope self-regulation after failure and ecological factors leading to successful slope restoration (other than soil-root reinforcement; e.g. Stokes et al. 2008) need further exploration (Restrepo et al. 2009). Moreover, there is a need to establish simple and effective protocols capable of capturing the biophysical diversity and self-organisation triggered by landslides while providing interpretable and useful information for slope restoration.

Post-landslide self-organisation processes depend on the emergent biophysical heterogeneity within slope ecosystems (Velázquez and Gómez-Sal 2007). Knowledge about landslide-derived biophysical diversity may improve the success of slope restoration actions using vegetation (Walker et al. 2009). For example, insight related to the post-disturbance emergence of distinct slope zones, or habitats (Velázquez and Gómez-Sal 2008; Neto et al. 2017), and the associated colonising plants may aid in the choice of different restoration strategies within the same slope. Insight related to the environmental factors governing post-slide plant diversity, succession and establishment may provide useful information to landslide engineers and restoration ecologists in order to achieve the goals of slope restoration actions (e.g. diverse plant cover, dense plant cover, etc.) or to monitor the subsequent slope evolution. However, plant interactions with the post-landslide abiotic slope features must be clarified to enhance the slope restoration success. These interactions are not yet entirely understood (Rajaniemi et al. 2003; Restrepo et al. 2009), and tools to adequately detect and interpret them are needed.

Plant-environment relationships can be complex. For example, plants need to compensate for dramatic differences in resource (i.e. energy, water and nutrients) availability across the environment in order to thrive (Chapin et al. 1987; Herbert et al. 2004). In particular, trade-offs between light and nitrogen seem to be the key controls in the establishment of plant communities after disturbance (Walters and Reich 1997; Myster and Walker 1997; Soto et al. 2017). Nitrogen-rich soils tend to foster the production of plant aboveground biomass (Urbano 1995). This may enhance the ability of vegetation to intercept light (Wilson and Tilman 1995; Walters and Reich 1997), but it also may promote plant competition for aboveground resources such as light (Rajaniemi et al. 2003; Herbert et al. 2004), leading to shifts in the composition and diversity of plant communities (e.g. Wilson and Tilman 1995). Light and nitrogen availability may be governed by landslide dynamics, but also by the landscape topography (e.g. Franklin 1995; Hood et al. 2003; Wang et al. 2015). For instance, slope orientation (i.e. aspect or azimuth) will determine the amount of incoming energy (e.g. Gallardo-Cruz et al. 2009), which in turn will affect photosynthesis, evapotranspiration, soil temperature and soil moisture content—all crucial variables for plant development. Slope surface curvature affects plant performance (e.g. Moeslund et al. 2013) by its influence on subsurface water flow, litter accumulation and soil erosion/deposition rates which, in turn, are related to soil depth and texture, water holding capacity and nutrient availability (Heimsath et al. 1997). A slope gradient greater than 45° will hinder the establishment of vegetation (Bochet and Garcia-Fayos 2004), in part not only because such slope will be prone to instability (i.e. more frequent disturbance; Velázquez and Gómez-Sal 2008; Lu and Godt 2013), but also because the nutrients will likely be washed down from steeper zones and accumulate in flatter areas (Heimsath et al. 1997; Hood et al. 2003; Bertoldi et al. 2006).

The study of the complex interactions between plants and their environment requires the use of advanced statistical tools (e.g. principal component analysis; Borcard et al. 2011). As indicated earlier, the challenge includes the need to clarify and interpret the results of statistical analyses to provide understandable and practical information to landslide engineers and restoration ecologists. The advent of machine learning techniques, such as regression trees (Breiman et al. 1984), may aid in the interpretation of complex interactions between plants and their environment (e.g. Kallimanis et al. 2007; Revermann et al. 2016). Regression trees have four major benefits: (1) They are more powerful than classical multiple regressions (as the relationship between response and predictor variables is not prespecified), (2) they can effectively model the change in direction of the effect of one factor in response to the levels of another factor, (3) they are more interpretable than the commonly used ordination techniques, and (4) they are able to depict the hierarchical structure by which ecosystems are built (Jørgensen 2007).

The goal of this study is to detect early traits of landslide-triggered ecosystem evolution, self-regulation and biophysical diversity in a small-scale landslide-prone coastal slope in Northeast Scotland. To achieve this, a novel protocol was defined for exploring the emergence of landslide-driven slope habitats. Plant diversity, species richness and plant biomass differences together with their interactions with certain soil and topographic attributes at three different slope strata are studied during two consecutive growing seasons following an assemblage of shallow landslide episodes. The protocol also assesses potential plant indicator species for the different landslide-driven slope habitats and studies soil properties indicative of slope habitat and landscape evolution. In the light of the findings and observations, recommendations related to slope restoration actions are proposed.

Materials and methods

Study site

Plant and biophysical diversity were studied on an accessible section of a landslide-prone slope during two consecutive growing seasons (2014 and 2015) following an assemblage of shallow landslide events. The slope is located adjacent to Catterline Bay, Northeast Scotland, UK (WGS84 Long −2.21 Lat 56.90; Fig. 1a), where multiple small-scale shallow landslides (ca. 100–300 m^2; Fig. 1a) occurred in April 2013 (Kincardineshire Observer 11/4/2013) as a result of a period of prolonged rainfall. Consequently, a number of failure zones appeared distributed close to each other (i.e. less than ca.10 m apart) across the studied slope (Fig. 1a); their scars were easily identifiable as exposed bare ground or areas of sparse vegetation. The surveyed section of the slope was mainly vegetated with early successional plant communities (i.e. herbs and grasses; Fig. 1b, c) surrounded by a mixture of areas dominated by herbaceous weeds and grasses, riparian trees and shrubs (e.g. willow, sycamore, ash, hawthorn) and agricultural crops of wheat and barley.

Figure 1. A Study site location and area of study (*dashed frame*). *White lines* indicate the location of shallow landslides occurring in April 2013. **B** Slope strata (i.e. crest, middle and toe) and sampling direction within each stratum (*white arrows*). **C** Section of the studied slope where the different slope strata are indicated. Aerial image source: GetMapping (2014)

The study site has a mean annual temperature of 8.0 °C and mean annual rainfall of 1232 mm (UK Met Office 2015), characteristic of a humid temperate climate site (Cfc: subpolar oceanic climate; Köppen 1884). The topography of the site is dominated by sloped (25–50°) terrain and cliffs running into the North Sea (Fig. 1), where often shallow (ca. <2 m), silty sand soils (sand 79.82%; silt 5.85%; clay 3.08%; BS 1377-2: 1990) can be found resting on sedimentary bedrock (i.e. conglomerate; BGS 1999). Overall, the studied slope is facing east, southeast and south directions (Fig. 1a).

Sampling approach

The studied slope was divided prior to sampling into three slope strata of similar width—i.e. crest (*c*), middle (*m*) and toe (*t*) (Fig. 1b, c). The criteria employed to stratify the slope were the accumulation of slope-forming materials following landslides and the recurrence of shallow landslide (i.e. disturbance) events. The feature used to delineate the strata was the presence or absence of landslide scars (Fig. 1c). The middle and toe strata (Fig. 1b, c) were delineated as the initiation and deposition zones, respectively, of the shallow landslides occurring in April 2013. The middle stratum was delimited from the toe for presenting visible landslide scars (Fig. 1c). The crest stratum, comprising the top part of the studied slope (Fig. 1c, d), remained stable during the 2013 landslide episode and has been disturbance-free for a longer time. This can be seen from the absence of landslip scars and from the establishment of dense patches of shrub species such as gorse, hawthorn and willow in some zones (Fig. 1b, c).

A stratified haphazard sampling design (e.g. Hall et al. 2013) was implemented to collect vegetation samples within each predefined slope stratum at the apices of the 2014 and 2015 growing seasons (i.e. mid-July). A 1-m buffer zone from each boundary was delimited within each stratum to avoid edge effects upon sampling. The vegetation samples were obtained by using a 0.5-m^2 (1.0 m × 0.5 m) aluminium frame (Fig. 2a). This was placed with the long side facing the top of the landslide (Myster and Walker 1997) on locations (i.e. sampling units or quadrats) chosen by throwing the aluminium frame aimlessly within each slope stratum. The quadrats were spaced approximately 10 m from each other along the direction of the contour lines within each slope stratum (Fig. 1b), ensuring an adequate spatial coverage of the studied slope. The whole slope section annotated in Fig. 1a was sampled, as landslide scars were observed throughout. All of the plant material found within the aluminium frame (i.e. quadrat) was clipped to the ground surface (USDA-NRCS 1997) using a sickle. The harvested vegetation was tied into bundles and transported to the laboratory for further processing (see "Plant identification, dry biomass determination and species abundance" section). In total, 59 quadrats were surveyed: 25 in 2014 ($c = 8$; $m = 8$; $t = 9$) and 34 in 2015 ($c = 12$; $m = 9$; $t = 13$).

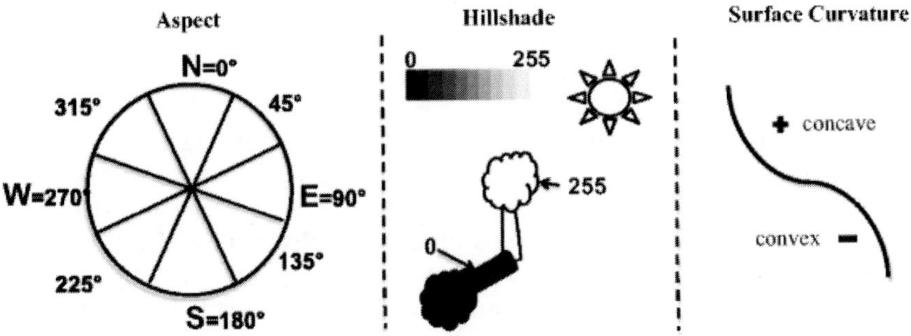

Figure 2. Explanatory illustration for the topographic attributes. Aspect or azimuth is the compass direction that the slope faces. Hillshade is derived from the aspect and determines which areas of the terrain are shaded in relation to a predefined position and inclination of the sun. A higher hillshade value refers to more sun exposure. The surface curvature is the amount by which the terrain surface deviates from being flat, acquiring a positive value when concave and negative when convex

The geographical position of each quadrat was recorded with a Garmin Monterra™ GPS. Soil samples were collected with a core auger from each quadrat at a depth ranging between 0 and 15 cm below the ground level (b.g.l.) for determination of soil organic matter (SOM; Lost on Ignition Method; Schulte 1996) and soil total Kjeldahl nitrogen (TKN; Bremner 1965). The soil resistance to penetration (SRP; kPa) was measured at each quadrat with a static cone penetrometer (60° vertex angled cone; 1.5-cm^2 basal cone area; BS 1377-4: 1990) at four different soil depths (i.e. 0, 15, 30 and 40 cm b.g.l.), from which the mean soil SRP was obtained. Each soil depth was reached by carefully removing the overlaying soil with a soil core auger, before pushing the cone into the soil up to its base and reading the stress (i.e. SRP) from the dial gauge.

The slope gradient of each quadrat was measured with a hand inclinometer while the topographic attributes aspect, hillshade (azimuth: east—i.e. morning sun position, when clear skies were more frequent during the sampling period; solar angle 45°—i.e. mean solar angle during the sampling period) and profile curvature (Fig. 2) were retrieved from a high-resolution (point spacing 2 × 2 m), photogrammetrically derived digital surface model (GetMapping 2014) using the recorded GPS coordinate points. The topographic attributes were used for analysis of the effects of topography on plant diversity metrics (see "Plant diversity metric determination" and "Environmental covariate effect on plant diversity, species richness and biomass" sections).

Plant identification, dry biomass determination and species abundance

The plant material (vascular plants only) harvested in each bundle (see "Sampling approach" section) was separated by hand into species for subsequent identification. The identification process was carried out to the family and species level using a plant identification key (Bonnier and De Layens 2002) and the Atlas of the British and Irish flora (BRC 2016). The vegetation remnants that could not be classified (e.g. remains of grass, herbs leaf and litter) were pooled together and labelled as mixed biomass. All plant material was placed on aluminium trays tagged with the species and location and oven-dried at 70 °C for 48 h, after which the vegetation dry biomass was recorded using a digital scale (±0.1 g). The species abundance (Ab; %) at each quadrat was then calculated as the percentage of dry biomass of a given species with respect to the total biomass within a given quadrant (i.e. sample unit), including the mixed biomass.

Plant diversity metric determination

Plant diversity was assessed by estimating the Shannon-Wiener index ($H'=-\sum_{i=1}^{S} p_i \ln p_i$; p_i: proportion of biomass belonging to the ith species; S: total number of species; Shannon and Weaver 1964) and Pielou's evenness ($J' = H'/H'_{max}$; Mulder et al. 2004) and by computing the cumulative and average species richness (i.e. α-diversity; Magurran 2004) for the three considered slope strata (i.e. crest, middle, toe). To do so, species presence-absence contingency tables (i.e. incidence matrix; e.g. Gotelli and Chao 2013) were built for computing H' and species richness with the packages 'vegan' (Oksanen et al. 2016) and 'rich' (Rossi 2011), respectively, in the statistical software R 3.2.1 (R Core Team 2015). The cumulative species richness (S) was computed as the cumulative richness encountered over the quadrats belonging to a given slope stratum and sampling year. Species richness per unit area (Sa) was obtained by dividing S over the total sampled area per slope stratum and year (i.e. number of quadrats × 0.5 m^2; see "Sampling approach" section). The average species richness (Sav) was estimated as the mean richness over the sampled quadrats per stratum and year. To explore stratum differences in terms of species richness and assess consistency of sampling effort, sample-based rarefaction curves (Gotelli and Colwell 2011; Chao and Jost 2012) were implemented by plotting cumulative randomised species richness against bootstrapped (Efron 1979) sampling intensity for the three evaluated strata with the R package rich (Rossi 2011).

Differences between slope strata

Slope strata differences in terms of plant diversity (i.e. H' and J') were quantified using Kruskal-Wallis tests at 95 and 99% confidence levels in R 3.2.1 (R Core Team 2015). The same procedure was used to evaluate strata differences in terms of Sa, SRP, SOM, TKN and total aboveground plant dry biomass. Sampling year differences (i.e. 2014 vs. 2015) in terms of the previous variables were evaluated in the same manner. Differences in cumulative species richness were assessed by a randomisation (i.e. permutation) test (e.g. Richardson and Richards 2008) with the R package rich (Rossi 2011); this test does not control for differences in sampling effort (i.e. number of samples or sample units) between the three strata but instead returns the probability of different species richness among the compared strata. Significant differences were considered when the probability was beyond 95%. Moreover, differences in terms of the species composition and relative abundance distribution (i.e. β-diversity; Anderson et al. 2011) between the three considered slope strata were evaluated using the Bray-Curtis (B-C) similarity index (i.e.BC=1−(2cij/si−sj)BC=1−(2cij/si−sj); c_{ij}: number of common species between two sites and s_i and s_j: total number of species counted at both sites; Somerfield 2008), which ranges between 0 and 1 (i.e. a B-C value of 0 denotes identical sites).

Indicator species analysis

Potential plant indicator species for the three considered slope strata (i.e. crest, middle, toe) were identified by estimating the indicator value (Dufrene and Legendre 1997) of each encountered plant species with the R package 'indicspecies' (De Caceres and Legendre 2009). Indicator value (IndVal) is an index related to the species abundance (Ab) that measures the association between species and a particular habitat (De Caceres 2013) or stratum. Thus, indicator value is the product of species specificity (*A*) and fidelity (*B*) (Dufrene and Legendre 1997), where *A* is the probability that the surveyed site (i.e. sampling unit) belongs to the target habitat given the fact that the species have been found there and *B* is the probability of finding the species in units belonging to a particular habitat. Hence, indicspecies' functions first estimate IndVal between the species and each habitat and then look for the habitat corresponding to the highest association value. The statistical significance of this relationship is tested using a randomisation (i.e. permutation) test (De Caceres 2013). Species preferences for a given slope habitat, or stratum, were also evaluated with the previous approach. Additionally, the extent that a given habitat is covered by a plant indicator was quantified by estimating coverage (De Caceres et al. 2012), which is the proportion of sampling locations belonging to a given habitat where one or another plant indicator (i.e. a species or a species combination) is found.

Environmental covariate effect on plant diversity, species richness and biomass

The effects of the environmental covariates (i.e. soil resistance to penetration (SRP), soil organic matter (SOM), soil total Kjeldahl nitrogen (TKN), slope gradient, hillshade, aspect and profile curvature) on plant diversity (H'), species richness and plant dry biomass were first evaluated using Pearson's correlation test in R 2.3.1. To infer more complex and hierarchical effects of the environmental covariates and include slope stratum (i.e. factor or qualitative variable) as a predictor variable, regression trees (Breiman et al. 1984; Prasad et al. 2006) were fitted between the three considered response variables (i.e. H', species richness and plant dry biomass) measured in each quadrat and the environmental covariates using the R package 'rpart' (Therneau et al. 2015). In total, six regression trees were implemented: two for each response variable, with one set including SRP as covariate and only considering information collected during the 2015 campaign and another set without SRP and considering information from both years (i.e. 2014 and 2015) since SRP was only measured in 2015 and may obscure the effect of other relevant attributes. Outcomes derived from regression trees were plotted as dendrograms. The position of each node (i.e. tree leaf) within the tree, or dendrogram, is the hierarchical contribution of a given predictor variable (i.e. environmental covariate) to the prediction of the response variable (i.e. H', species richness or plant dry biomass). The terminal nodes of the dendrogram contain the mean response. The regression trees were pruned using a complexity parameter of 0.06 to avoid overfitting, reduce the number of final nodes (e.g. Prasad et al. 2006) and enhance the clearness of the outcomes. The goodness of fit from the regression trees was evaluated with a random holdback method (i.e. the model was fitted with 70% of the original data and then validated with the 30% remaining) by estimating the coefficient of determination (R^2) and residual mean square error (RMSE) and then plotting the objective functions between observed and predicted values (e.g. Malone 2013).

RESULTS

Plant diversity, species richness and slope strata

The dominant vegetation found on the studied landslide-prone slope (Fig. 1a) comprised herbs and grasses. The undisturbed portions of the slope (i.e. stable during the last landslide event and not surveyed herein) generally presented woody vegetation species such as willow, hawthorn, sycamore and ash.

Plant diversity (Table 1; Fig. 3a), evaluated in terms of H', ranged from 1.19 to 1.41 with a relatively high species evenness in all cases ($J' > 0.6$; Table 1) and did not present significant differences between the slope strata (H': $\chi^2 = 0.52$, $df = 2$, $p = 0.77$; J': $\chi^2 = 0.52$, $df = 2$, $p = 0.77$) or the years ($\chi^2 = 0.04$, $df = 1$, $p = 0.84$). However, the cumulative species richness (S; i.e. α-diversity; Table 1; Fig. 3b) showed an increasing trend down the slope strata gradient, with S being significantly higher ($p < 0.05$) at the toe (t) (Table 1; Fig. 3b) when compared to the crest (c) for the 2014 campaign (Table 1; Fig. 3b). The species richness per unit area (Sa; Table 1) did not present statistically significant differences between the strata ($\chi^2 = 2.57$, $df = 2$, $p = 0.28$), although the middle stratum showed the highest value for both sampling campaigns (Table 1). Overall, the toe stratum had more plant species in both years, particularly not only compared to the crest stratum (11 and 6 in total, 2014 and 2015, respectively) but also compared to the middle stratum (2 and 1 in total, 2014 and 2015, respectively). In the 2015 campaign, the middle stratum had in total seven species more than the crest. The species composition differences (i.e. β-diversity) between the slope strata tended to be more similar for the 2014 campaign (B-$C_{c\text{-}m}$ 0.24; B-$C_{c\text{-}t}$ 0.28; B-$C_{m\text{-}t}$ 0.39) than for the 2015 (B-$C_{c\text{-}m}$ 0.40; B-$C_{c\text{-}t}$ 0.52; B-$C_{m\text{-}t}$ 0.59), with the toe stratum being the most different in both cases. On the basis of the rarefaction curves (Fig. 3b), the sampling intensity was fair for the toe and middle strata; the rarefaction curves just started to reach the asymptote. For the crest stratum, however, the asymptote was reached, indicating that species pool was well covered with the sampling intensity employed (Fig. 3b).

Table 1. Mean values ± standard deviation for the assessed plant diversity metrics and soil environmental covariates for the studied slope strata and field campaigns

Habitat	Season	H'	J'	S	Sa (species m⁻²)	Sav	Biomass (g m⁻²)	SRP (kPa)	OM (%)	TKN (mg g⁻¹)
Crest	2014	1.35 ± 0.27	0.68 ± 0.13	18	4.5	1.3 ± 2.7	460 ± 211	–	4.03 ± 1.70	0.07 ± 0.05
	2015	1.19 ± 0.29	0.59 ± 0.14	20	3.3	1.82 ± 2.81	736 ± 182*	26.39 ± 7.15	3.91 ± 1.78	0.05 ± 0.04
Middle	2014	1.22 ± 0.47	0.61 ± 0.24	27	6.75	1.59 ± 3.56	931 ± 817	–	4.55 ± 0.45	0.12 ± 0.02
	2015	1.41 ± 0.27	0.70 ± 0.14	28	6.2	1.87 ± 3.51	535 ± 442	29.30 ± 9.02	4.27 ± 1.28	0.08 ± 0.03
Toe	2014	1.32 ± 0.55	0.66 ± 0.28	29*	6.4	2.01 ± 4.06	834 ± 639	–	4.14 ± 1.04	0.11 ± 0.04
	2015	1.32 ± 0.40	0.66 ± 0.20	27	4.15	2.16 ± 3.34	292 ± 249	36.14 ± 6.93**	3.77 ± 1.14	0.11 ± 0.04**

H' Shannon index, *J'* species evenness, *S* cumulative species richness over the total sampled area per stratum, *Sa* species richness per unit area of ground, *Sav* average species richness over the sampling units, *Biomass* aboveground plant dry biomass, *Penetrability* soil penetrability, *OM* soil organic matter, *TKN* soil total Kjeldahl nitrogen

*Statistically different at 95% significance; **statistically different at 99% significance

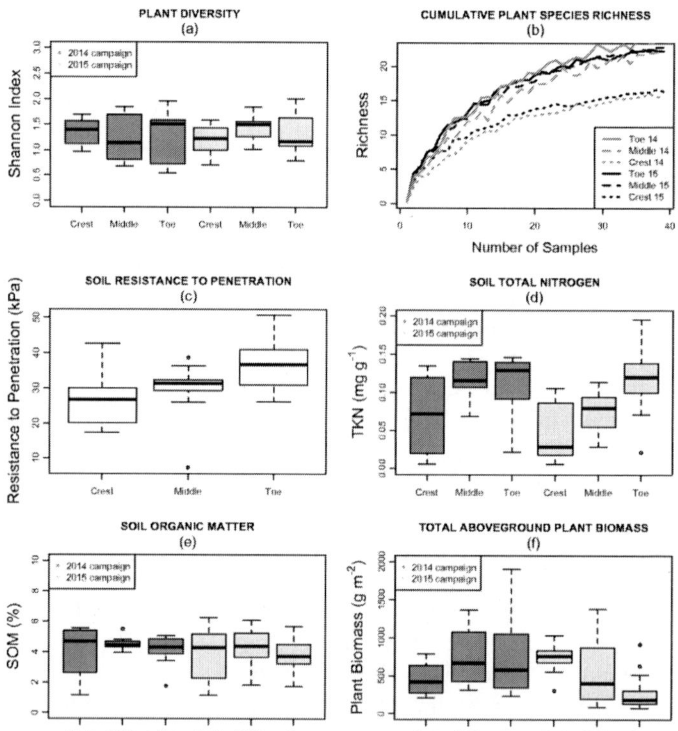

Figure 3. **A** Plant diversity boxplots indicating the variation in terms of the Shannon index (H') across the studied slope strata and field campaigns. **B** Rarefaction curves indicating the cumulative species richness for the studied slope habitats and field campaigns. **C** Soil resistance to penetration (SRP; kPa) boxplots indicating its variation between the studied slope strata for the 2015 field campaign. **D** Soil total Kjeldahl nitrogen (TKN; mg g^{-1}) boxplots indicating its variability across the studied slope habitats and field campaigns. **E** Soil organic matter (SOM; %) boxplots for the studied slope habitats and field campaigns. **F** Total aboveground plant dry biomass (g m^{-2}) boxplots for the studied slope habitats and field campaigns

Consistent stratum differences were also encountered in terms of SRP (Table 1; Fig. 3c), which was statistically higher (i.e. higher SRP) at the toe ($\chi^2 = 9.59$, $df = 2$, $p < 0.01$). The total soil nitrogen (TKN; Table 1; Fig. 3d) was statistically higher at the toe ($\chi^2 = 12.16$, $df = 2$, $p < 0.01$) in 2015, although it remained relatively unchanged during the studied years ($\chi^2 = 3.64$, $df = 1$, $p = 0.06$). Similarly, the total plant dry biomass (Table 1; Fig. 3f) was statistically higher ($\chi^2 = 11.96$, $df = 2$, $p < 0.05$) at the crest for 2015, although no differences were found between the sampled years ($\chi^2 = 2.08$, $df = 1$, $p = 0.15$). Conversely, the SOM did not show differences among the studied strata (SOM: $\chi^2 = 2.05$, $df = 5$, $p = 0.84$) or the years ($\chi^2 = 0.84$, $df = 1$, $p = 0.36$), although, in general, it showed higher levels at the crest (Fig. 3e).

Plant indicator species for slope habitats

The indicator species analysis detected six potential plant indicators (Table 2) out of 42 encountered plant species (i.e. γ-diversity) across 22 botanical families with different habitat (or stratum) preferences (Table 3). The crest and toe presented a single species each with high specificity (i.e. *A*) and moderate and low fidelity (i.e. *B*) and high and moderate coverage, respectively (Table 2). However, the middle habitat showed four potential indicator species with very high specificity but low fidelity and, in combination, relatively high coverage (Table 2). The toe habitat presented the highest number of species (i.e. 9) that were found only within it, while the middle and crest habitats only had two specific plant species each (Table 3). The two plant species with the highest encountered abundance over the entire study site (Fig. 1a) were *Arrhenatherum elatius* (Ab = 5.39%) and *Rumex obtusifolius* (Ab = 5.32%).

Table 2. Plant species indicators for the different studied slope strata or habitats. *A* species specificity, *B* species fidelity, *IndVal* species indicator value, *p value* statistical significance level, *Coverage* proportion of sampled quadrats belonging to a given habitat where one or another plant indicator (i.e. species or a species combination) was found. For the middle habitat, the combined coverage is shown

Habitat	Indicator	*A*	*B*	IndVal	*p* value	Coverage (%)
Crest	*Galium aparine*	0.83	0.65	0.74	0.001	80.00
Middle	*Lythrum salicaria*	0.76	0.65	0.70	0.001	76.47
	Chamerion angustifolium	1.00	0.18	0.42	0.019	
	Geranium robertianum	1.00	0.18	0.42	0.015	
	Dryopteris pseudodisjuncta	0.99	0.18	0.42	0.038	
Toe	*Plantago lanceolata*	0.96	0.27	0.51	0.007	59.09

Table 3. List of plant species found within the study area and their respective botanical families, absolute abundance in the study area (Ab), total biomass (TB), habitat preference (i.e. 1: presence; 0: absence; *c*: crest; *m*: middle; *t*: toe), species indicator value (IndVal), statistical significance as indicator species (*p* value) and presence (✓) or absence (X) in the field campaigns carried out (C14: 2014 campaign; C15: 2015 campaign)

Species	Family	Ab (%)	TB (g)	*c*	*m*	*t*	IndVal	*p* value	C14	C15
Arrhenatherum elatius[a]	Poaceae	5.39	3062.6	1	1	0	0.69	0.04	✓	✓
Rumex obtusifolius	Polygonaceae	5.32	4465.3	1	1	1	0.65	NA	✓	✓
Viburnum acerifolium	Aboxaceae	3.67	2926.5	1	1	0	0.49	0.05	✓	✓
Lythrum salicaria[b]	Lythraceae	3.50	3242.2	0	1	0	0.70	0.001	✓	✓
Onopordum acanthium	Asteraceae	3.44	1904.6	1	1	1	0.66	NA	✓	✓
Silene dioica	Caryophyllaceae	2.98	1704.9	1	1	1	0.74	NA	✓	✓
Daucus carota	Apiaceae	2.86	2215.2	1	1	0	0.50	0.13	✓	✓
Erigeron acris	Asteraceae	2.63	968.0	1	1	1	0.49	NA	✓	✓
Poa pratensis[a]	Poaceae	2.42	605.3	0	1	1	0.64	0.01	✓	✓
Eleocharis lacustris	Cyperaceae	2.35	1294.5	0	1	1	0.42	0.24	✓	✓
Heracleum maximum	Apiaceae	1.67	954.6	1	0	1	0.34	0.41	X	✓
Galium aparine[b]	Rubiaceae	1.64	1170.9	1	0	0	0.73	0.001	✓	✓
Lathyrus pratensis	Fabaceae	1.42	1672.3	1	1	1	0.52	NA	✓	✓
Anthriscus sylvestris	Apiaceae	1.18	627.7	1	1	0	0.40	0.41	✓	✓
Bromus lepidus	Poaceae	0.97	379.3	1	1	1	0.39	NA	✓	✓
Dryopteris pseudodisjuncta[a]	Pteridophyte	0.76	291.8	0	1	0	0.42	0.04	✓	✓
Plantago lanceolata[b]	Plantaginaceae	0.73	103.6	0	0	1	0.51	0.01	✓	✓
Trifolium dubium	Fabaceae	0.71	90.4	0	0	1	0.42	0.34	✓	✓
Phalaris truncata	Poaceae	0.68	62.2	0	1	0	0.24	0.29	✓	✓
Pteridium aquilinum	Dennstaedtiaceae	0.61	171.8	0	0	1	0.21	1.00	X	✓
Convolvulus arvensis	Convolvulaceae	0.55	299.7	1	0	1	0.50	0.17	✓	✓
Cynodon dactylon[a]	Poaceae	0.46	81.0	0	1	1	0.57	0.04	✓	✓
Avena brevis	Poaceae	0.43	327.6	1	1	0	0.37	0.22	✓	✓
Descurainia sophia	Brassicaceae	0.33	248.7	1	1	1	0.32	NA	✓	✓
Sonchus oleraceous	Asteraceae	0.24	194.6	1	0	0	0.22	0.63	X	✓
Urtica dioica	Urticaceae	0.24	162.4	1	1	0	0.46	0.05	✓	✓
Hydrocotyle vulgaris	Apiaceae	0.21	17.0	0	1	1	0.23	0.75	X	✓
Aurina saxatilis	Brassicaceae	0.20	108.4	1	1	1	0.34	NA	X	✓
Trifolium repens	Fabaceae	0.20	35.4	0	0	1	0.37	0.14	✓	✓
Poa annua	Poaceae	0.12	41.7	0	1	1	0.36	0.24	✓	✓
Agrimonia eupatoria	Rosaceae	0.11	36.2	1	0	0	0.38	0.06	✓	X
Vicia cracca	Fabaceae	0.11	23.0	0	0	1	0.37	0.09	✓	✓
Lolium perenne	Poaceae	0.09	16.4	0	0	1	0.21	1.00	✓	✓
Crepis biennis	Asteraceae	0.07	27.8	1	0	1	0.26	0.94	X	✓
Chamerion angustifolium[a]	Onagraceae	0.06	41.3	0	1	0	0.42	0.02	✓	✓
Bromus mollis	Poaceae	0.05	14.6	0	0	1	0.21	1.00	✓	✓
Orchid	Orchideaceae	0.03	4.8	0	0	1	0.21	1.00	X	✓
Geranium robertianum[a]	Geraniaceae	0.02	13.7	0	1	0	0.42	0.01	X	✓
Vicia bithynica	Fabaceae	0.02	20.1	0	0	1	0.21	1.00	✓	✓
Papaver rhoeas	Papaveraceae	0.01	4.8	1	0	0	0.22	0.63	X	✓
Salvia pratensis	Lamiaceae	0.002	2.8	0	1	0	0.24	0.29	✓	X
Erodium ciconium	Geraniaceae	0.001	1.5	0	0	1	0.21	1.00	✓	X

p value = NA: plant species occur in all sites and cannot be statistically tested

[a] Significant at 95% confidence level: plant species indicator on the basis of the statistical significance of IndVal

[b] Significant at 99% confidence level

Effect of environmental covariates on plant diversity, species richness and biomass

Correlation tests

Results derived from Pearson's correlation tests (Fig. 4a, b) showed a strong negative correlation between plant dry biomass and SRP ($r = -0.68$; Fig. 4b) and a low correlation with the rest of the evaluated environmental covariates. Plant diversity (H') exhibited a moderate positive correlation with SRP, SOM and TKN (Fig. 4a, b); a moderate negative correlation with the plant dry biomass ($r = -0.31$; Fig. 4a); and no correlation with the topographic attributes (i.e. slope, shade, aspect, curvature). The cumulative species richness showed a low positive correlation ($r = 0.28$; Fig. 4a) with TKN and did not show a clear relationship with the topographic attributes or plant biomass.

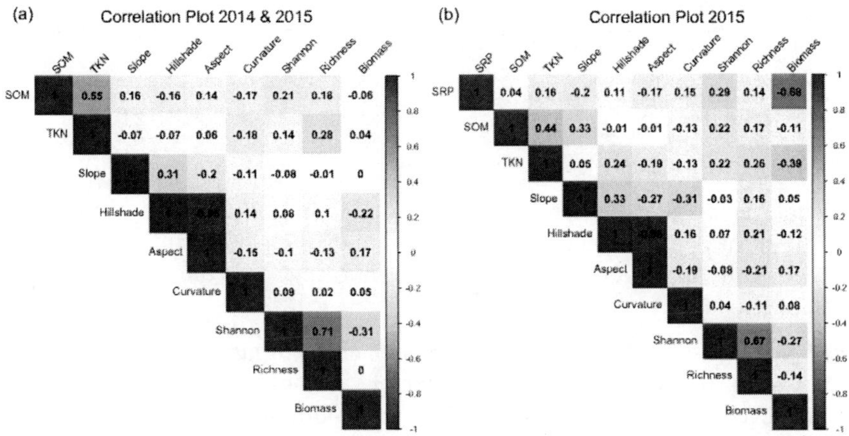

Figure 4. Correlation (i.e. Pearson's correlation coefficient) between the three considered plant diversity metrics (i.e. Shannon index, species richness and aboveground plant dry biomass) and the considered environmental covariates for **A** the combination of 2014–2015 field campaigns and **B** the 2015 field campaign. *SRP* soil resistance to penetration, *SOM* soil organic matter, *TKN* soil total Kjeldahl nitrogen, *Slope* slope gradient, *Hillshade* topographic hillshade, *Aspect* topographic azimuth, *Curvature* topographic surface curvature. The *darker* the *shade* the higher the correlation. *Blue colour*: positive correlation; *red colour*: negative correlation

Regression trees

The fitted regression trees (Fig. 5a–f) show a significant goodness of fit in all cases (Fig. 5a–f; Table 1). The importance of each predictor variable within the regression trees after pruning is shown in Table 4. It is worth noting that the factor variable 'slope stratum' was treated as a surrogate of nitrogen differences across the slope strata for the final interpretation of the regression trees (see "Environmental covariate effect and useful recommendations for slope restoration" section). This approach was adopted to provide a quantitative meaning to the factor slope stratum, and it is supported by the strong strata differences in terms of TKN (Fig. 3d; Wilcke et al. 2003).

Table 4. Regression tree outcomes showing the coefficient of determination (R^2), residual mean square error (RMSE) and variable importance expressed in %

Response variable	Model	R^2	RMSE	Slope gradient	Aspect	Hillshade	Surface curvature	TKN	SOM	SRP	Stratum
Plant diversity	A	0.62	0.22	10	24	26	4	13	23	–	0
	B	0.86	0.11	36	4	4	15	16	12	10	3
Species richness	A	0.77	1.21	3	9	10	26	34	6	–	13
	B	0.78	0.73	34	5	7	5	2	0	11	35
Plant biomass	A	0.71	271.93	19	29	18	0	20	9	–	5
	B	0.82	143.85	0	16	0	13	10	0	42	18

Model A includes information from 2014 and 2015 field campaigns without the soil resistance to penetration; model includes soil resistance to penetration as a predictor variable and information from the 2015 field campaign only

TKN soil total Kjeldahl nitrogen, *SOM* soil organic matter, *SRP* soil resistance to penetration, *Stratum* factor variable including the three considered slope strata in the study (i.e. crest, middle and toe)

Model A includes information from 2014 and 2015 field campaigns without the soil resistance to penetration; model includes soil resistance to penetration as a predictor variable and information from the 2015 field campaign only

TKN soil total Kjeldahl nitrogen, *SOM* soil organic matter, *SRP* soil resistance to penetration, *Stratum* factor variable including the three considered slope strata in the study (i.e. crest, middle and toe)

Plant diversity

Slope aspect and hillshade were the main controlling variables of plant diversity (H') when data from both field campaigns were combined (Table 4; Fig. 5a). Aspect (Fig. 2) was the unique predictor (Fig. 5a) for plant diversity when the slope faced southerly (i.e. aspect = 180; Fig. 2) directions. For easterly facing directions (i.e. 90 < aspect < 170; Fig. 2), however, other variables were more limiting for plant diversity, in particular TKN (Fig. 5a). Additionally, soil organic matter (SOM) gained relevance when the slope was not fully insolated (i.e. hillshade <255; Fig. 2) (Fig. 5a).

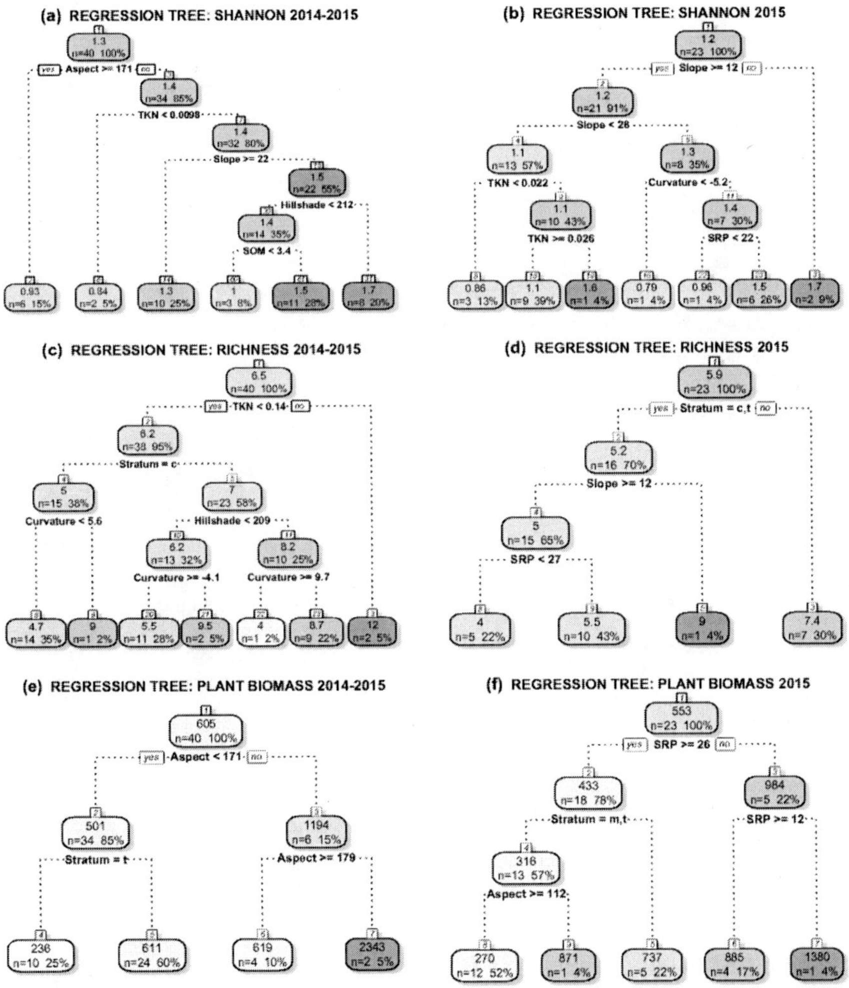

Figure 5. Regression tree dendrograms for the different fitted models. **A** Shannon diversity for the 2014 and 2015 campaigns. **B** Shannon diversity for the 2015 campaign and including SRP. **C** Species richness for the 2014 and 2015 campaigns. **D** Species richness for the 2015 campaign including SRP. **E** Aboveground plant dry biomass for the 2014 and 2015 campaigns. **F** Aboveground plant dry biomass for the 2015 campaign including SRP. Each tree leaf (i.e. *box*) indicates the mean response, number and percentage of observations. Interpretation of the topographic attribute (i.e. aspect, hillshade, curvature) values is shown in Fig. 3

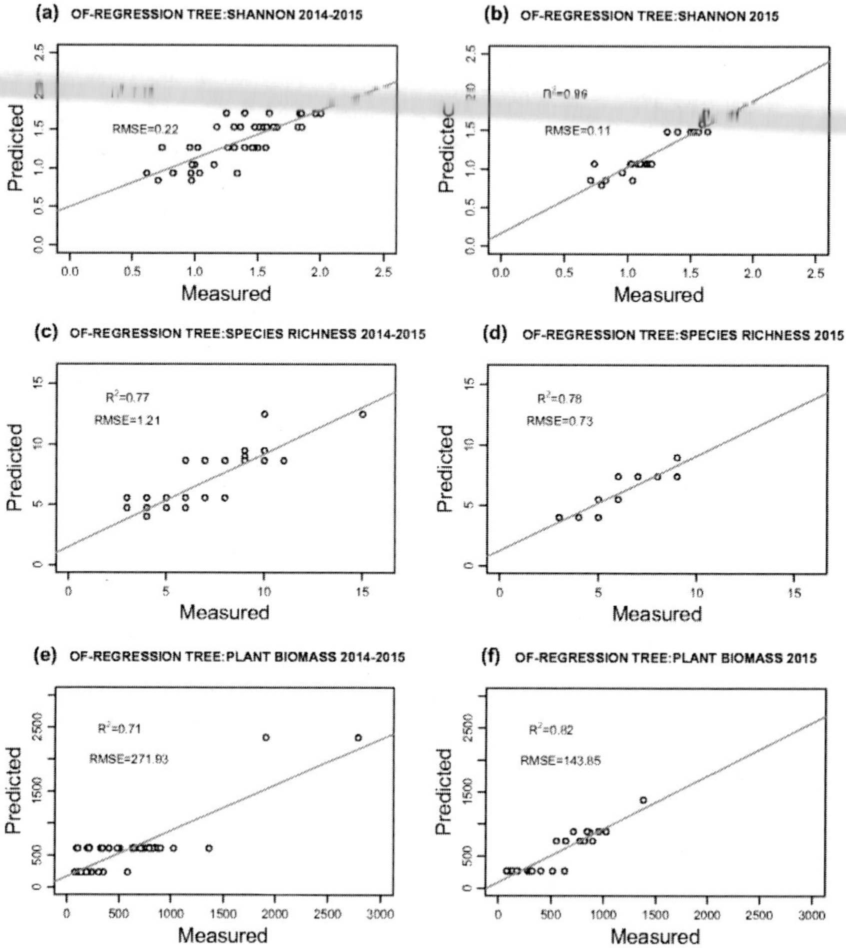

Figure 6. Objective functions (OF) for the different fitted regression trees. **A** Shannon diversity for the 2014 and 2015 campaigns excluding SRP. **B** Shannon diversity for the 2015 campaign including SRP. **C** Species richness for the 2014 and 2015 campaigns excluding SRP. **D** Species richness for the 2015 campaign including SRP. **E** Aboveground plant dry biomass for the 2014 and 2015 campaigns excluding SRP. **F** Aboveground plant dry biomass for the 2015 campaign including SRP. *SRP* soil resistance to penetration

Slope gradient, TKN and curvature were the most important predictor variables for plant diversity when data from the 2015 campaign were used (Table 4; Fig. 5b). TKN was a more relevant predictor when the slope gradient was lower than 28° (Fig. 5b). However, curvature was a better predictor when slope gradient was greater than 28° (Fig. 5b), and the more convex the curvature (i.e. more negative in value; Fig. 2), the more this attribute controlled plant diversity.

Species richness

TKN and slope stratum were the most relevant predictors for species richness (Table 4; Fig. 5c, d). The fitted trees for species richness clearly identified the crest stratum as a distinct habitat when both years (i.e. 2014 and 2015) were combined (Fig. 5c) and the middle slope stratum with respect to the toe and crest strata when 2015 data were used (Fig. 5d). As for the toe and middle habitats, when the combined set of measurements was used (Fig. 5c), hillshade value hierarchically determined the curvature effect. When shading was greater (i.e. hillshade lower than 255; Fig. 2), a convex curvature (i.e. negative in value; Fig. 2) was more limiting to species richness. Conversely, when insolation was higher (i.e. higher value of hillshade; Fig. 2), a concave curvature (i.e. positive in value; Fig. 2) was shown to affect the species richness.

Plant biomass

Slope aspect, stratum and soil resistance to penetration (SRP) were the most relevant covariates for plant biomass (Table 4; Fig. 5e, f). As slope orientation became less favourable for irradiation with respect to southerly orientations (i.e. easterly directions; 90 < aspect < 170; Fig. 2), the stratum became more important (Fig. 5e) for plant biomass. A lower SRP was directly related to a higher plant biomass (Fig. 5f).

DISCUSSION

Signs of landslide-driven slope ecosystem dynamics

Landslide-derived geomorphological processes led to the differentiation of the slope ecosystem into three clearly defined habitats (see "Plant diversity, species richness and slope strata" section; Velázquez and Gómez-Sal 2008; Neto et al. 2017). This differentiation supported the division of the studied slope into three predefined strata (see "Sampling approach" section) and resulted from an increase in biophysical diversity (Geertsema and Pojar 2007).

The increase in biophysical diversity was evident in terms of the encountered differences in SRP (Fig. 3c). The toe habitat showed a much higher SRP than the crest (Fig. 3c). This outcome may be related to landslide-driven changes in the local soil texture, porosity and density (Coutrema and Polar 2007) which, in turn, are linked to the geotechnical properties of the slope-forming materials (see "Study site" section) exposed after the shallow landslide episodes. Settlement and consolidation processes fostered by wetting and drying meteorological cycles after the landslide events could have increased the soil density at the toe (Craig 2004; Head and Epps 2011). The plastic nature of the bare soil, related to the soil fine fraction (see "Study site" section), may have contributed to an increase in soil firmness at the middle stratum following wetting and drying cycles (Consentini and Foti 2014). As a result, SRP increased in the exposed strata subject to slope failure (i.e. middle and toe). Alternatively, SRP differences (Fig. 3c) could be due to other landscape evolution processes triggered by landslides (see in the following; Walker et al. 2009).

The differences in species richness between the slope strata (Table 1; Fig. 3b) could be interpreted as a sign of landscape evolution (Dale et al. 2005; Velázquez and Gómez-Sal 2008). For example, species richness was lower at the crest, which has been stable for a longer period than the other two studied slope zones (Fig. 3b). The disturbance-free periods could have facilitated the establishment and succession of plant communities (del Moral and Jones 2002). Once plant communities were established, competitive exclusion may have arisen (Wilson and Tilman 1995), decreasing species richness (Herbert et al. 2004). The absence of disturbance would have facilitated the formation of slope soil (Myster and Walker 1997; Smale et al. 1997), fostered by the incorporation of plant-derived decaying organic matter (Fig. 3e), thus improving the soil structure (Bronick and Lal 2005) and reducing the SRP (Fig. 3c; Franzluebbers 2002). Nonetheless, we believe that the temporal expansion of this study could contribute towards the elucidation on whether our observations can be interpreted as signs of hillslope evolution.

Conversely, landslide recurrence may have led to an increase in species richness (Velázquez and Gómez-Sal 2009) in the middle and toe strata (Table 1; Fig. 3b; Neto et al. 2017). Lower levels of plant competition found in the nutrient-rich gaps produced by shallow landslides support this observation (Walker et al. 1996). In particular, the lack of herbaceous plant competition for light with higher-order plants (e.g. woody species) could be the reason for the higher comparative species richness found after disturbance in landslide-prone zones (e.g. Myster and Walker 1997; Rajaniemi et al. 2003; Soto et al. 2017).

The active role of shallow landslides on slope ecosystem differentiation was also shown by the accumulation of nitrogen at the lowest slope stratum (i.e. toe; Fig. 3d; Wilcke et al. 2003). It is highly likely that this was a result of the downward movement of nutrient-rich soil volumes following slope failure, albeit nitrogen leaching from the crest soil after rains could be possible as well (e.g. Hood et al. 2003). Based on this result, nitrogen could be the key abiotic factor that facilitated the establishment of vegetation within the gaps opened up by landslides (Walker and del Moral 2003; Shiels et al. 2008) and one of the main factors triggering further ecosystem dynamics (e.g. Herbert et al. 2004). The higher nitrogen levels observed in the two lower slope strata in 2014 (Fig. 3d) could explain the higher plant biomass observed there (Fig. 3f; Gross and Collins 2000; Herbert et al. 2004) if adequate nitrogen mineralisation is assumed (Urbano 1995). TKN differences between the two sampling campaigns (Fig. 3d), supported by correlated plant biomass differences (Fig. 3f), may reflect consumption of nitrogen by the plant community established on the slope after disturbance: the higher levels of soil nitrogen allowed more plant biomass development during 2014 (Dalling and Tanner 1995; Rowe et al. 2006). As a result, the levels of soil nitrogen decreased in the subsequent season (Fig. 3d; Niklaus et al. 2001) and led to lower biomass production (Fig. 3f). Nonetheless, soil nitrogen could have also led to an increase in belowground plant competition and, consequently, in reduction of plant aboveground biomass (Rajaniemi et al. 2003; Herbert et al. 2004), as seen in the results from 2015 (Fig. 3d, f) and supported by the correlation tests (Fig. 4a). These processes may have led to a decrease in plant diversity (Gross and Collins 2000; Rajaniemi et al. 2003), hinted at in the results from the toe stratum in 2015 when compared to 2014 (Fig. 3a, b).

Plant diversity metrics as signs of slope ecosystem self-organisation

Species richness differences between the slope strata (see "Plant diversity, species richness and slope strata" section; Fig. 3b) may result from the self-regulating capacity of the slope ecosystem after disturbance (Neto et al. 2017). The downward movement of nutrient-rich materials and the way that plant communities respond (e.g. through competing for light and nutrients) played a key role in the self-organisation process (Walker et al. 1996), as discussed earlier (see 4.1). As a result, species richness showed evident differences between the slope strata (Table 1; Fig. 3b). These were consistent with the findings of other authors (e.g. Armesto and Pickett 1985; Myster and Walker 1997; Walker et al. 2006; Elias and Dias 2009; Velázquez and Gómez-Sal 2009; Neto et al. 2017) and with the intermediate disturbance hypothesis (Connell 1978).

In terms of species composition, the Bray-Curtis similarity index (see "Plant diversity, species richness and slope strata" section) highlighted a distinct two habitat during both sampling campaigns and showed clear differences between 2014 and 2015 (see "Plant diversity, species richness and slope strata" section). This may suggest an increasing system complexity as the landscape starts recovering from disturbance (Odum 1969; Myster and Walker 1997; Velázquez and Gómez-Sal 2009). The relatively high species evenness (*J'*) encountered (Table 1) indicates an absence of dominant species and, thus, may suggest an early successional stage within the plant community of our study site (Mulder et al. 2004).

The plant diversity (*H'*) on the study site (Table 1; Fig. 3a) was below or within the range reported by other diversity studies focusing on grasslands (e.g. Bochet and Garcia-Fayos 2004; Kahmen et al. 2005; Spehn et al. 2005; Pohl et al. 2009). Although plant diversity (*H'*) did not differ between slope strata (Table 1; Fig. 3a), the observed *H'* values with respect to the ones provided in the literature could suggest an adequate plant colonisation and establishment on the studied slope. Many plant species found (Table 3) were common weeds (BRC 2016), usually present on disturbed grounds, indicating that most of them entered the community from the adjacent landscape (Velázquez and Gómez-Sal 2008). Overall, the species assemblage was relatively similar between the studied slope strata (Table 3) indicating that many species could have been dormant in the displaced soil materials as seeds or propagules (e.g. Myster and Walker 1997).

Relating these observations to the self-regulation of a slope after disturbance, some ideas can be drawn regarding the early stages of human-driven restoration actions on slopes, which usually include planting a single or few plant species. More than 40 plant species were identified in the initial stages of slope colonisation in this study (Table 3), suggesting that planting with a higher species diversity would be more ecologically relevant. In some instances, landslide-derived gaps can be overwhelmed by alien species that make the most of the low levels of plant competition in these patches to the detriment of native species (Catford et al. 2012). In this regard, two alien species were found within the studied slope with a relative high abundance (Table 3): *Heracleum maximum* and *Viburnum acerifolium.* This highlights an important issue to consider when planning planting strategies.

Plant species as indicators of landslide-driven biophysical diversity

Plant indicator species results (Table 2) corroborated the differentiation of the slope into distinct habitats and provided further support to the notion that landslides may increase biophysical diversity. For instance, the principal indicator for the middle slope habitat, *Lythrum salicaria*, is a herbaceous species belonging to wet habitats and slow-flowing creeks (BRC 2016). As the slope failed, the phreatic zone was exposed in some areas within the slope's middle zone, creating an entire foreign aquatic habitat that did not exist before within our slope ecosystem. Likewise, the indicator for the crest habitat, *Galium aparine*, might have thrived as a consequence of small mammals, such as rabbits, dispersing its seeds over the crest habitat (BRC 2016). Despite the former speculation, the crest of the slope remained stable during the last slope failure episode (see "Study site" section), allowing the prolonged stability to host mammals (i.e. rabbits, cats and mice), as noticed by the authors during the sampling campaigns. Moreover, the plant indicator species identified for the toe habitat, *Plantago lanceolata*, may suggest modifications to the slope geochemistry, as many of the species identified within the slope (Table 3) are generally associated with calcareous or basic (i.e. pH >7) soils (BRC 2016), while *Plantago lanceolata* generally thrives on more acidic soils (i.e. pH <7). The outcomes derived from the plant indicator analysis (Tables 2 and 3), however, should be taken with caution as the fidelity (i.e. B; Table 2) for most of the identified indicators was low, and the coverage (Table 2) did not reach 100%. Nonetheless, the use of the plant indicator species analysis (Dufrene and Legendre 1997; De Caceres and Legendre 2009) presented here opens up an exciting possibility to explore landslide-derived habitats and establish different slope restoration and management strategies in accordance with the features of a given slope habitat.

Environmental covariate effect and useful recommendations for slope restoration

Our results are consistent with the hypothesis that slope topography has a significant effect on establishment and dynamics of vegetation (e.g. Bochet and Garcia-Fayos 2004; Gallardo-Cruz et al. 2009; Moeslund et al. 2013; Stein et al. 2014) after landslides. Terrain attributes such as slope gradient, aspect, hillshade and curvature (Fig. 2) play a vital role in the establishment of vegetation as they are intimately related to the availability of natural resources (e.g. nutrients, light, water) to plants (e.g. Fu et al. 2004; Gallardo-Cruz et al. 2009; Yu et al. 2015). These effects were seen after fitting regression tree models (Table 4; Figs. 5a–f and 6a–f), which showed hierarchies of importance among topographical attributes and with their associated environmental covariates, such as light and nitrogen (see "Regression trees" section). The effect of the evaluated environmental covariates varied depending on the vegetation metric to be predicted (Table 4), the direction of change of one factor in response to the level of another factor and the amount of information employed to fit the regression tree (i.e. 2014–2015: Fig. 5a, c, e; 2015: Fig. 5b, d, e). It is worth noting that the factor variable slope stratum was treated as a surrogate of nitrogen differences across the slope strata for the final interpretation of the regression trees, as indicated in "Regression trees" section. Overall, five main aspects related to post-landslip establishment of plant communities on slopes were confirmed with the regression trees:

1.Soil nitrogen and light availability have a strong effect on plant diversity (H'; Fig. 5a, b), species richness (Fig. 5c, d) and plant biomass (Fig. 5e, f) as noted previously in the literature (e.g. Myster and Walker 1997; Rajaniemi et al. 2003; Herbert et al. 2004; Pykälä et al. 2005; Shiels et al. 2008; Gallardo-Cruz et al. 2009; Soto et al. 2017).

2.The amount of incoming light with respect to the level of soil nitrogen can regulate plant diversity (Fig. 5a, c) and biomass (Fig. 5e). Thus, trade-offs may occur between light and nitrogen (Fig. 5a, c, e; e.g. Walters and Reich 1997; Rajaniemi et al. 2003; Herbert et al. 2004; Soto et al. 2017) and their effect upon plant communities must be evaluated together.

3.Slope topography governs the availability of both nitrogen (Weintraub et al. 2015) and light (Gallardo-Cruz et al. 2009) (Fig. 5a, b, e, f). The spatial availability of nitrogen can be limited by the topographical exposure (i.e. slope gradient and curvature; Chapman 2000), where a higher slope gradient and a more convex curvature (i.e. negative in value; Fig. 2) will constrain the formation of soil and the accumulation of nutrients (Fig. 5b; Heimsath et al. 1997; Bertoldi et al. 2006; Weintraub et al. 2015). On a landslide-prone slope, however, landslide dynamics will exert a greater control over the availability of nitrogen (e.g. Guariguata 1990; Walker et al. 1996; Myster and Walker 1997; Velázquez and Gómez-Sal 2008) than the topographical exposure, supported here by the differentiation of the studied slope into habitats (Figs. 3d and 5c, e, f; Walker et al. 1996; Velázquez and Gómez-Sal 2008; Neto et al. 2017). With regard to light availability, we used the hillshade parameter in our approach, knowing that it is intended as a cartographical representation option to enhance the graphical representation of the study site topography (e.g. Zu 2016) and, therefore, carried little ecological meaning. Although its use provided relevant results, for the future, it would be advisable to employ indices aiming at estimating solar radiation input in order to better clarify the effect of light on plant communities (e.g. topographic radiation index; Piedallu and Gegout 2008).

4.Topographical exposure may also control plant diversity (i.e. H' and richness; Table 1; Fig. 5b, d). High slope gradients and convexity (Fig. 2) may have a negative impact on plant establishment (Bochet and Garcia-Fayos 2004). As a result, it is likely that plant communities will develop less in zones of high topographical exposure (i.e. more prone to shallow landslides; Lu and Godt 2013), where the diversity of early successional species will be maintained at relatively high levels (Velázquez and Gómez-Sal 2008) compared to more stable zones undergoing plant succession (e.g. crest of the studied slope and undisturbed sections of the slope).

5.The SRP is a viable and readily measurable indicator of slope habitat (Fig. 3c) and plant biomass (Fig. 5f). Soil penetrability is related to soil structure (Franzluebbers 2002): a lower SRP (i.e. higher penetrability) will be indicative of better soil physical conditions for plant establishment (Bengough and Mullins 1990) and, therefore, for more plant biomass development. It is worth noting that plant biomass development is of high interest in terms of slope stabilisation with plants (e.g. Gonzalez-Ollauri and Mickovski 2016).

Overall, the earlier outcomes indicate that regression trees are a viable tool for evaluating and understanding the hierarchical complexity of slope ecosystems, which may not be fully detected with Pearson's correlation test (Fig. 4a, b). The information obtained from the regression trees (Fig. 5a–f) can be directly applied in the implementation of slope restoration strategies against shallow landslides involving the use of vegetation (e.g. Norris et al. 2008; Mickovski 2014; Tardio and González-Ollauri 2016). For example, a balanced application of soil amendments (e.g. nitrogen and OM-rich compost) on shaded zones of the slope could foster the establishment of a dense and more diverse early-stage plant community (e.g. Walters and Reich 1997; Elmarsdottir et al. 2003; Rowe et al. 2006) or root systems (Mickovski and Ennos 2003; Gonzalez-Ollauri and Mickovski 2016). Care must be taken on slope zones presenting high topographical exposure to obtain adequate vegetation establishment. On these zones, human-driven revegetation strategies, such as facilitation (e.g. Brooker et al. 2008), could be implemented. Information related to SRP could be used to implement soil tillage on certain slope zones to improve the physical properties of soil (Bronick and Lal 2005), which, combined with the application of soil amendments, can facilitate the establishment of a dense plant cover (e.g. Bengough and Mullins 1990; Wilson and Tilman 1995) on the slope.

CONCLUSIONS

In the light of our results and observations, we can conclude that

- Shallow landslides are dynamic geomorphological processes contributing to biophysical diversity and able to distinguish the landscape into different slope habitats even at small spatial scales.
- The slope habitats can be distinct in terms of plant species richness and compositional differences, as well as in terms of soil properties such as nitrogen content and resistance to penetration.
- Differences in plant species richness throughout the slope habitats are a direct consequence of the slope ecosystem self-regulating capacity after disturbance.
- Soil nitrogen and light availability have a strong effect on plant establishment (i.e. plant diversity, richness and biomass) after disturbance.
- Landscape topography and landslide dynamics play an important role on soil nitrogen and light availability and, hence, on landscape evolution and its self-regulating capacity.
- Certain plant species may be good indicators of landslide-driven biophysical diversity.
- The effects of the environmental covariates on plant diversity metrics can be effectively appraised with regression tree models.

A comprehensive novel approach has been presented here to detect traits of ecosystem evolution, self-regulation and biophysical diversity triggered by shallow landslides. We acknowledge that the results presented herein refer to a particular site with specific environmental conditions and to a limited monitoring time. Further research along the lines presented in this study and, over longer periods of time, larger spatial scales is needed to shed more light on how shallow landslide-prone ecosystems evolve after slope disturbance. Undoubtedly, the approach presented herein provides a good basis to explore shallow landslides within an ecological perspective and to apply the resulting knowledge into the sustainable restoration of landslide-prone slopes.

ACKNOWLEDGMENTS

The authors want to thank to Catterline Brae Action Group (CBAG) for site access and logistical support. The help of students funded by national and international programmes: Victoria Thompson (Nuffield Foundation), Leonardo Gentile, Gabriel Marsalis (Science Without Borders), Anabel Perdomo and Julie Thiriat (Erasmus +) is greatly acknowledged. We are grateful to Elizabeth Mittell for language and style editing. We also acknowledge the useful comments and suggestions from the editor and two anonymous referees that helped us to enhance the manuscript. This research project was funded by a PhD scholarship awarded by the School of Engineering and Built Environment at the Glasgow Caledonian University (S1340554).

References

1. Anderson MJ, Crist TO, Chase JM, Vellend M, Inouye BD, Freestone AL, Sanders NJ, Cornell HV, Comita LS, Davies KF, Harrison SP, Kraft NJB, Segen JC, Swenson NG (2011) Navigating the multiple meanings of β diversity: a roadmap for the practicing ecologist. Ecol Lett 14:19–28
2. Armesto JJ, Pickett STA (1985) Experiments on disturbance in old-field plant communities: impact on species richness and abundance. Ecology 66(1):230–240
3. Bengough AG, Mullins CE (1990) Mechanical impedance to root growth: a review of experimental techniques and root growth responses. J Soil Sci 41:341–358
4. Bertoldi G, Rigon R, Over T (2006) Impact of watershed geomorphic characteristics on the energy and water budgets. J Hydrometeorol 7:389–403
5. BGS (1999) British Geological Survey Rock Classification Scheme. Vol. 3: classification of sediments and sedimentary rocks. Research Report No RR 99-03. BGS, Nottingham, UK
6. Bochet E, Garcia-Fayos P (2004) Factors controlling vegetation establishment and water erosion on motorway slopes in Valencia, Spain. Restor Ecol 12(2):166–174
7. Bonnier G, De Layens G (2002) Claves para la determinacion de plantas vasculares, 4th edn. Omega Eds, Barcelona
8. Borcard, D, Gillet, F and Legendre, P (2011) Numerical ecology with R. Series: use R! Springer.
9. BRC, British Records Centre (2016) Online atlas of the British and Irish flora. Retrieved on 4/7/2016 http://www.brc.ac.uk/plantatlas/
10. Breiman L, Freidman J, Olshen R, Stone C (1984) Classification and regression trees. Wadsworth, Belmont, CA
11. Bremner JM (1965) Organic forms of nitrogen. Agronomy 9:1238–1255
12. Bronick CJ, Lal R (2005) Soil structure and management: a review. Geoderma 14:3–22
13. Brooker RW, Maestrem FT, Callaway RM, Lortie CL, Cavieres LA, Kunstler G, Liancourt P, Tielbörger K, Travis JMJ et al (2008) Facilitation in plant communities: the past, the present, and the future. J Ecol 96:18–34
14. BS 1377-2 (1990) Methods of test for soils for civil engineering purposes. Classification tests. British Standards Institution, London
15. BS 1377-4 (1990) Methods of test for soils for civil engineering purposes. Compaction related tests. British Standards Institution, London
16. Catford JA, Daehler CC, Murphy HT, Sheppard AW, Hardesty BD, Westcott DA, Rejmanek M, Bellingham PJ, Pergl J, Horvitz CC, Hulme PE (2012) The intermediate

disturbance hypothesis and plant invasions: implications for species richness and management. Perspectives in Plant Ecology Evolution and Systematics 14:231–241
17. Cendrero A, Dramis F (1996) The contribution of landslides to landscape evolution in Europe. Geomorphology 15:191–211
18. Chao A, Jost L (2012) Coverage-based rarefaction curves and extrapolation: standardizing samples by completeness rather than size. Ecology 93:2533–2547
19. Chapin FS, Bloom AJ, Field CB, Waring RH (1987) Plant responses to multiple environmental factors. Bioscience 37(1):49–57
20. Chapman L (2000) Assessing topographic exposure. Meteorol Appl 7:335–340
21. Connell JH (1978) Diversity in tropical rain forests and coral reefs. Science 199:1302–1310
22. Consentini RM, Foti S (2014) Evaluation of porosity and degree of saturation from seismic and electrical data. Geotechnique 64(4):278–286
23. Craig R (2004) Craig's soil mechanics, 7th edn. E and FN Spon, London
24. Dale, VH, Campbell, DR, Adams, WM, Crisafulli, CM, Dains, VI, Frenzen, PM, Holland, RF (2005) Plant succession on the Mount St. Helens debris avalanche deposit. In: Dale VH, Swanson, FJ, Crisafulli, CM eds. Ecological responses to the 1980 eruption of Mount St. Helens, pp 59–73. Springer, Dordrecht, The Netherlands
25. Dalling JW, Tanner EVJ (1995) An experimental study of regeneration on landslides in montane rain forests in Jamaica. The Journal of Ecology 83(1):55–64
26. De Caceres, M and Legendre, P (2009) Associations between species and groups of sites: indices and statistical inference. Ecology, http://sites.google.com /site/miqueldecaceres/
27. De Caceres M, Legendre P, Wiser SK, Brotons L (2012) Using species combinations in indicator value analyses. Methods Ecol Evol 3:973–982
28. De Caceres, M (2013) How to use the indicspecies package (ver. 1.7.1). https://cran.r-project.org/web/packages/indicspecies/vignettes/indicspeciesTutorial.pdf
29. del Moral R, Jones C (2002) Vegetation development on pumice at Mount St. Helens, USA. Plant Ecol 162(1):9–22
30. Dufrene M, Legendre P (1997) Species assemblages and indicator species: the need for a flexible asymmetrical approach. Ecol Monogr 67(3):345–366
31. Efron B (1979) Bootstrap methods: another look at the jackknife. Ann Statist 1:1–26
32. Elias RB, Dias E (2009) Effects of landslides on the mountain vegetation of Flores Island. Azores J Veg Sci 20:706–717
33. Elmarsdottir A, Aradottir AL, Trlica MJ (2003) Microsite availability and establishment of native species on degraded and reclaimed sites. J Appl Ecol 40:815–823
34. Franklin J (1995) Predictive vegetation mapping: geographic modelling of biospatial patterns in relation to environmental gradients. Prog Phys Geogr 19(4):474–499
35. Franzluebbers AJ (2002) Water infiltration and soil structure related to organic matter and its stratification with depth. Soil Tillage Res 66:197–205
36. Fu BJ, Liu SL, Ma KM, Zhu YG (2004) Relationships between soil characteristics, topography and plant diversity in a heterogeneous deciduous broad-leaved forest near Beijing, China. Plant Soil 261:47–54
37. Gallardo-Cruz JA, Perez-Garcia EA, Meave JA (2009) β-Diversity and vegetation structure as influenced by slope aspect and altitude in a seasonally dry tropical landscape. Landsc Ecol 24:473–482
38. Geertsema M, Pojar JJ (2007) Influence of landslides on biophysical diversity—a perspective from British Columbia. Geomorphology 89:55–69
39. GetMapping (2014) GetMapping 2 m resolution digital surface model (DSM) for Scotland and Wales. NERC Earth Observation Data Centre. Retrieved from http://catalogue.ceda.uk/uuid/4b0ed418e30819e4448dc89a27dc8388
40. Gonzalez-Ollauri A, Mickovski SB (2016) Using the root spread information of pioneer plants to quantify their mitigation potential against shallow landslides and erosion in temperate humid climates. Ecol Eng 95:302–315

41. Gotelli NJ, Chao A (2013) Measuring and estimating species richness, species diversity, and biotic similarity form sampling data. Encyclopedia of Biodiversity, Volume 5:195–210
42. Gotelli, NJ and Colwell, RK (2011) Estimating species richness. In: Magurran, AE and McGill, BJ, 2011. Frontiers in measuring biodiversity (pp. 39–53). Oxford University Press, New York, US. Gough, L, Osenberg, CW
43. Gross KL, Collins SL (2000) Fertilization effects on species density and primary productivity in herbaceous plant communities. Oikos 89:428–439
44. Guariguata MR (1990) Landslide disturbance and forest regeneration in the Upper Luquillo Mountains of Puerto Rico. J Ecol 78(3):814–832
45. Hall TW, Higson AW, Pierce BJ, Price KH, Skousen CJ (2013) Haphazard sampling: selection biases and the estimation consequences of these biases. Current Issues in Auditing 7(2):16–22
46. Head KH, Epps RJ (2011) Manual of soil laboratory testing: permeability, shear strength and compressibility tests, vol Vol. 2. CRC Press, Boca Raton
47. Heimsath AM, Dietrich WE, Nishiizumi K, Finkel RC (1997) The soil production function and landscape equilibrium. Nature 388:358–361
48. Herbert DA, Rastetter EB, Gough L, Shaver GR (2004) Species diversity across nutrient gradients: an analysis of resource competition in model ecosystems. Ecosystems 7:296–310
49. Hood EW, Williams MW, Caine N (2003) Landscape controls on organic and inorganic nitrogen leaching across alpine/subalpine ecotone, Green Lakes Valley, Colorado Front Range. Ecosystems 6:31–45
50. Jørgensen, S (2007) An integrated ecosystem theory. Ann.-Eur. Acad. Sci. (2006–2007):19–33.
51. Kahmen A, Perner J, Buchmann N (2005) Diversity-dependent productivity in semi-natural grasslands following climate perturbations. Funct Ecol 19:594–601
52. Kallimanis AS, Ragia V, Sgardelis SP, Pantis JD (2007) Using regression trees to predict alpha diversity based upon geographical and habitat characteristics. Biodivers Conserv 16:3863–3876
53. Kincardineshire Observer(2013), 11/4/2013. Retrieved on 3/7/2016 from http://www.Kincardineshireobserver.co.uk/news/catterline-villagers-pull-together-to-clear-road-1-2890185
54. Köppen W (1884) The thermal zones of the Earth according to the duration of hot, moderate and cold periods and the impact of heat on the organic world. Meteorol Z 1:215–226
55. Lu N, Godt J (2013) Hillslope hydrology and stability. Cambridge University Press, New York, US
56. Malone B (2013) Use R for digital soil mapping. Soil Security Laboratory, The University of Sidney, Australia
57. Magurran A (2004) Measuring biological diversity. Blackwell Science, Oxford
58. Mickovski SB, Ennos AR (2003) Anchorage and asymmetry in the root system of Pinus peuce. Silva Fennica 37(2):161–173
59. Mickovski, SB (2014) Resilient design of landslip prevention measures: a case study. Forensic Engineering 10.1680/feng.14.00001
60. Moeslund JE, Arge L, Bøcher PK, Dalgaard T, Ejrnæs R, Odgaard MV, Svenning J-C (2013) Topographically controlled soil moisture drives plant diversity patterns within grasslands. Biodivers Conserv 22:2151–2166
61. Mulder CPH, Bazeley-White E, Dimitrakopoulus PG, Hector A, Scherer-Lorenzen M, Schmid B (2004) Species evenness and productivity in experimental plant communities. Oikos 107:50–63
62. Myster RW, Walker LR (1997) Plant successional pathways on Puerto Rican landslides. J Trop Ecol 13:165–173
63. Neto C, Cardigos P, Cruz Oliveira S, Zezere JL (2017) Floristic and vegetation successional process within landslides in a Mediterranean environment. Sci Total Environ 574:969–981

64. Niklaus PA, Kandeler E, Leadlley PW, Schmis B, Tscherko D, Korner C (2001) A link between plant diversity and, elevated CO2 and soil nitrate. Oecologia 127:540–548
65. Norris J, Stokes A, Mickovski SB, Cameraat E, Van Beek R, Nicoll B, Achim A (2008) Slope stability and erosion control: ecotechnological solutions. Springer, Dordrecht
66. Odum EP (1969) The strategy of ecosystem development. Science, New Series 164(3877):262–270
67. Oksanen J, Blanchet FG, Kindt R, Legendre P, Minchin PR, O'Hara RB, Simpson GL, Solymos P, Stevens MHH, Wagner H (2016) Vegan: community ecology package. R package:v2.3–v2.5 http://CRAN.R-project.org/package=vegan
68. Piedalu C, Gegout JC (2008) Efficient assessment of topographic solar radiation to improve plant distribution models. Agric For Meteorol 148:1696–1706
69. Prasad AM, Iverson LR, Liaw A (2006) New classification and regression techniques: bagging and random forest for ecological protection. Ecosystems 9:191–199
70. Pohl M, Alig D, Körner C, Rixen C (2009) Higher plant diversity enhances soil stability in disturbed alpine ecosystems. Plant Soil 324:91–102
71. Pykälä J, Luoto M, Heikkinen RK, Kontula T (2005) Plant species richness and persistence of rare plants in abandoned semi-natural grasslands in northern Europe. Basic and Applied Ecology 6:25–33
72. R Core Team (2015) R: a language and environment for statistical computing. Vienna, Austria. https://cran.r-project.org/
73. Rajaniemi TK, Allison VJ, Goldberg DE (2003) Root competition can cause decline in diversity with increased productivity. J Ecol 91:407–416
74. Restrepo C, Walker LR, Shiels AB, Bussmann R, Claessens L, Fisch S, Lozano P, Negi G, Paolini L, Poveda G, Ramos-Scharron C, Richter M, Velázquez E (2009) Landsliding and its multiscale influence on mountainscapes. Bioscience 59(8):685–698
75. Revermann, R, Finckh, M, Stellmes, M, Strohbach, BJ, Frantz, D and Oldeland, J (2016) Linking land surface phenology and vegetation-plot databases to model terrestrial plant α-diversity of the Okavango Basin. Remote Sens. 8 (370);
76. Richardson JML, Richards MH (2008) A randomisation program to compare species-richness values. Insect Conservation and Diversity 1:135–141
77. Rossi JP (2011) Rich: an R package to analyse species richness. Diversity 3(1):112–120
78. Rowe EC, Healey JR, Edwards-Jone G, Hills J, Howells M, Jones DL (2006) Fertilizer application during primary succession changes the structure of plant and herbivore communities. Biol Conserv 131:510–522
79. Schulte, E and Hopkins, BG (1996) Estimation of soil organic matter by weight loss-on-ignition. In Magdoff, F, Tabatabai, MA and Hanlon, EA (eds.) 1996. Soil organic matter: analysis and interpretation. Soil Sci. Soc. Am., Madison, US, pp. 21–31
80. Shannon CE, Weaver W (1964) The mathematical theory of communication. The University of Illinois Press, Urbana, US
81. Shiels AB, West CA, Weiss L, Klawinski PD, Walker LR (2008) Soil factors predict initial plant colonization on Puerto Rican landslides. Plant Ecol 195:165–178
82. Smale MC, McLeod M, Smale PN (1997) Vegetation and soil recovery on shallow landslide scars in Tertiary Hill Country, East Cape region, New Zealand. N Z J Ecol 21(1):31–41
83. Somerfield PJ (2008) Identification of the Bray-Curtis similarity index: comment on Yoshioka (2008). Mar Ecol Prog Ser 372:303–306
84. Soto DP, Jacobs D, Salas C, Donoso PJ, Fuentes C, Puettmann KJ (2017) Light and nitrogen interact to influence regeneration in old-growth Nothofagus-dominated forests in south-central Chile. For Ecol Manag 384:303–313
85. Spehn EM, Hector A, Joshi J, Scherer-Lorenzen M, Schmid B, Bazeley-White E, Beierkuhlein C, Caldeira MC, Diemer M, Dimitrakopoulus PG, Finn JA, Freitas H et al (2005) Ecosystem effects of biodiversity manipulations in European grasslands. Ecol Monogr 75(1):37–63

86. Stein A, Gerstner K, Kreft H (2014) Environmental heterogeneity as a universal driver of species richness across taxa, biomes and spatial scales. Ecol Lett 17:866–880
87. Stokes, A, Norris, J, van Beek, L, Bogaard, T, Cammeraat, E, Mickovski, SB, Jenner, A, Di Iorio A, Fourcaud, T (2008) How vegetation reinforces soil on slopes. In: Norris, J, Stokes, A, Mickovski, SB, Cameraat, E, Van Beek, R, Nicoll, B, Achim, A (2008) Slope stability and erosion control: ecotechnological solutions (pp. 65–116). Springer, Dordrecht, The Netherlands
88. Tardio G, González-Ollauri A, Mickovski SB (2016) A non-invasive preferential root distribution analysis methodology from a slope stability approach. Ecol Eng 97:46–57
89. Therneau T, Atkinson B, Ripley B (2015) Rpart: recursive partitioning and regression trees. R package version 4:1–9 http://CRAN.R-project.org/package=rpart
90. UK Met Office (2015) MIDAS Land Surface Stations data, 1853-current. Retrieved from http://badc.nerc.ac.uk/view/badc.nerc.ac.uk_ATOM_dataent_ukmo-midas
91. Urbano P (1995) Tratado de Fitotecnia General. Mundi-Prensa, Madrid
92. USDA-NRCS (1997) National Grazing Lands Handbook. USDA-NRCS, Washington DC, US
93. Velázquez E, Gómez-Sal A (2007) Environmental controls of early succession on a large landslide in a tropical dry ecosystem (Casita volcano, Nicaragua). Biotropica 35(5):601–609
94. Velázquez E, Gómez-Sal A (2008) Landslide early succession in a neotropical dry forest. Plant Ecol 199:295–308
95. Velázquez E, Gómez-Sal A (2009) Changes in the herbaceous communities on the landslide of the Casita volcano, Nicaragua, during early succession. Folia Geobot 44:1–18
96. Walker LR, Bellingham PJ, Peltzer DA (2006) Plant characteristics are poor predictors of microsite colonization during the first two years of primary succession. J Veg Sci 17:397–406
97. Walker LR, del Moral R (2003) Primary succession and ecosystem rehabilitation. Cambridge University Press, Cambridge
98. Walker LR, Landau FH, Velázquez E, Shiels A, Sparrow AD (2010) Early successional woody plants facilitate and ferns inhibit forest development on Puerto Rican landslides. J Ecol 98:625–635
99. Walker LR, Shield A (2013) Landslide ecology. Cambridge University Press, Cambridge
100. Walker LR, Velazquez E, Shiels AB (2009) Applying lessons from ecological succession to the restoration of landslides. Plant Soil 324:157–168
101. Walker LR, Zarin DJ, Fetcher N, Myster RW, Johnson AH (1996) Ecosystem development and plant succession on landslides in the Caribbean. Biotropica 28(4a):566–576
102. Walters MB, Reich PB (1997) Growth of Acer saccharum seedlings in deeply shaded understories of northern Wisconsin: effects of nitrogen and water availability. Can J For Res 27:237–247
103. Wang B, Zhang G, Duan J (2015) Relationship between topography and the distribution of understory vegetation in a Pinus massoniana forest in Southern China. International Soil and Water Conservation Research 3:291–304
104. Weintraub SR, Taylor PG, Porder S, Cleveland CC, Asner GP, Townsend AR (2015) Topographical controls on soil nitrogen availability in a lowland tropical forest. Ecology 96(6):1561–1574
105. Wilcke W, Valladarez H, Stoyan R, Yasin S, Valarezo C, Zech W (2003) Soil properties on a chronosequence of landslides in montane rain forest, Ecuador. Catena 53:79–95
106. Wilson SD, Tilman D (1995) Competitive responses of eight old-field plant species in four environments. Ecology 76(4):1169–1180

107. Yu F, Wang T, Groen TA, Skidmore AK, Yang X, Geng Y, Ma K (2015) Multi-scale comparison of topographic complexity indices in relation to plant species richness. Ecol Complex 22:93–101

108. Zu X (2016) GIS for environmental applications: a practical approach. Routledge, Oxford

CHAPTER3

Assessment of Landslide Pre-Failure Monitoring and Forecasting Using Satellite SAR Interferometry

Serena Moretto [1,2,*], Francesca Bozzano [1,2], Carlo Esposito [1], Paolo Mazzanti [1,2] and Alfredo Rocca [2]

[1]*Department of Earth Sciences, Sapienza University of Rome, Rome 00185, Italy*

[2]*NHAZCA S.r.l., Spin-off of Sapienza, University of Rome, Rome 00185, Italy*

ABSTRACT

In this work, the ability of advanced satellite interferometry to monitor pre-failure landslide behaviours and the potential application of this technique to Failure Forecasting Methods (FFMs) are analysed. Several limits affect the ability of the technique to monitor a landslide process, especially during the pre-failure phase (tertiary creep). In this study, two of the major limitations affecting the technique have been explored: (1) the low data sampling frequency and (2) the phase ambiguity constraints. We explored the time series of displacements for 56 monitored landslides inferred from the scientific literature and from different in situ and remote monitoring instruments (i.e., extensometers, inclinometers, distometers, Ground Base InSAR, and total station). Furthermore, four different forecasting techniques have been applied to the monitoring data of the selected landslides. To analyse the reliability of the FFMs based on the InSAR satellite data, the 56 time series have been sampled based on different satellite features, simulating the satellite revisit time and the phase ambiguity constraints. Our analysis shows that the satellite InSAR technique could be successful in monitoring the landslide's tertiary creep phase and, in some cases, for forecasting the corresponding time of failure using FFMs. However, the low data sampling frequency of the present satellite systems do not capture the necessary detail for the application of FFMs in actual risk management problems or for early warning purposes.

Keywords: A-DInSAR; InSAR; landslide; failure forecasting methods; landslide forecasting

1. INTRODUCTION

The scientific community started to explore landslide behaviour during the first half of the twentieth century [1,2]. In 1950, Terzaghi recognized the connection between landslide evolution and creep theory, laying the foundation for landslide prediction. Landslide forecasting remains a current issue and a significant challenge in natural hazard risk mitigation. Since the beginning of the 1960s, many authors have been studying this topic [3,4,5,6,7,8,9,10,11,12,13,14]. Saito and Uezawa initiated this field of research, and the first successful landslide prediction was performed by Saito in 1965 for the Dosan Line landslide [4]. They defined semi-empirical methods to estimate the timing of slope failure using displacement monitoring data acquired via different techniques.

Subsequently, different semi-empirical methods for landslide prediction were developed. In 1985, a new equation based on laboratory experiments was derived by Fukuzono [6] and later validated by Voight [8]:

$$\Omega'' = A\Omega'^{\alpha}, \qquad (1)$$

This equation describes the relationship between velocity (Ω') and acceleration (Ω'') of the surface displacement during the pre-failure phase under constant stress and temperature conditions using two empirical constants A and α.

Based on Equation (1), failure forecasting methods (FFMs) that estimate the time of failure (TF) were developed by Cornelius and Voight [11].

The methods are based on the creep theory [2], which describes the time-dependent deformational behaviour of the slope under constant stress conditions. Specifically, the final creep phase known as tertiary creep (TC), characterized by an acceleration of the deformation affecting a slope, is modelled using semi-empirical functions to predict the TF. Pre-failure monitoring data are the basic requirement for the application of FFMs. Many successful predictions are described in the literature [15,16,17,18,19,20,21] that were developed with data acquired from different monitoring techniques such as extensometers, inclinometers, GB-InSAR, etc.

In this paper, the potential of satellite interferometry to monitor the pre-failure landslide behaviour for the application of FFMs is assessed.

The advanced techniques of satellite SAR image processing (A-DInSAR Advanced Differential Synthetic Aperture Radar Interferometry, [22,23,24,25]) can provide the time series of displacements for stable radar reflectors in the investigation area. The processing algorithms developed during the last few decades have made A-DInSAR a powerful technique for landslide risk management. It currently allows for landslide detection, mapping, characterization, and monitoring at different scales, both for large area and site specific analysis. However, the present features of the technique influence its suitability for landslide forecasting and early warning purposes, even if, in some cases, satellite InSAR can be successfully applied to disaster risk reduction by identifying unstable slope indicators [26].

The limitations that affect the capability of the technique to monitor a landslide process during the pre-failure phase can be summarized as follows [27]: (i) radar distortions; (ii) presence of backscatter; (iii) availability of data; (vi) areal extension of the landslide process; and (v) deformational behaviour (length of tertiary creep and rate of deformations). In this study, the site-specific feasibility of satellite InSAR monitoring has been deliberately neglected, and the focus is on the constraints related to the slope's deformational behaviour. Two major limitations of the technique for pre-failure landslide monitoring were accounted for: (i) the long satellite revisit time (RT, i.e., the sampling period of the data acquisition) and (ii) the phase ambiguity constraints.

The satellite RT is one of the major limitations affecting the A-DInSAR technique for forecasting purposes because it reduces the applicability of the technique to landslide processes with a TC long enough to be captured by the satellites. The phase ambiguity constraint influences the ability of the technique to monitor rapid displacements, which characterize the pre-failure stage. Furthermore, there is a strong dependence between RT and phase ambiguity. Specifically, if no prior information is available, it is not possible to achieve A-DInSAR results if the displacements are greater than $\lambda/4$ (where λ is the sensor's wavelength) between two consecutive SAR acquisitions and two adjacent pixels [28,29].

Considering these limitations, in this paper the potential of satellite interferometry for landslide forecasting has been explored by back-analysing 56 case histories.

2. MATERIALS AND METHODS

Is it possible to monitor a landslide during TC with satellite interferometry? Is it possible to apply landslide forecasting methods using the InSAR time series of displacement?

In this study, we try to answer these questions by exploring two of the primary limitations affecting satellite InSAR: (i) the low satellite data sampling frequency and (ii) the ambiguity phase constraints.

To assess the satellite InSAR potential for landslide pre-failure monitoring and forecasting, the following approach has been adopted:

1. A database collection composed of 56 past landslides, including pre-failure displacement data inferred from the scientific literature;
2. The digitization of the pre-failure datasets from graphs;
3. The sampling of 56 time series of displacements based on different satellites' revisit time to reproduce the satellite data acquisition;
4. Simulation of the phase ambiguity constraint based on the different satellites' features (wavelength and revisit time);
5. The application of four FFMs on three different datasets: (i) digitized datasets, which represent the actual monitoring datasets; (ii) sampled datasets (based on the revisit time constraint); and (iii) simulated datasets (considering the phase ambiguity limit);
6. A comparison between the forecast results obtained using actual monitoring data and simulated satellite data.

The pre-failure monitoring data of 56 past landslides have been explored by observing their evolution from the beginning of the acceleration phase to the occurrence of failure.

The collected monitoring data show different characteristics; they were acquired using different monitoring instruments (i.e., extensometers, inclinometers, distometers, GB-InSAR, and total station) and were displayed in the instruments using different formats, including displacement vs. time, velocity vs. time, and inverse velocity vs. time. Furthermore, the data was displayed in different units such as millimetres, centimetres, metres, mm/day, mm/hour, and cm/day. To homogenize the information and to compare the results, all the datasets are reported in mm/day.

The 56 landslide datasets collected (Table 1) show different features in terms of: triggering factor, mechanism, volume, material, and deformational behaviour.

For the objective of this work, the most important feature is the deformational behaviour of the slope before failure; specifically, the TC duration and the deformation rates occurring during the TC. The beginning of TC has been manually defined on the strain rate vs. time plot as the point at which the velocity begins to increase just before the final collapse. As shown in Table 1, the length of the TC period is extremely variable, ranging from several minutes to more than 1000 days. Additionally, the average velocity during the pre-failure phase varies substantially from 0.01 m/day to more than 4 m per day.

Monitoring data presented graphically in the scientific literature have been digitized using the MATLAB® code "grabit". An example of the digitized data for the Vajont landslide is shown in Figure 1. To simulate the satellite data acquisition, the monitoring datasets of the 56 landslides have been sampled in the velocity time space based on different satellites' RT values (Figure 2). The features of nine satellites in the past, present, and future have been considered (Table 2).

Furthermore, according to the satellites' wavelengths and RT values, time series have been customized to consider the phase ambiguity limits. Because the phase is a periodic function and the wavelength number in the two travel paths is unknown, it follows that displacements greater than $\lambda/2$ between two consecutive acquisitions may remain unsolved or underestimated. Furthermore, for Advanced DInSAR (i.e., Persistent Scatterers Interferometry (PSI), Small Baseline Subset (SBAS), and similar techniques), the maximum amount of displacement that can be measured between two consecutive acquisitions is $\lambda/4$ [28]. To simulate this limit, we assumed that satellite InSAR is not able to monitor displacements greater than $\lambda/4$ between two consecutive acquisitions. For example, considering the Sentinel features, measurement points with displacements greater than 14 mm in 6 days have not been considered (Figure 3). Because the datasets are in the velocity–time space, the

Table 1. Landslide database collected from the scientific literature.

Landslide	Reference	ID	Tertiary Creep (TC) Length (days)	Average Velocity (mm/day)
Jitsukiyama	[30]	1	0.01	0.04 (mm/s)
Moriwaki Experiment	[31]	2	0.02	0.04 (mm/min)
Unnamed Mufundirawa	[32]	3	0.1	0.87 (mm/s)
Letlhakane Mine	[33]	4	0.2	68.60 (mm/h)
Dosan Line	[34]	5	1.2	664.79
La Saxe	[35]	6	1.6	4541.68
Betz Post 2 mc	[15]	7	3.2	63.42
Asamushi	[34]	8	3.4	163.76
Rock Dump	[36]	9	4.1	1011.21
Cavallerizzo	[37]	10	4.3	26.32
Roesgrenda B	[38]	11	6	171.93
Tuckabianna West	[39]	12	6.1	39.55
Afton	[40]	13	9.4	591.07
Excavation A	[40]	14	9.7	14.22
Preonzo	[41]	15	10.6	91.30
Silla Montecchi	[42]	16	10.8	80.31
Eskihisar Coal Mine	[43]	17	11.5	6.72
Ryan Call Slide1	[44]	18	12.8	1564.10
Crack Gauge	[45]	19	13	11.98
Ruahihi	[46]	20	14.1	54.58
Es.2 Rose & Hungr2007	[15]	21	15	52.16
Roesgrenda A	[38]	22	20.1	2.00
Abbotsford	[46]	23	20.4	106.26
New Tredegar	[47]	24	22.5	36.22
Victorious East Pit	[40]	25	22.7	4.56
Haveluck	[40]	26	25.8	1.94
Bohemia	[48]	27	30.3	0.51
Val Pola	[49]	28	30.5	51.33
Luscar_51B2	[40]	29	31.6	76.73
Ooigawa Railroad	[34]	30	38.9	12.76
Telfer	[40]	31	39.4	2.73
Liberty Pit Mine	[15]	32	40.2	4.73
Selbourn	[50]	33	41.5	100.54
Betz Post 18mc	[15]	34	44	6.58
Lijiaxia	[51]	35	44.7	4.47
Takabayama	[52]	36	44.9	145.78
Kennekott#1	[53]	37	46.9	57.19
Chuquicamata Mine2	[52]	38	47.6	224.24
Ryan Call Slide2	[44]	39	53.7	21.35
Vajont	[13]	40	54.6	30.41
Luscar 50A2	[40]	41	60.9	1272.11
Hogart	[54]	42	77.8	63.85
Roberts	[55]	43	105	57.46
Nevis Bluff	[56]	44	110.4	2.98
Monte Beni	[16]	45	132.2	19.67
Xintan	[57]	46	138.9	459.60
Azimi 1988	[7]	47	143.8	3.39
Threatening Rock	[58]	48	153.8	2.49
Bomba	[59]	49	167.9	10.86
Smoky River	[40]	50	174.2	1.59
Scalate	[55]	51	183.7	0.76
Brada River	[60]	52	187.8	0.10
Excavation C	[40]	53	211.3	1.22
Delabole Quarry	[61]	54	380.7	2.89
Saleshan	[62]	55	631.3	1.88
Puigcercos Scarp	[63]	56	1013	0.39

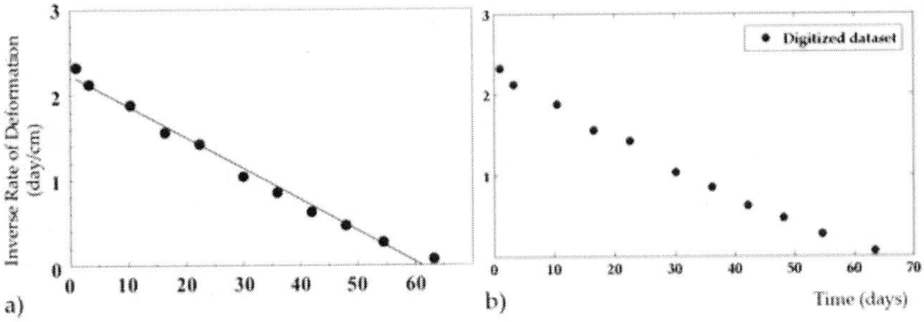

Figure 1. An example of the digitized monitoring data for the Vajont landslide (9 October 1963). (**a**) Monitoring data available in the literature ([12] after Muller [64] and Voight [8]) referencing the inverse rates (days/cm) of horizontal slope movement registered before the Mt. Toc catastrophic collapse. Velocity measurements were obtained from stations near the toe of the moving mass [12]; (**b**) Digitized data.

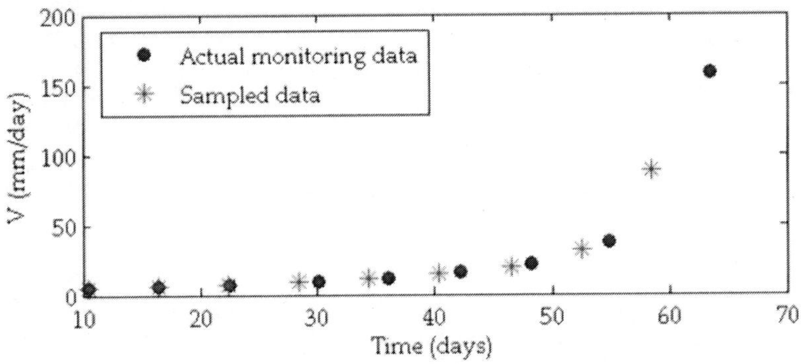

Figure 2. Example of sampled datasets (red stars) based on the Sentinel 1A + 1B revisit time of 6 days. Blue circles represent the pre-failure monitoring data of the Vajont landslide.

measurement points with a velocity greater than the velocity of ambiguity ($V_{\text{ambiguity}}$) have been removed, where $V_{\text{ambiguity}}$ is defined as:

$$V_{\text{ambiguity}} = (\lambda/4)/\text{RT}, \qquad (2)$$

Four FFMs have been used to obtain the TF estimation [65]:

i. Inverse velocity method (INV, [6]);

ii. Log rate vs. log acceleration technique (LOG, [11]);

iii. Linearized least-square technique (LSM, [11]);

iv. Non-linear least-square technique (NL, [11]).

Table 2. Satellites considered in this study and their principal features (Revisit Time (RT) and wavelength (λ)).

Period	Satellite	Revisit Time (days)	Band	λ (mm)
Past	ALOS PALSAR 1	46	L	236
	J-ERS	44	L	235
	ERS 1/2-Envisat	35	C	56
Present	COSMO-SkyMed	4	X	31
		8	X	31
		16	X	31
	Sentinel-1A	12	C	56
	Sentinel 1A + 1B	6	C	56
	ALOS PALSAR 2	14	L	236
	TerraSAR-X	11	X	31
Future	Radarsat Constellation	4	C	56

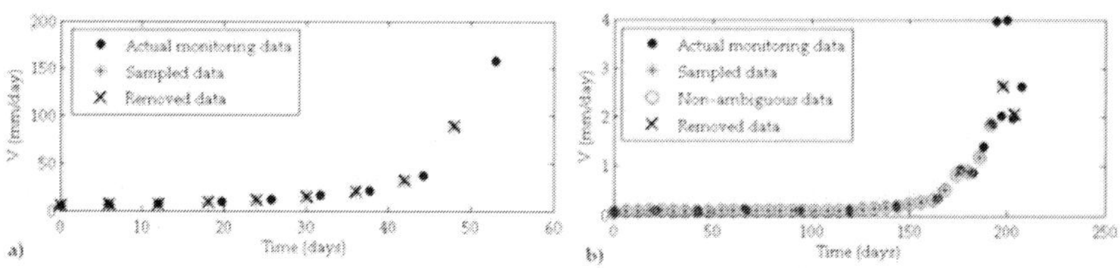

Figure 3. An example of the ambiguity phase simulation based on the Sentinel 1A+1B features (RT = 6 days and Vambiguity = 2.33 mm/day). Removed data (black X) represent the measurements with a velocity greater than the velocity of ambiguity, while the green circles represent data with $V <$ Vambiguity. (**a**) Vajont's datasets; in this case, the deformations occurring during the pre-failure stage were too fast to be monitored with satellite interferometry; (**b**) Excavation C datasets. The raw data consist of surface displacement data acquired using a prism monitoring system.

The INV method is a graphical technique based on the linear approximation of the inverse velocity data, while the other methods are numerical techniques derived via the manipulation of Equation (1). The INV method is the most utilized due to its simple implementation in actual emergency situations. It works well for linearly distributed data (α = 2); however, the other methods allow data characterized by non-linear trends to be fit based on the estimation of the two constants A and α.

The four forecasting methods have been applied to all the landslide time series using the following datasets:

i. the actual monitoring data, i.e., the digitized datasets;

ii. the sampled time series, i.e., the sampled time series based on the satellite's revisit time;

iii. the simulated time series, which considers the phase ambiguity constraint.

Using this approach, it was possible to assess the suitability of the satellite InSAR technique for monitoring the landslide process during the tertiary creep phase and for the application of FFMs based on the revisit time and phase ambiguity constraints of different satellite sensors.

3. RESULTS

The FFM results are reported in the following subsections according to the datasets used: (i) the actual monitoring data (Section 3.1); (ii) the sampled data (Section 3.2) and (iii) the simulated time series (Section 3.3).

3.1. Actual Monitoring Data

The FFM results for the actual monitoring data are compared to the results obtained for the sampled and simulated time series. The forecasting methods have been applied to 56 past landslides considering all the tertiary creep phases from the beginning to the last available datum. For each landslide, the actual time of the slope failure (actual failure) is known, allowing us to estimate the prediction error. The prediction error obtained for each landslide according to each forecasting method is shown in Figure 4. The error (E) has been defined as the difference in days between the actual failure and the computed time of failure (Tf):

$$E = \text{Actual Failure} - \text{Tf}, \quad (3)$$

Based on this equation, positive error values would represent safe predictions in a priori analysis [29]; in contrast, negative error values are related to unsafe predictions (Tf is delayed with respect to the actual failure).

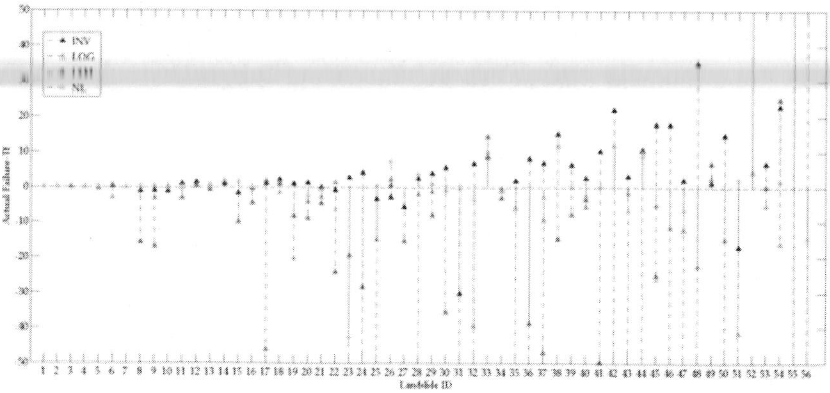

Figure 4. Error diagram obtained for the actual monitoring datasets. The colours identify the different failure forecasting methods (FFMs).

Landslide ID values are reported on the *x*-axis, while the prediction error is on the *y*-axis. ID values have been assigned according to an increasing TC length. Additionally, a decrease in data sampling frequency is contained in our database toward the higher ID values.

Figure 4 shows that the prediction error increases towards the higher ID values. Predictions close to the actual failure have been obtained for the low ID values (from ID 1 to ID 21), i.e., based on the datasets with higher sampling frequency and a short TC phase.

Overall, the semi-empirical methods returned predictions with an error less than 3 days for: 30 examples using the graphical technique, 28 using the NL technique, 24 using the LOG technique, and 15 using the LSM technique.

3.2. Sampled Datasets

For the application of forecasting methods, a minimum of three monitoring points within the accelerating phase is required. Under these conditions, we can build a function that can provide information on the increasing displacement velocity and, thus, we are able to apply semi-empirical models. Based on the TC duration and on the revisit time of the different satellites, the landslides with at least three data points within the TC phase have been defined as compatible for the application of the forecasting methods (Table 3).

Table 3. Analysis results for the sampled datasets related to different RTs. The number of compatible landslides and the number of satisfactory predictions (error lower than ±5 days) for each satellite are reported. The compatible landslides include at least three data samples in the TC phase. For each forecasting method (inverse velocity method—INV, log rate vs. log acceleration technique—LOG, linearized least-square technique—LSM, and non-linear least-square technique—NL), the number of landslides with a prediction error lower than ±5 days is reported. The best performances obtained in terms of prediction accuracy are underlined.

Period	Satellite	RT (Days)	Band	λ (mm)	Compatible Landslides (Revisit Time)	E < |5| Days INV	E < |5| Days LOG	E < |5| Days LSM	E < |5| Days NL
					Number of Landslides				
Past	ALOS PALSAR 1	46	L	236	14	1	2	1	1
	J-ERS	44	L	235	14	2	1	0	1
	ERS 1/2-Envisat	35	C	56	14	1	0	1	0
Present	COSMO-SkyMed	4	X	31	43	17	11	11	11
		8	X	31	35	8	5	10	10
		16	X	31	26	4	2	7	7
	Sentinel-1A	12	C	56	30	7	2	7	4
	Sentinel 1A + 1B	6	C	56	38	13	3	5	9
	ALOS PALSAR 2	14	L	236	29	4	6	3	6
	TerraSAR-X	11	X	31	30	6	2	4	3
Future	Radarsat Constellation	4	C	56	43	17	11	11	11

Considering the past satellites (ALOS PALSAR, J-ERS, ERS, and Envisat) characterized by the longer revisit times ranging from 35 to 46 days, 14 compatible landslides (i.e., those that can be monitored using satellite InSAR) have been identified out of the 56 collected in the database. For the present and the future satellite generations, a maximum of 43 compatible landslides have been identified based on the revisit time of the Sentinel 1A-1B, Cosmo Sky-Med (theoretical RT of 4 days), and Radarsat Constellation.

Considering the present satellites' features, the number of compatible landslides ranges from 26 to 43, i.e., from 25% to 86% of the landslides collected in the database. This means that, at present, considering only the revisit time constraint, we could observe a change in the displacement trend via satellite with at least three data points for more than 25% of the collected landslides, reaching 86% of the landslides considering the Sentinel-1 characteristics.

The four FFMs have been applied to every compatible landslide based on each satellite's revisit time. Figure 5 shows the analysis performed considering the shorter revisit time (namely, 4 days), belonging to the future Radarsat Constellation and, theoretically, to the COSMO-SkyMed Constellation. For the forecasting analysis applied to the actual monitoring data, the prediction error increases toward the higher ID values. Specifically, the errors derived from the INV application are small (<±5 days) up to ID 25 and start to increase after ID 25.

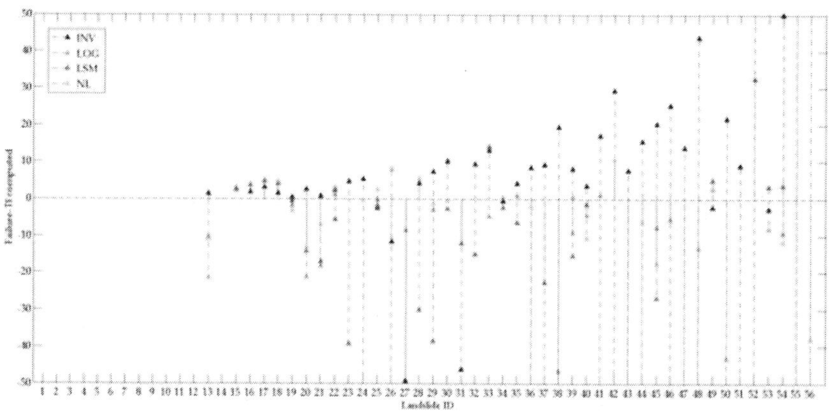

Figure 5. Diagram of the prediction error for the compatible landslides based on a RT of 4 days.

Table 3 and Figure 6 show that the number of satisfactory predictions increases toward the shorter revisit times. A prediction is considered satisfactory if the prediction error is within ±5 days. The INV method shows the strongest relationship between the number of successful forecasts and the RT (Figure 6b), because a shorter RT results in a higher number of accurate predictions.

The analysis shows that the INV method results in a better accuracy for the TF estimation. The INV technique has a prediction error of less than 5 days for 17 landslides based on an RT of 4 days, and for 13 landslides based on the RT of the Sentinel-1 satellite.

3.3. Simulated Time Series

Considering the phase ambiguity constraints, a maximum of 24 compatible landslides out of the 56 have been identified based on the Radarsat Constellation features and characterized by the best association between the revisit time and the sensor's wavelength. Currently, the Sentinel-1 constellation would have observed 17 landslides with at least three measurement points inside the acceleration phase preceding the slope collapse. By introducing the phase ambiguity, the suitability of the failure forecasting methods decrease. For the simulated satellite datasets, only a maximum of four satisfactory predictions have been obtained using the inverse velocity method (Table 4).

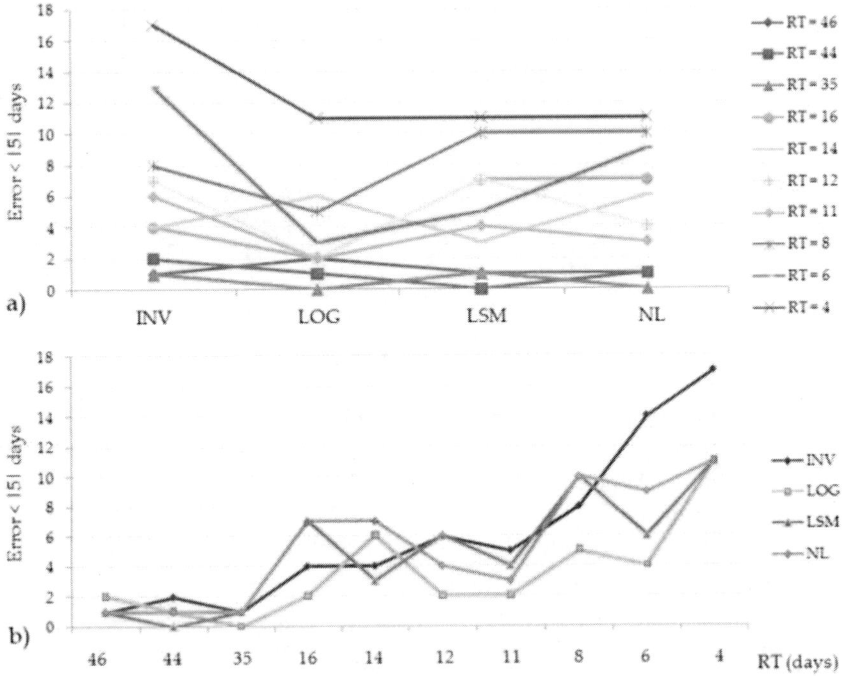

Figure 6. (**a**) Relationship between the number of successful forecasts and the forecasting method considering the different revisit times; (**b**) Relationship between the revisit time and the number of satisfactory predictions based on each FFM.

4. DISCUSSION

The forecasting methods have great potential for landslide hazard management [16,18,19,20]. By applying FFMs to actual monitoring data, we have obtained a prediction error lower than 3 days in 53.7% of the examples using the INV method, 51.7% using the LOG technique, 48.2% using the NL technique, and 27.8% using the LSM technique. The percentage of successful predictions increases when considering

Table 4. Analysis results introducing the phase ambiguity constraint based on the different satellite's features. "Compatible Landslides (Phase ambiguity)" represents the number of landslides with at least three data points in the TC phase accounting for the phase ambiguity limit. For each forecasting method, the number of landslides with a prediction error lower than 5 days is reported.

Period	Satellite	RT (days)	Band	λ (mm)	Compatible Landslides (RT)	Compatible Landslides (Phase Ambiguity)	E < |5| days INV	E < |5| days LOG	E < |5| days LSM	E < |5| days NL
					Number of Landslides					
Past	ALOS PALSAR 1	46	L	236	14	7	0	1	0	0
	J-ERS	44	L	235	14	8	0	0	0	0
	ERS $\frac{1}{2}$-Envisat	35	C	56	14	4	1	0	1	0
Present	COSMO-SkyMed	4	X	31	43	19	3	0	3	1
		8	X	31	35	11	0	0	2	0
		16	X	31	26	9	0	0	0	0
	Sentinel-1A	12	C	56	30	11	1	0	0	0
	Sentinel 1A + 1B	6	C	56	38	17	4	3	2	1
	ALOS PALSAR 2	14	L	236	29	14	1	2	1	1
	TerraSAR-X	11	X	31	30	9	0	0	0	0
Future	Radarsat Constellation	4	C	56	43	24	4	1	4	2

landslides with shorter TC phases and datasets characterized by a higher sampling frequency, namely, the landslide database presented here with lower ID values. When considering only the first 21 landslides (TC < 15 days), an error lower than 3 days has been obtained in 100% of the examples using the INV method, 85.7% using the LOG technique, 76.2% using the NL technique, and 57.1% using the LSM method.

For the sampled and simulated datasets, the number of successful predictions decreases (Table 3 and Table 4). Only a maximum of four satisfactory predictions have been obtained with the inverse velocity method including the ambiguity phase constraints. The principal reason is because the FFMs have been applied to only the landslides with a TC phase long enough to be observed with the satellite's RT (high ID values). The predictions obtained using the actual and simulated datasets are similar (Figure 7) when the FFMs are accurate using the actual monitoring data (ID 23, 27, 35, 49, 53). A reliable prediction can also be obtained using the simulated datasets; however, if the prediction error is high using the actual time series, the FFMs fail using the simulated satellite data as well.

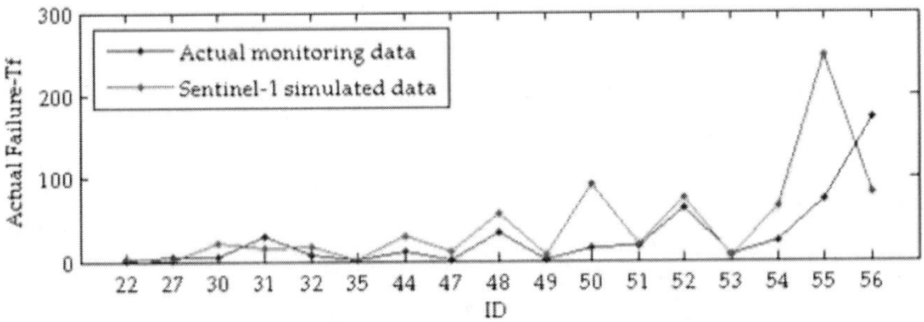

Figure 7. Comparison between the INV method applied to the actual monitoring data and to the simulated datasets. The results of the simulated datasets are referenced to the Sentinel 1A + 1B features.

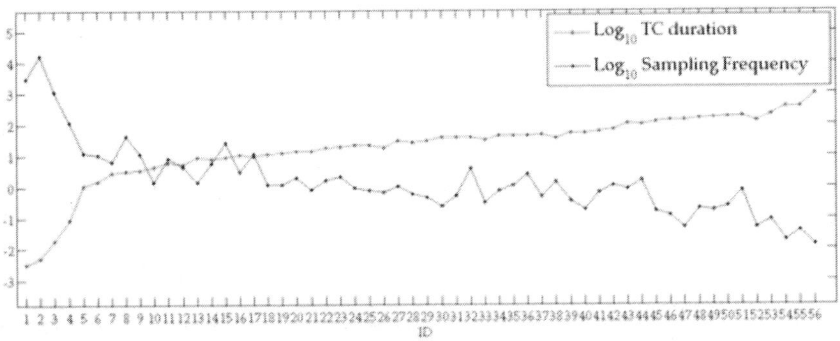

Figure 8. Inverse correlation between the TC's length and data sampling frequency.

The error growth towards higher ID values could be related to both the decrease of the data sampling frequency and to the increase of the TC phase duration. Because of the inverse relationship between the TC length and data sampling frequency (Figure 8), we are not able to clearly determine which parameter primarily affects the success or the failure of the forecasting analyses. A higher sampling frequency produces more robust datasets for the application of forecasting methods, especially near the time of slope collapse; however, a very long TC phase can lead to complex deformational behaviours due to the variation of the stress conditions, which has a negative impact on the FFM stability. The difference in the prediction errors obtained with the actual and simulated time series data is related to the loss of the last monitoring data caused by the sampling and ambiguity phase reproduction. It is assumed that the last monitoring data before the collapse significantly

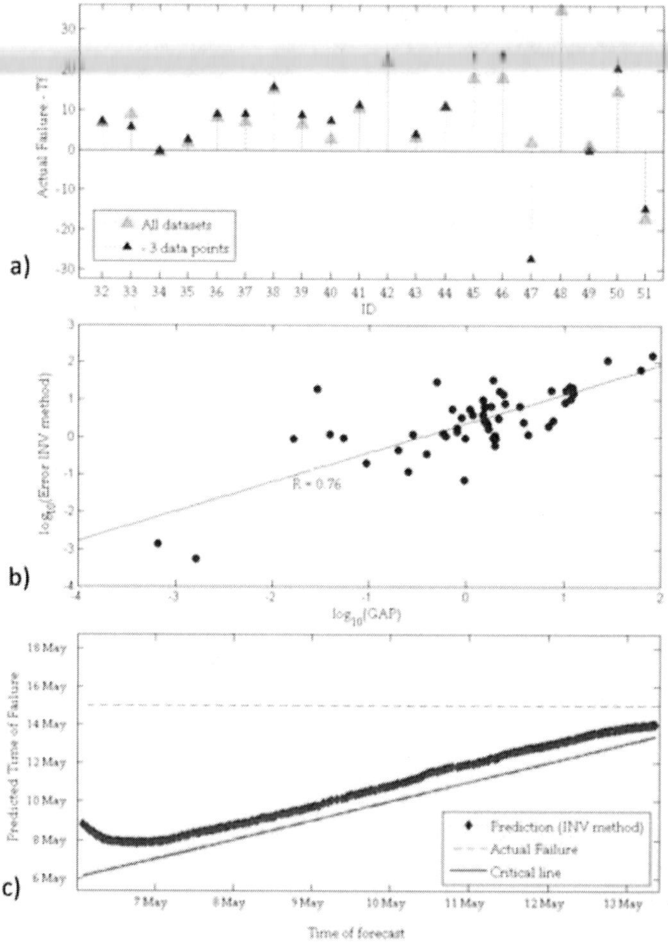

Figure 9. Prediction errors. (**a**) Errors using the INV method to: (i) the entire available time series (green predictions) and (ii) removing the last three measurement points (blue predictions); (**b**) Prediction error vs. GAP (time interval between the last datum and the actual failure); (**c**) INV iterative forecasting analysis for the Preonzo Landslide (ID 12) (modified after [66]). The predictions have been updated by simulating the data acquisition over time. The time of the actual failure (15 May) is displayed as a red dashed line. The critical line (red line) corresponds to the day the forecast was made. Consequently, it represents the lower boundary of a reasonable prediction. The analysis shows that the closer the last available datum is to the actual failure, the more accurate the prediction. When approaching the actual failure, the prediction error decreases, in this case, following a linear trend for the INV methods.

affects the forecasting analyses results because they include the highest velocity of the phenomenon. Under equal conditions, the closer the monitoring data are to the actual slope failure time, the more accurate the forecast. This can be proven by comparing the predictions obtained using the entire time series (datasets from the beginning of TC until the last monitoring datum) and only a part of the time series (i.e., removing the last measurement points) (Figure 9). Based on our results, there is a connection between the time intervals from the last available measurement point to the actual failure (GAP) and the accuracy of the forecasting methods. The relationship between the GAP and the prediction accuracy can be observed for the majority of the landslides collected (Figure 9a,b). In Figure 9c, an example is shown where the prediction accuracy increases when approaching the actual failure based on the Preonzo landslide monitoring data.

The GAP reduction capability is related to the sampling frequency; the higher the sampling frequency, the greater the ability to obtain measurements closer to the actual failure. Figure 10 shows the relationship between the GAP and the sampling frequency for the time series of our database.

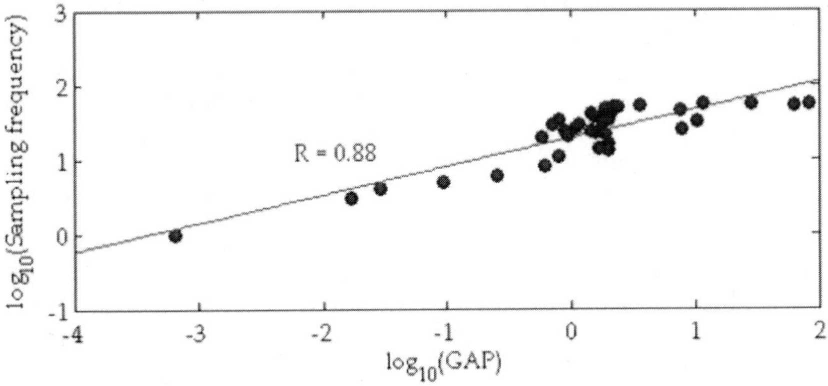

Figure 10. Relationship between the data sampling frequency and the GAP for the collected landslide database.

5. CONCLUSIONS

From our analyses, the following conclusions can be drawn:

1. Monitoring sampling frequency is important for the application of forecasting methods. For some events, an accurate prediction using FFMs on simulated satellite data has been obtained.

However, satellite interferometry cannot provide, at present, the sampling frequency required for a reliable FFM for early warning purposes.

2. Satellite interferometry has potential for monitoring landslide precursors and for detecting the slopes affected by a change in their deformational behaviour. According to the deformational behaviours of our landslide database, at present, 30% of the events could have been monitored during the acceleration phase with at least three data points, even considering the phase ambiguity constraints. Thus, based on the Sentinel 1A + 1B features, it is possible to monitor 17/56 landslides using satellite SAR interferometry.

3. Considering both the RT and ambiguity phase limits, it would have been possible to forecast a maximum of 4 landslides with an error smaller than 5 days, based on the Sentinel-1 features. This is due to the strong constraints related to the phase ambiguity, which does not allow the monitoring of the high displacement rate occurring during the tertiary creep phase. Notably, it is sometimes possible to monitor consecutive displacements greater than $\lambda/4$. If one direction of motion can be assumed ('upward' or 'downward', as for the majority of landslides), the maximum measurable displacement between a pair of scenarios becomes one-half of the wavelength [28,29]. However, considering $\lambda/2$ as the maximum detectable displacement between two consecutive acquisitions, we achieved a similar result: six successful forecasts were obtained based on the Sentinel-1 features.

4. If a landslide process is monitored using a suitable technique (high spatial and temporal resolution), forecasting methods can be a powerful tool for risk management and for early-warning systems.

ACKNOWLEDGMENTS

The present work has been carried out in the framework of the project *"Grandi Frane in roccia e Frane Superficiali a Cinematica Rapida in Aree Montane: Metodi Per La Previsione Temporale e Spaziale (Prediction and Susceptibility)"*, supported by the Dipartimento per gli Affari Regionali e le Autonomie della Presidenza del Consiglio dei Ministri and the Università di Roma «La Sapienza». We would like to thank the two anonymous reviewers and the Editor for their helpful and constructive commonts.

AUTHOR CONTRIBUTIONS

Francesca Bozzano and Paolo Mazzanti conceived the study. Francesca Bozzano, Carlo Esposito and Paolo Mazzanti coordinated the research and contributed to the article organization. Serena Moretto and Alfredo Rocca built the landslides database. Serena Moretto designed and performed the forecasting analyses. Serena Moretto and Francesca Bozzano drafted the manuscript which was revised by all Authors.

CONFLICTS OF INTEREST

The authors declare no conflict of interest.

REFERENCES

1. Heim, A. Bergsturz und Menschenleben; Fretz. & Wasmuth: Zürich, Switzerland, 1932. (In Germany)
2. Terzaghi, K. Mechanism of landslides. In Application of Geology to Engineering Practice (Berkeley Volume); Geological Society of America: Washington, DC, USA, 1950; pp. 83–123.
3. Saito, M.; Uezawa, H. Failure of Soil due to Creep. In Proceedings of the 5th International Conference on Soil Mechanics and Foundation Engineering (ICSMFE), Paris, France, 17–22 July 1961; Volume 1, pp. 315–318.
4. Saito, M. Forecasting the Time of Occurrence of a Slope Failure. In Proceedings of the 6th International Conference on Soil Mechanics and Foundatin Engineering, Montreal, QC, Canada, 8–15 September 1965; pp. 537–541.
5. Voight, B.; Kennedy, B.A. Slope Failure of 1967–1969, Chuquicamata Mine, Chile. Dev. Geotech. Eng. 1979, 4, 595–632.
6. Fukuzono, T. A new method for predicting the failure time of a slope. In Proceedings of the 4th International Conference and Field Workshop on Landslides, Tokyo, Japan, 23–31 August 1985; pp. 145–150.
7. Azimi, C.; Biarez, J.; Desvarreux, P.; Keime, F. Forecasting time of failure for a rockslide in gypsum. In Proceedings of the 5th international symposium on landslides, Lausanne, Switherland, 10–15 July 1988; pp. 531–536.
8. Voight, B. A relation to describe rate-dependent material failure. Science 1989, 243, 200–203. [PubMed]
9. Voight, B. A method for prediction of volcanic eruptions. Nature 1988, 332, 125–130.
10. Fukuzono, T. A simple method for predicting the failure time of slope using reciprocal of velocity. Technol. Disaster Prev. Sci. Technol. Agency Jpn. Int. Coop. Agency 1989, 13, 111–128.

11. Cornelius, R.R.; Voight, B. Graphical and PC-software analysis of volcano eruption precursors according to the materials failure forecast method (FFMS). J. Volcanol. Geotherm. Res. 1995, 64, 295–320.
12. Kilburn, C.R.; Petley, D.N. Forecasting giant, catastrophic slope collapse: Lessons from Vajont, Northern Italy. Geomorphology 2003, 54, 21–32.
13. Petley, D.N.; Higuchi, T.; Petley, D.J.; Bulmer, M.H.; Carey, J. Development of progressive landslide failure in cohesive materials. Geology 2005, 33, 201–204.
14. Petley, D.N.; Petley, D.J.; Allison, R.J. Temporal prediction in landslides. Understanding the Saito effect. In Landslides and Engineered Slopes: From the Past to the Future; Taylor & Francis: Xi'an, China, 2008; pp. 794–800.
15. Rose, N.D.; Hungr, O. Forecasting potential rock slope failure in open pit mines using the inversevelocity method. Int. J. Rock Mech. Min. Sci. 2007, 44, 308–320.
16. Gigli, G.; Fanti, R.; Canuti, P.; Casagli, N. Integration of advanced monitoring and numerical modeling techniques for the complete risk scenario analysis of rockslides: The case of Mt. Beni (Florence, Italy). Eng. Geol. 2011, 120, 48–59.
17. Bozzano, F.; Cipriani, I.; Mazzanti, P.; Prestininzi, A. A field experiment for calibrating landslide time-of-failure prediction functions. Int. J. Rock Mech. Min. Sci. 2014, 67, 69–77.
18. Mazzanti, P.; Bozzano, F.; Cipriani, I.; Prestininzi, A. New insights into the temporal prediction of landslides by a terrestrial SAR interferometry monitoring case study. Landslides 2015, 12, 55.
19. Manconi, A.; Giordan, D. Landslide early warning based on failure forecast models: The example of the Mt. de La Saxe rockslide, northern Italy. Nat. Hazards Earth Syst. Sci. 2015, 15, 1639–1644.
20. Sättele, M.; Krautblatter, M.; Bründl, M.; Straub, D. Forecasting rock slope failure: How reliable and effective are warning systems? Landslide 2016, 13, 737.
21. Loew, S.; Gschwind, S.; Keller-Signer, A.; Valenti, G. Monitoring and Early Warning of the 2012 Preonzo Catastrophic Rockslope Failure. Landslides 2017, 14, 141.
22. Ferretti, A.; Prati, C.; Rocca, F. Nonlinear subsidence rate estimation using permanent scatterers in differential SAR interferometry. IEEE Trans. Geosci. Remote Sens. 2000, 38, 2202–2212.
23. Ferretti, A.; Prati, C.; Rocca, F. Permanent scatterers in SAR interferometry. IEEE Trans. Geosc. Remote Sens. 2001, 39, 8–20.
24. Berardino, P.; Fornaro, G.; Lanari, R.; Sansosti, E. A new algorithm for surface deformation monitoring based on small baseline differeential SAR interferograms. IEEE Trans. Geosc. Remote Sens. 2002, 40, 2375–2383.
25. Perissin, D.; Wang, T. Repeat-Pass SAR Interferometry with Partially Coherent Targets. IEEE Trans. Geosci. Remote Sens. 2012, 50, 271–280.
26. Early Warning from Space of Homes on the Slide. Available online: http://www.esa.int/Our_Activities/Observing_the_Earth/Early_warning_from_space_of_homes_on_the_slide (accessed on 9 May 2017).
27. Mazzanti, P.; Rocca, A.; Bozzano, F.; Cossu, R.; Floris, M. Landslides Forecasting Analysis by Time Series Displacement Derived from Satellite InSAR Data: Preliminary Results. In Proceedings of the Fringe 2011, Frascati, Italy, 19–23 September 2011.
28. Ferretti, A.; Prati, C.; Rocca, F.; Nicola, C.; Farina, P.; Young, B. Permanent scatterers technology: A powerful state of the art tool for historic and future monitoring of landslides and other terrain instability phenomena. In International Conference on Landslide Risk Management; Balkema: Vancouver, BC, Canada, 2005.
29. Crosetto, M.; Monserrat, O.; Iglesias, R.; Crippa, B. Persistent scatterer interferometry: Potential, limits and initial cand X-band comparison. Photogramm. Eng. Remote Sens. 2010, 76, 1061–1069.
30. Fukuzono, T. Experimental Study of Slope Failure Caused by Heavy Rainfall. IAHS-AISH Publ. 1987, 165, 133–134.

31. Moriwaki, H.; Inokuchi, T.; Hattanji, T.; Sassa, K.; Ochiai, H.; Wang, G. Failure processes in a full-scale landslide experiment using a rainfall simulator. Landslides 2012, 1, 277–288.
32. Mufundirwa, A.; Fujii, Y.; Kodama, J. A new practical method for prediction of geomechanical failure-time. Int. J. Rock Mech. Min. Sci. 2010, 47, 1079–1090.
33. Kayesa, G. Prediction of slope failure at Letlhakane mine with the geomos slope monitoring system. In International Symposium on Stability of Rock Slopes in Open Pit Mining and Civil Engineering; The South African Institute of Mining and Metallurgy: Johannesburg, South Africa, 2006; pp. 606–622.
34. Saito, M. Forecasting Time of Slope Failure by Tertiary Creep. In Proceedings of the 7th International Conference on Soil Mechanics and Foundation Engineering; Sociedad Mexicana de Mecanica: Mexico City, Mexico, 1969.; pp. 677–683.
35. Manconi, A.; Giordan, D. Landslide failure forecast in near-real-time. Geomat. Nat. Hazard. Risk 2014, 7, 639–648.
36. Kent, A. Coal mine waste dumps in British Columbia stability issues and recent developments. Int. Mine Waste Manag. 1992, 10–18.
37. Iovine, G.; Petrucci, O.; Rizzo, V.; Tansi, C. The march 7th 2005 Cavallerizzo (Cerzeto) Landslide in Calabria—Southern Italy. In Proceedings of the 10th IAEG Congress, Nottingham, UK, 6–10 September 2006; pp. 1–12.
38. Okamoto, T.; Larsen, J.O.; Matsuura, S.; Asano, S.; Takeuchi, Y.; Grande, L. Displacement properties of landslide masses at the initiation of failure in quick clay deposits and the effects of meteorological and hydrological factors. Eng. Geol. 2004, 72, 233–251.
39. Thompson, P.W. Confirmation of a Failure Mechanism Using Open Pit Monitoring Methods. In Proceedings of the Australian Conference on Geotechnical Instrumentation and Monitoring in Open Pit and Underground Mining, Kalgoorlie, Australia, 21–23 June 1993; pp. 483–489.
40. Glastonbury, B.; Fell, R. Report on the Analysis of the Deformation Behaviour of Excavated Rock Slopes; School of Civil & Environmental Engineering, University of New South Wales: Sydney, Australia, 2002.
41. GEOPRAEVENT. Available online: www.geopraevent.ch (accessed on10 May 2017).
42. Metodi Innovativi per il Controllo e la Gestione dei Fenomeni Franosi. Available online: http://www.eventi.saie.bolognafiere.it/media/saie/presentazioni%20sba/sala%20A/Matteo_Berti.pdf (accessed on 10 May 2017). (In Italian)
43. Ulusay, R.; Aksoy, H. Assessment of the failure mechanism of a highwall slope under spoil pile loadings at a coal mine. Eng. Geol. 1994, 38, 117–134.
44. Ryan, T.M.; Call, R.D. Applications of Rock Mass Monitoring for Stability Assessment of Pit Slope Failure. In Proceedings of the 33th U.S. Symposium on Rock Mechanics (USRMS), Santa Fe, NM, USA, 3–5 June 1992; Balkema: Rotterdam, The Netherlands, 1992.
45. Cahill, J.; Lee, M. Ground control at Leinster nickel operations. In International Symposium on Stability of Rock Slopes in Open Pit Mining and Civil. Engineering; The South African Institute of Mining and Metallurgy: Johannesburg, South Africa, 2006; pp. 321–334.
46. Salt, G. Landslide Mobility and Remedial Measures. In Proceedings of the 5 International Symposium on Landslides, Rotterdam, The Netherlands, 1988; pp. 757–762.
47. Carey, J.M.; Moore, R.; Petley, D.N.; Siddle, H.J. Pre-failure behaviour of slope materials and their significance in the progressive failure of landslides. In Landslides and Climate Change: Challenges and Solutions; Taylor & Francis Group: London, UK, 2007; pp. 207–215.
48. Urcioli, G.; Picarelli, L. Interaction between landslides and man-made works. In Landslides and Engineered Slopes; Taylor & Francis Group: London, UK, 2008.
49. Borsetto, M.; Frassoni, A.; La Barbera, G.; Fanelli, M.; Giuseppetti, G.; Mazzà, G. An application of Voight empirical model for the prediction of soil and rock instabilities.

In Proceedings of the 7th International Symposium on Landslides; Bell, D.H., Ed.; Balkema: Rotterdam, The Netherlands; Christchurch, New Zealand, 1991; pp. 335–341.
50. Petley, D.N. The evolution of slope failures: Mechanisms of rupture propagation. Nat. Hazard. Earth Sys. Sci. 2004, 4, 147–152.
51. Bai, J.G.; Lu, S.D.; Han, J.S. Importance of study of creep sliding mechanism to prevention and treatment of reservoir landslide. In Landslides and Engineered Slopes: From the Past to the Future; Taylor & Francis Group: Xi'an, China, 2008; pp. 1071–1076.
52. Bhandari, R.K. Special lecture—Some practical lessons in the investigation and field monitoring of landslides. Landslides 1988, 1–3, 1435–1457.
53. Zavodni, Z.M.; Broadbent, C.D. Slope failure kinematics. CIM Bull. 1980, 69–74.
54. Brawner, C.O.; Stacey, P.F. Hogarth pit slope failure, Ontario, Canada. In Rockslides and Avalanches, B; Elsevier: Amsterdam, The Netherlands, 1979; pp. 691–707.
55. Glastonbury, B.; Fell, R. Report On the Analysis of Rapid Natural Rock Slope. R-390; School of Civil & Environmental, University of New South Wales (UNSW): Sydney, Australia, 2001.
56. Brown, I.; Hittinger, M.; Goodman, R. Finite element study of the Nevis Bluff (New Zealand) rock slope failure. Rock Mech. 1980, 12, 231–245.
57. Keqiang, H.; Sijing, W. Double-parameter threshold and its formation mechanism of the colluvial landslide: Xintan landslide, China. Environ. Geol. 2006, 49, 696–707.
58. Schumm, S.A.; Chorley, R.J. The fall of Threatening Rock. Am. J. Sci. 1964, 262, 1041–1054.
59. D'Elia, B.; Picarelli, L.; Leroueil, S.; Vaunat, J. Geotechnical characterisation of slope movements in structurally complex clay soils and stiff jointed clays. Riv. Ital. Geotec. 1998, 3, 5–47.
60. Zabusky, L.; Swidzinski, W.; Kulczykowski, M.; Mrozek, T.; Laskowicz, I. Monitoring of landslides in the Brada river valley in Koronowo (Polish Lowlands). Environ. Earth Sci. 2015, 73, 8609–8619.
61. Boyd, J.M.; Hinds, D.V.; Moy, D.; Rogers, C. Two simple devices for monitoring movements in rock slopes. Quart. J. Eng. Geol. Hydrogeol. 1973, 6, 295–302.
62. Siqing, Q.; Sijing, W. A homomorphic model for identifying abrupt abnormalities of landslide forerunners. Eng. Geol.2000, 57, 163–168.
63. Royán, M.J.; Abellán, A.; Vilaplana, J.M. Progressive failure leading to the 3 December 2013 rockfall at Puigcercós scarp (Catalonia, Spain). Landslides 2015, 12, 585–595.
64. Muller, L. The rock slide in the Vaiont valley. Felsmech. Ing. 1964, 42, 148–212.
65. Bozzano, F.; Mazzanti, P.; Esposito, C.; Moretto, S.; Rocca, A. Potential of satellite InSAR monitoring for landslide Failure Forecasting. In Landslides and Engineered Slopes. Experience, Theory and Practice; CRC Press: Boca Raton, FL, USA, 2016; Volume 2, pp. 523–530.
66. Moretto, S.; Bozzano, F.; Esposito, C.; Mazzanti, P. Lesson learned from the pre-collapse time series of displacement of the Preonzo landslide (Switzerland). Rend. Online Soc. Geol. Ital. 2016, 41, 247–250.

CHAPTER 4

Do Landslides Follow Landslides? Insights in Path Dependency from a Multi-Temporal Landslide Inventory

Jalal Samia[1,2,]Arnaud Temme[23],Arnold Bregt[1],Jakob Wallinga[2],Fausto Guzzetti[4],Francesca Ardizzone[4],Mauro Rossi[4]

[1].*Wageningen URLaboratory of Geo-information Science and Remote SensingWageningenThe Netherlands*

[2].*Wageningen URSoil Geography and LandscapeWageningenThe Netherlands*

[3].*Institute of Arctic and Alpine ResearchUniversity of ColoradoBoulderUSA*

4.CNR IRPIPerugiaItaly

ABSTRACT

Landslides are a major category of natural disasters, causing loss of lives, livelihoods and property. The critical roles played by triggering (such as extreme rainfall and earthquakes), and intrinsic factors (such as slope steepness, soil properties and lithology) have previously successfully been recognized and quantified using a variety of qualitative, quantitative and hybrid methods in a wide range of study sites. However, available data typically do not allow to investigate the effect that earlier landslides have on intrinsic factors and hence on follow-up landslides. Therefore, existing methods cannot account for the potentially complex susceptibility changes caused by landslide events. In this study, we used a substantially different alternative approach to shed light on the potential effect of earlier landslides using a multi-temporal dataset of landslide occurrence containing 17 time slices. Spatial overlap and the time interval between landslides play key roles in our work. We quantified the degree to which landslides preferentially occur in locations where landslides occurred previously, how long such an effect is noticeable, and how landslides are spatially associated over time. We also investigated whether overlap with previous landslides causes differences in landslide geometric properties. We found that overlap among landslides demonstrates a clear legacy effect (path dependency) that has influence on the landslide affected area. Landslides appear to cause greater susceptibility for follow-up landslides over a period of about 10 years. Follow-up landslides are on average larger and rounder than landslides that do not follow earlier slides. The effect of earlier landslides on follow-up landslides has implications for understanding of the landslides evolution and the assessment of landslide susceptibility.

Keywords: Path dependency Follow-up landslides Spatial association Roundness Landslide susceptibility

INTRODUCTION

Existing landslide research recognizes the critical role that is played by external triggers (e.g. extreme rainfall events, earthquakes and human interferences) along with intrinsic attributes of site (e.g. slope and lithology) that contribute to landslide occurrence (Carrara et al. 1999; Crozier 1986; Guzzetti et al. 2008). Much work has been done on the modelling of landslide susceptibility and hazard. By definition, landslide susceptibility is a non-temporal concept that refers to locations where landslides preferentially occur (Guzzetti et al. 2005; Varnes 1984), whereas landslide hazard describes the likelihood of landslide occurrence in time and space (Guzzetti et al. 1999; Varnes 1984) along with the magnitude of landslide occurrence (Guzzetti et al. 1999). Landslide susceptibility and hazard have been studied using qualitative (Barredo et al. 2000; Ruff and Czurda 2008; Van Westen et al. 2003) and quantitative (Godt et al. 2008; Guzzetti et al. 2005; Lan et al. 2004; Remondo et al. 2003; Van Westen et al. 1997; Van Westen and Terlien 1996; Yeon et al. 2010) approaches. Qualitative approaches emphasize the role of experience and expert knowledge in determining landslide susceptibility (Van Westen et al. 2003). Quantitative approaches assume that conditions that lead to landslide occurrence in the past and present are likely to cause landslides in the future, thus, the probability of occurrence of future landslides is determined using correlations among various conditioning factors and landslide inventories by statistical methods (Tien Bui et al. 2016; Van Westen and Terlien 1996). Deterministic quantitative approaches use detailed geotechnical and hydrological data in combination with statistical models to estimate the probability of slope failure (Aleotti and Chowdhury 1999; Van Westen and Terlien 1996).

Empirical landslide inventories that document the location and sometimes the date, shape and type of landslides play an important role in assessing landslide susceptibility and hazard (Cardinali et al. 2006; Guzzetti et al. 2005, 2012; van Westen et al. 2006, 2008). Most of the landslide inventories are prepared through the interpretation of stereoscopic aerial photographs (Guzzetti et al. 2012). In addition, field mapping (Brunsden 1993), analysis of surface morphology through high-resolution DEMs and interpretation of satellite images (Guzzetti et al. 2012) are used for mapping landslide inventories. Overall statistics of landslide inventories have been studied, and it was documented that the frequency–area distribution of landslides follows a power-law distribution for medium and large landslides with an exponential rollover for small landslides (Guzzetti et al. 2002, 2009; Malamud et al. 2004; Stark and Hovius 2001; Turcotte et al. 2006; Wood et al. 2015).

Although a vast amount of research has been done on the prediction of landslide occurrence, there has, to our knowledge, been no empirical attention for the effect of earlier landslides on future landslides. This effect is nonetheless expected because landslides typically change the surface morphology (Schuster and Highland 2003), the sediment or regolith properties (Chen 2009), the vegetation (Singh et al. 2014) and the slope angle (van Westen et al. 2006), which are all factors that change landslide susceptibility. If true, such importance of landslide history for landslides susceptibility would be a form of path dependency (a concept from complexity theory (Phillips 2006; Temme et al. 2015))—indicating that the history of the landsliding process affects its future through one or more legacy effects. A likely reason for the lack of attention for quantifying the effect of earlier landslides on future landslides is that multi-temporal landslide inventories are very difficult to obtain (Atkinson and Massari 1998; Brenning 2005) and high-resolution multi-temporal datasets of intrinsic properties are virtually absent.

In this paper, we explore the possible effects of earlier landslides on future landslides. For this, we will use a rich multi-temporal landslide inventory from the Collazzone area in central Umbria, Italy (Ardizzone et al. 2013; Galli et al. 2008; Guzzetti et al. 2006).

Our main objective is to investigate whether earlier landslides determine the susceptibility for future landslides, i.e. whether there is path dependency in landslide occurrence. We test two hypotheses. The first hypothesis is that landslides follow landslides; i.e., that landslides increase the likelihood of another (in our terms: follow up) landslide occurring in the same place. We operationalize this hypothesis without attention for the role of intrinsic attributes (e.g. slope and geology) to offer a clear contrast with existing approaches. The second hypothesis is that follow-up landslides differ from other landslides in terms of their shape, size and frequency–area distribution.

STUDY AREA AND DATA

The Collazzone area extends for 78.9 km^2 in central Umbria, Italy (Fig. 1). Elevation in the area ranges from 145 to 634 m above sea level, and the slope, computed from a 10 × 10 m Digital Terrain Model (DTM), ranges from 0° to 63.7° with a mean value of 9.9°. The terrain is hilly with asymmetrical valleys, with lithology and attitude of bedding controlling the slopes. Only sedimentary rocks crop out in the area and include (1) alluvial deposits, Holocene in age; (2) travertine, Pleistocene in age; (3) continental deposits (gravel, sand and clay), Plio–Pleistocene in age (Guzzetti et al. 2006); (4) layered sandstone and marl, Miocene in age and (5) thinly layered limestone, Lias to Oligocene in age. The land use is characterized by cropland, forests, urban areas, pastures, vineyards, orchards and water. Farming in the area favours the development of slope failures and erosion. Soils range in thickness from a few decimetres to more than 1 m; they have a fine or medium texture (Fiorucci et al. 2015). Soils have a xeric moisture regime characterized by cold and moist winters and dry summers. Precipitation is most abundant in October and November; with a mean annual rainfall of 841 mm in the period from 1951 to 2013. Snow falls in the area on average every 2–3 years. Landslides are abundant and range in age, type, morphology and volume from relict—partly eroded—large and deep-seated landslides, to young, mostly shallow landslides involving the soil mantle. Landslides are triggered predominantly by meteorological events, including intense and prolonged rainfall and rapid snowmelt.

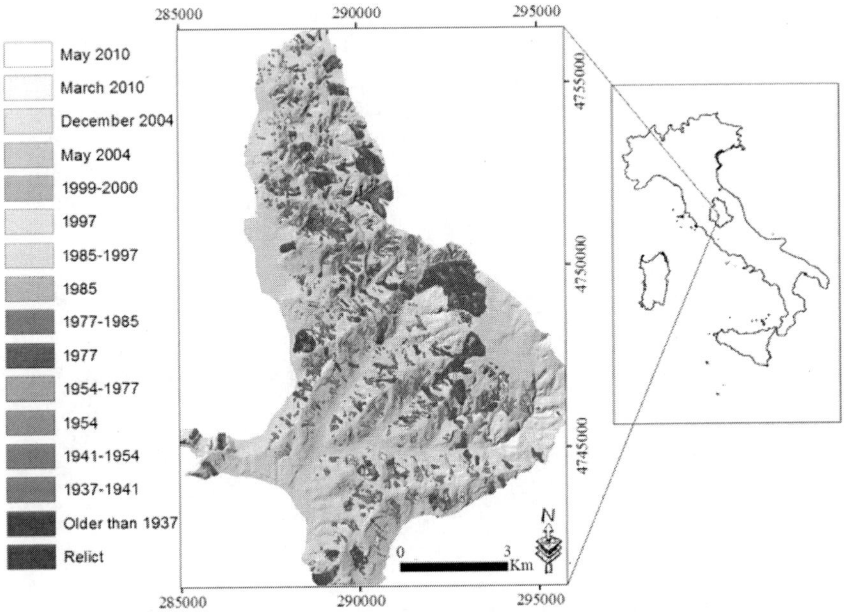

Figure 1. Location of Umbria region and of the Collazzone study area (right map). Multi-temporal landslide inventory overlying a shaded relief image (left map). The coordinate system of the relief image is EPSG:32633

Description of the multi-temporal landslide inventory

A detailed multi-temporal landslide inventory is available for the Collazzone area (Fig. 1 and Table 1). The inventory was originally prepared at 1:10,000 scale through the visual interpretation of five sets of stereoscopic aerial photographs taken unsystematically in the period 1941–1997 at scales ranging from 1:13,000 to 1:33,000. The landslide inventory was continued in the period from 1999 to December 2005 through field surveys carried out after periods of prolonged rainfall and in March and May 2010 using stereo satellite images (Ardizzone et al. 2013).

Landslides ages were estimated from the date of the aerial photographs and the morphological appearance of the landslide. In each of the five sets of stereoscopic aerial photographs used to prepare the multi-temporal inventory, landslides that appeared 'fresh' on the aerial photographs were separated from the other landslides that had occurred since the previous photograph. The date (i.e. year) of the aerial photographs used to identify the landslides was assigned to the 'fresh' slope failures. The other slope failures (i.e. the 'non-fresh' landslides) were attributed to the period between the date of the aerial photograph where they were identified and the date of the previous aerial photographs.

Table 1. Properties of landslides in different time slices of the multi-temporal landslide inventory for the Collazzone area, Umbria, Central Italy. Period covered, number and density of landslide per period, total affected area and inventory type per period. Inventory type: P, obtained from the systematic interpretation of stereoscopic aerial photographs; F, obtained through direct field mapping; S, obtained from the interpretation of stereoscopic satellite images. Landslide periods are estimated from the date of the aerial photographs and the morphological appearance of the landslide

Period covered	Number of landslides(N)	Density(N)/(Km^2)	Total Area of landslides(Km^2)	Inventory type
Relict	27	0.34	5.72	P
Older than 1937	269	3.41	6.55	P
1937–1941	706	8.86	4.09	P
1941–1954	63	0.79	0.42	P
1954	97	1.23	0.71	P
1954–1977	409	5.18	1.49	P
1977	252	3.19	0.69	P
1978–1985	105	1.33	0.62	P
1985	135	1.71	0.45	P
1986–1997	63	0.79	0.27	P
1997	413	5.23	0.78	P
1999–2000	17	0.21	0.07	F
May 2004	71	0.89	0.27	F
December 2004	154	1.94	0.38	F
December 2005	62	0.90	0.18	F
March 2010	158	2.00	0.29	S
May 2010	55	0.69	0.085	S

Galli et al. (2008) demonstrated the high quality of the multi-temporal landslide inventory map for the Collazzone study area. Nonetheless, we acknowledge that the mapping and age attribution are not perfect. Errors are probably largest in time slices that were obtained through field mapping due to the difficulty in translating lateral assessments onto vertical photographs, but even here we expect the uncertainty in determining a landslide boundary to be less than 3 m. For other time slices, the aerial photographs and satellite images that were used have resolutions of 1 m or better, and hence, uncertainty is of the same order. With an average size of landslides of over 5000 m^2, it was expected that the uncertainty in mapping did not substantially affect our results.

METHODS

In addition to our use of the term path dependency (indicating that previous process activity affects future process activity through legacy effects), it is useful to clarify some other terms that are important for our analysis. The mono-temporal inventories that together make up our multi-temporal inventory are separately called 'time-slices', whereas subdivisions of the multi-temporal inventory based on topology (see below) are called 'sub-inventories'. We use 'earlier' and 'follow-up' to describe landslides based on their relative order in the multi-temporal inventory. Note that follow-up landslides are not reactivated landslides. We consider a landslide a reactivated landslide when all or most of the landslide moved down again, under the same general condition as the first landslide. Instead, follow-up landslides are new landslides that have different size and shape than the pre-existing landslide. The particular topological relations of interest between landslides in earlier and follow-up time slices were called 'spatial association'.

To test our hypotheses, two sets of analyses were done on the multi-temporal landslide inventory. The first set of analyses focuses on the effect of overlap between landslides on the total area affected by landslides (Fig. 2 (a)), on the degree of overlap between landslides from different time slices (Fig. 2 (b)) and on the number of overlaps between landslides over the entire inventory (Fig. 2 (c)). The second set of analyses focuses on the properties of classes of landslides that vary in their spatial association with landslides from the earlier periods (Fig. 2(d)).

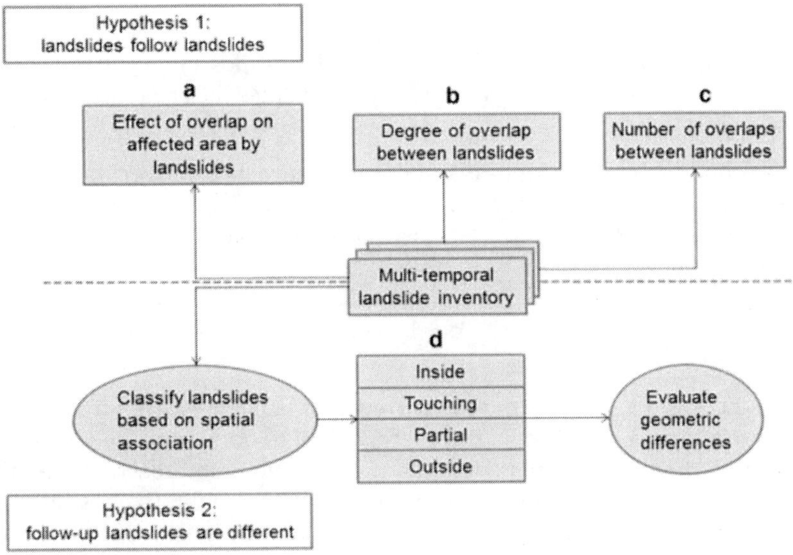

Figure 2. Overview of the analyses performed on the multi-temporal landslide inventory to test hypothesis 1 (**A, B, C**) and hypothesis 2 (**D**)

Degree of overlap

We propose three complementary methods for this purpose to express the importance of overlap between landslides: (1) unaffected area, (2) overlap index and (3) number of overlaps (Fig. 2).

The first method measures the cumulative effect that landslide overlapping has on the total area affected by landslides, by comparing a theoretically unaffected area with the actually unaffected area. This method allows assessment of the amount of reduction in affected area caused by (potentially followed-up) overlapping landslides. The theoretically unaffected relative area (–, where – denotes that this is a dimensionless parameter) if no overlap between landslides would occur is defined as:

$$\text{Theoretically unaffected area}_t = \left(\sum_{i=1}^{t} AL_{Ti}\right)/AS \quad (1)$$

where AS is the area of interest (m^2) and AL_{Ti} is the total area (m^2) of all landslides in the time slices *i* of an inventory. The relative area that is actually unaffected by landslides over the entire period is calculated as:

$$\text{Actually unaffected area}_t = \left(\bigcup_{i=1}^{t} AL_{Ti}\right)/AS \quad (2)$$

where ∪ denotes the spatial union among landslides in consecutive time slices, i.e. the total area covered by landslides after accounting for overlaps.

The second method provides a standardized measure of overlap between landslides in consecutive time slices. This overlap index (–), earlier termed relative area overlap (Maruca and Jacquez 2002), is calculated as:

$$\text{Overlap index}_{t-t+n} = \frac{AL_t \cap AL_{t+n}}{AL_t + AL_{t+n} - (AL_t \cap AL_{t+n})} \quad (3)$$

where *t* is the assumed date of occurrence of landslides from a first time slice (see below), and *t* + *n* is the average date of occurrence of landslides in a later time slice, and ∩ denotes the geometric intersection (i.e. the overlapping area) between two time slices. This is qualitatively similar to the error index that has been used to calculate the positional mismatch for pairs of corresponding landslides in the two inventories (Ardizzone et al. 2002; Carrara et al. 1992; Santangelo et al. 2015). The overlap index is non dimensional, ranges from zero (no overlap) to unity (perfect overlap) and is not a function of the size of the study area. The overlap index was calculated for consecutive time slices and for pairs of time slices that are two or three time slices apart. The value for the overlap index was then related to the time that passed between time slices.

As explained above, in some time slices, there is more uncertainty about the date of occurrence of a landslide than in others. Uncertainty is larger in the time slices that describe longer periods, such as the time slice describing landslides that occurred between 1954 and 1977 (see Table 1). In some other cases, the date of occurrence is relatively well known, for instance in the time slice made of landslides that occurred after the 01-01-1997 rapid snowmelt event. These uncertainties, large or small, propagate into uncertainty about the time elapsed between landslides. We assumed uniform probability distributions of landslides in the time slices that describe periods—meaning that landslides may have happened at any moment during the considered time slice, with equal probability. This is despite the fact that there may have been significant rain, snowfall, and snowmelt events in these periods that may have caused landslides to be clustered in time. No information about this intra-period variability was available. Using the uniform probability distributions, we performed stochastic simulation, randomly placing 10,000 pairs of landslides in the periods of two time slices and then recording the time passed between each pair. The median time passed between time slices ($t - t + n$) was used as the time passed between time slices, and the first and third quartile were recorded to express uncertainty of the time between both time slices. This approach is unbiased and therefore no bias in consequent analyses is expected.

The third method quantifies the relation between area and number of overlaps. This method was used to quantify repeated overlapping over multiple time slices of the multi-temporal inventory. Every time slice of the inventory was rasterized at 1 m resolution, and the cells involved in landsliding in each inventory were given a value of 1. Then, all raster were summed, resulting in a raster where the values correspond to the number of overlaps over the entire multi temporal inventory. The analysis was performed using standard raster operations in ArcGIS. This measure does not change with the order of time slices, but is affected by the area involved in landsliding in the multi-temporal inventory and by the number of time slices in it. Therefore, it was compared with the number of overlaps from a null model in which the observed area of landsliding per time slice was assigned randomly to the study area, regardless of slope or lithology. Since spatial information was not needed for this calculation, the null model was calculated in Microsoft Excel.

Effect of different spatial association

Spatial associations between landslides from a given time slice and the immediately preceding time slice were used (Fig. 2 (d)). Four classes of spatial association were defined (Fig. 3). The 'inside' class contains landslides that are completely inside landslides from the earlier time slice. The 'partial' class contains landslides that partially overlap landslides from the earlier time slice. The 'touching' class contains landslides that are outside landslides from the earlier time slice, but that touch landslides from this time slice, and the 'outside' class contains landslides that neither overlap nor touch landslides from the earlier time slice. This classification was chosen based on the assumption that different spatial associations between landslides relate to different mechanisms in which landslides affect the probability of follow-up landslides. Landslides that are 'inside' earlier landslides, for instance, may particularly point to saturation of materials on top of an earlier landslide (Igwe and Fukuoka 2015), and landslides that are 'touching' an earlier landslide may point to a slope-effect because slopes are most strongly changed along the borders of earlier landslides (van Westen et al. 2006). The 'outside' class was considered likely unrelated to earlier landslides, by any mechanism.

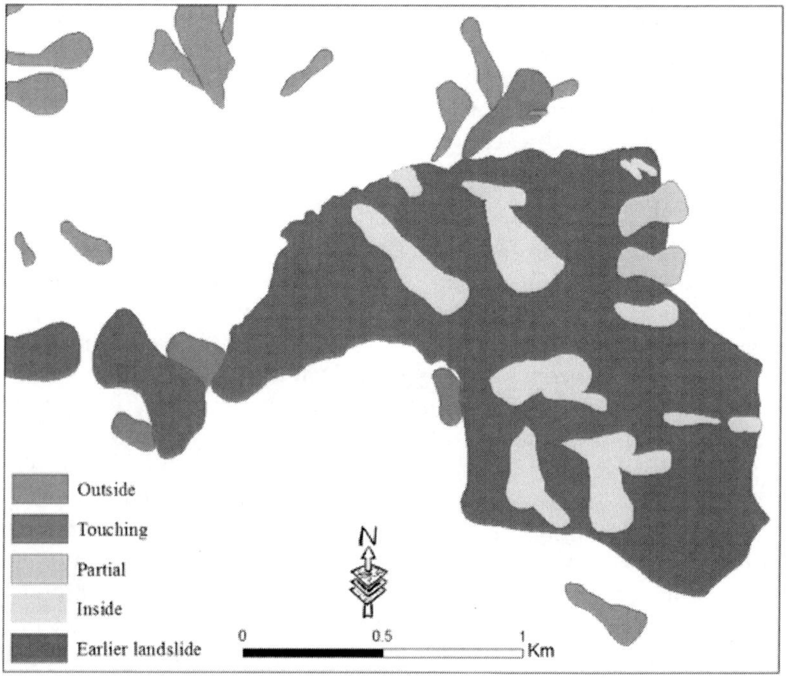

Figure 3. Examples of spatial association of landslides with landslides from the previous time slice

Standard GIS tools were used to separate landslides from each time slice of the multi-temporal inventory, except the first time slice, into the four classes of association. The sub-classes of association (sub-inventories) were then merged over the different time slices. Then, the following geometrical properties were calculated for landslides in each merged sub-inventory: minimum, mean and maximum area, roundness (a measure of shape) and the three parameters of the inverse gamma distribution that are often used in landslide research to describe frequency–area relationships (Malamud et al. 2004), see below. This allowed us to test our second hypothesis that landslides following up on earlier landslides have different geometric properties.

Minimum, mean, median and maximum sizes of landslides were calculated with standard GIS tools. Roundness (–) is introduced here as a simple measure of shape:

$$\text{Roundness} = \frac{\text{theoretical circular perimeter}}{\text{actually measured perimeter}} \quad (4)$$

where the theoretical circular perimeter (*m*) is the perimeter of the landslide, if it would have been perfectly round with the same area:

$$\text{Theoretical circular perimeter} = 2\pi\sqrt{\left(\frac{A_L}{\pi}\right)} \quad (5)$$

where A_L is the area of the landslide (m^2). Roundness values closer to unity indicate round shapes, values close to zero indicate more elongated shapes (Fig. 4). In earlier research, a landslide geometry generating algorithm was used to approximate the shape of landslides using geometric features such as upper and lower length and lateral boundary of the landslide scar (Chiang 2015). Taylor et al. (2015) approximated landslide shapes by ellipses and then used the length to width ratio of these ellipses to characterize shape. We argue that the proposed roundness is a simpler measure that makes no assumptions about landslide shape that nonetheless captures the most important difference between landslide shapes.

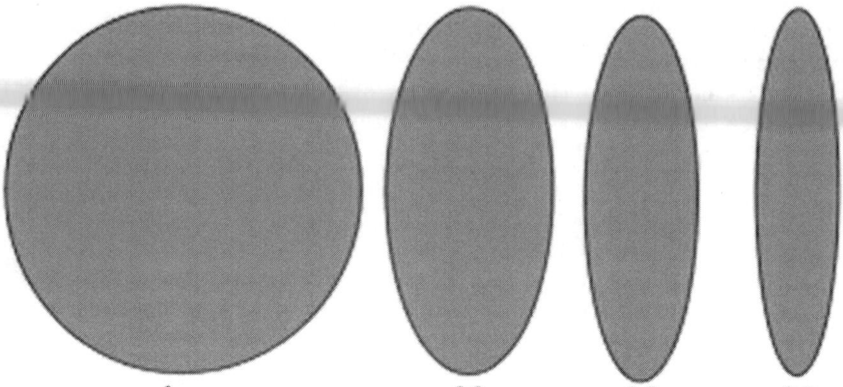

Figure 4. Example shapes illustrating different degrees of roundness

To estimate power-law parameters, we fitted the three-parameter inverse gamma distribution (Malamud et al. 2004) to the four sub-inventories, using LANDSTAT software (version 9) (Rossi and Malamud 2014). The three-parameter inverse gamma distribution is:

$$pdf(A_l|\alpha,\eta,\lambda) = \left[\frac{\lambda^{2\alpha}}{\Gamma(\alpha)}\right]\left[\left(\frac{1}{x+\eta^2}\right)^{(\alpha+1)}\right] exp\left[-\frac{\lambda^2}{x+\eta^2}\right] \quad (6)$$

Parameter α controls the steepness of the right tail of the probability density function. Parameter η controls the steepness of the left tail of the probability density function. Parameter λ controls the position of the rollover. From these parameters, the rollover, the most likely size of a landslide, can be calculated. Maximum likelihood estimation was used and uncertainties for the three parameters and the rollover were calculated using bootstrapping with 250 repetitions.

We used standard Analysis of Variance to test whether differences in landslide properties between sub-inventories are significant, complemented with T-tests for normally distributed properties (Student 1908) and U-tests for non-normally distributed properties (Mann and Whitney 1947).

RESULTS

Degree of overlap

Over time, significant differences occur between the theoretically unaffected area (Eq. 1) and the actually unaffected area (Eq. 2) (Fig. 5). After the last time slice in the inventory, from May 2010, 7.8 % of the area has not experienced landsliding during the entire period of observation because of overlapping landslides. Clearly, there is a significant amount of overlap between landslides.

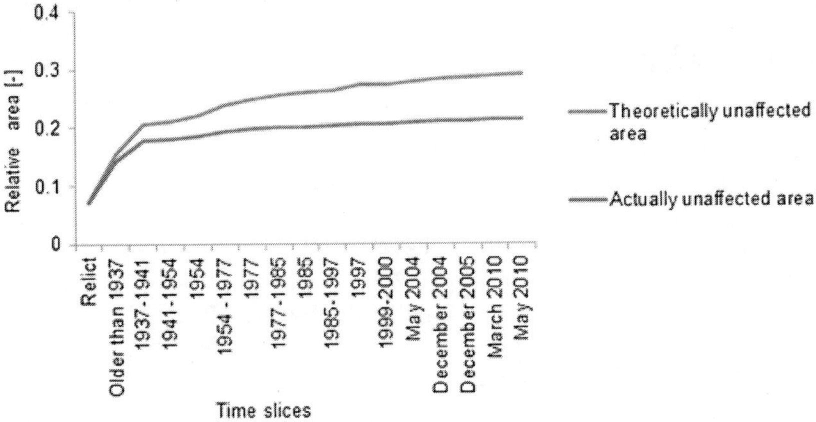

Figure 5. Theoretically and actually relative unaffected area by landslides in the Collazzone study area

The overlap index (Eq. 3) appears to be negatively correlated with the time passed between time slices, although variation in overlap index is large, especially when observations are close together (Fig. 6). The overlap of landslides with those from an earlier time slice appears to decrease substantially over a period of about 10 years. This suggests that the cause of new landslides is not completely external. If the cause of overlap was completely external (for instance through repeated landsliding at particularly dangerous locations), then no relation between overlap index and the amount of time passed between slides should exist.

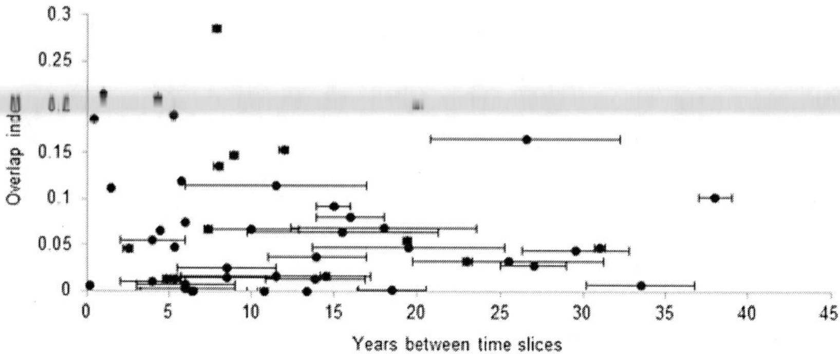

Figure 6. Overlap index as a function of time between time slices of the multi-temporal landslide inventory; the data shows a higher overlap index for time intervals up to approximately 10 years

Instead, earlier landslides themselves appear to affect the probability of reoccurrence to a substantial degree—probably through legacy effects. Importantly, this effect decreases over time. This means that if a landslide happened longer ago, it is less likely to be overlapped by a future landslide. However, this is not always the case: some time slices that are close together in time, have low overlap index. Apparently, not all landslides affect their environment such that there is a larger probability for future landslides.

About 200 m^2 of the study area has been affected seven times by landslides in the 17 time slices of the multi-temporal inventory (Fig. 7). In the null model in which landslide cells are placed randomly in the study area, seven overlaps do not occur. Overall, much more of the study area has been affected by three or more overlapping landslides than in the random model, whereas the area where no overlapping occurs, or where only one or two overlaps occur, is less than in the random model. The relation between number of overlaps and the logarithm of area is about twice as steep in the random model than in the dataset, again a clear sign of spatial overlap between follow-up landslides.

Effect of spatial association

Overall, 28 % of landslides are in the classes inside (9 %), partial (10 %) and touching (9 %). The remaining 72 % of landslides belong to the 'outside' class.

The mean size of the landslides with spatial association 'inside', 'partial' and 'touching' is larger than those with spatial association 'outside'

(Table 2, $p < 0.001$). The subtle decrease in mean area from 'inside' to 'partial' and 'touching' is not statistically significant ($p > 0.05$).

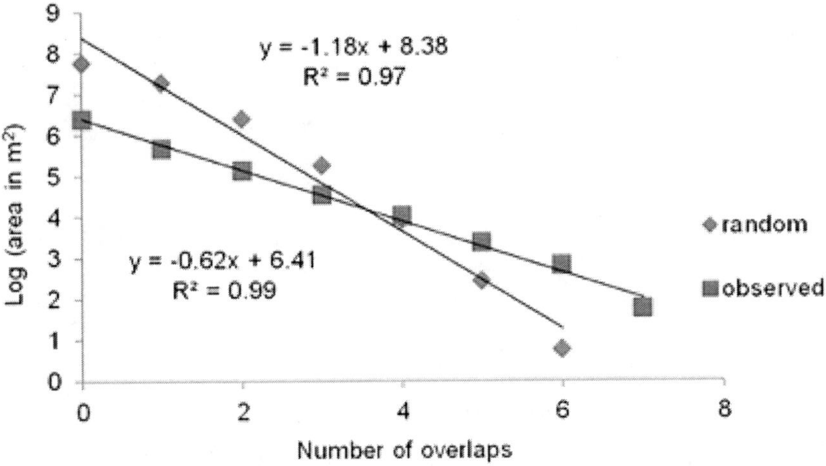

Figure 7. Area versus number of overlapping landslides. The area experiencing more than four overlaps is larger for the observed dataset than for the random dataset

Table 2. Geometrical and area–frequency distribution properties of landslides depending on spatial association

Inventories		Inside	Partial	Touching	Outside
Area (ha)	n	165	199	185	2480
	Min	0.03	0.01	0.02	0.002
	Mean ± SD	0.91 ± 0.013	0.89 ± 0.010	0.82 ± 0.011	0.51 ± 0.010
	Median	0.38	0.46	0.46	0.21
	Max	6.90	5.70	7.50	1.73
Roundness	Mean ± SD	0.79 ± 0.10	0.78 ± 0.12	0.78 ± 0.12	0.83 ± 0.11
Area–frequency statistics	α ± SD	1.04 ± 0.08	1.28 ± 0.50	1.78 ± 0.25	1.32 ± 0.04
	η ± SD	16.2 ± 4.3	28.9 ± 2.6	32.8 ± 4.2	19.3 ± 0.5
	λ ± SD	52 ± 6	69 ± 5	90 ± 11	49 ± 1
	Rollover ± SD (m2)	1073 ± 157	1317 ± 158	1858 ± 241	670 ± 25

Shape factor

Landslides that are not spatially associated with landslides from the immediately preceding time slice appear to be more round than other landslides (Fig. 8). The type of spatial association is a significant determinant of shape (Table 2, $p < 0.001$). This is because of the higher roundness of the 'outside' type—all other types of association are not significantly different from each other ($p > 0.05$).

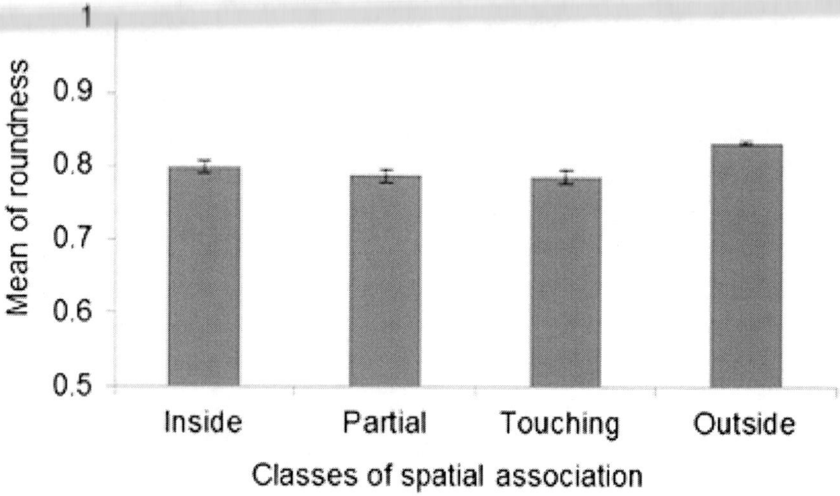

Figure 8. Roundness as a measure of shape for different classes of spatial association. Error bars indicate the standard error

Frequency–area statistics

The frequency–area distributions for the four sub-inventories representing different spatial association exhibit power-law scaling for large and medium landslides with a rollover for small landslides (Fig. 9a–d). The three parameters of the inverse gamma distribution, and the resulting rollover, differ between sub-inventories (Table 2). When the strength of spatial association between landslides decreases (from 'inside' to 'partial' and 'touching'), the exponent of the inverse power-law (α) increases. The average value of (α) in the spatially associated sub-inventories is greater than in the 'outside' spatially un-associated sub-inventory. Interestingly, the value of the rollover is much smaller for the un-associated 'outside' sub-inventory than for the three other sub-inventories. The rollover also decreases with decreasing strength of spatial association. All differences between sub-inventories were significantly different ($p < 0.001$).

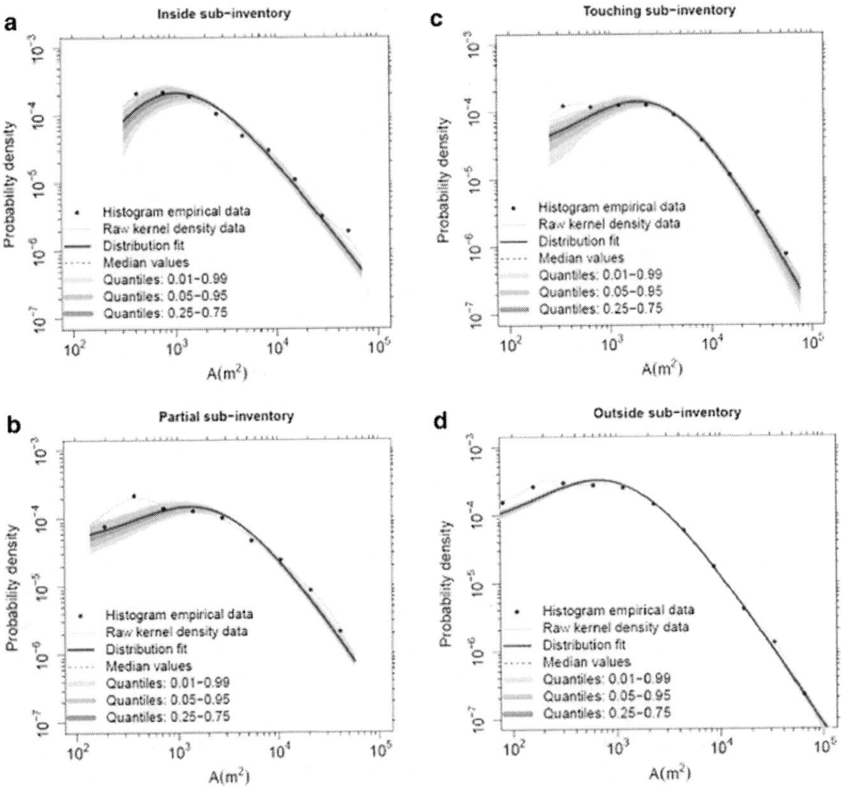

Figure 9. Comparison of the inverse gamma probability density function fit to frequency densities from four landslide sub-inventories representing different spatial association types

DISCUSSION

In this discussion, we will first focus on our findings relating to the two hypotheses, before considering wider implications.

Do landslides follow landslides?

We found that the multi-temporal landslide inventory of our study site recorded substantial overlap between landslides and an associated reduction in the fraction of the study area affected by landslides (Fig. 5). The overlap is more for landslides happening sooner after an earlier landslide and decreases over a timescale of about 10 years (Fig. 6). This indicates that landslides in our study site do follow landslides. Importantly, it also suggests an internal control on the landslide system next to the range of intrinsic controls such as lithology, slope steepness and vegetation, which are commonly the focus of landslide susceptibility studies (Guzzetti et al. 2005; Van Westen et al. 1997). We propose that the mechanisms that can explain the nature of this internal control (i.e. a positive landslide-landslide effect that decreases over time) fit into two categories. First, the probability of landslides overlapping earlier landslides may decrease over time because the deposits of an earlier landslide stabilize due to, e.g. a more stable slope geometry, the regrowth of vegetation and the repair of soil structure and cohesion. Second, the probability can decrease over time because follow-up landslides have already occurred, erasing topographic instabilities that remained after an earlier landslide. Targeted field observations with high temporal resolution could quantify the relative importance of these two reasons by focussing on vegetation regrowth and detailed topographic evolution.

Regardless of the mechanism, our findings suggest that there is path dependency among landslides: older landslides act in some way or other as initiators for follow-up landslides for a certain period. This is also consistent with the number of landslide overlaps over the entire period captured by the multi-temporal inventory (Fig. 7). This number of overlaps is larger than in the null model where landslides occur randomly in the study area—although when considered separately from Fig. 6, this observation can also be explained by spatial differences in landslide susceptibility. Such landslide self-organization (Turcotte et al. 2002) into emergent patterns is not captured in traditional cause-effect studies of landslide susceptibility. It also adds an important consideration to the discussion about landslide self-organized critical behaviour (Guzzetti et al. 2002; Hergarten 2003; Turcotte et al. 2002). Apparently, not all landsliding potential ('metastable regions', Guzzetti et al. 2002, p171) is removed by a landslide—instead first landslides appear in some cases to increase the potential for follow-up landslides. The interplay between the traditional cause-effect approach and path dependency in a self-organization approach should be explored and quantified in follow-up studies.

The period over which the positive landslide-landslide effect is observed—about 10 years in our study area—is very likely specific for the local settings and other areas with similar climate, geology, topography, soils and vegetation. We maintain that this period is a measure of landslide path dependency that reflects the rate of processes that reduce landslide susceptibility after a first landslide, or the rate at which follow-up landslides occur in our study site (see above). In areas with more frequent landsliding, or in sites with more rapid vegetation growth that restores stability, it may be shorter (all other factors being equal).

Are follow-up landslides different?

We also hypothesised that follow-up landslides have different properties in terms of size, shape and frequency–area statistics. This is to some extent the case. Mainly, the mean size of landslides is lower for landslides that are not spatially associated with earlier landslides, than for those that are. Apparently, on average, follow-up landslides are larger (although there are also less of them, Table 2). In addition, our empirical data suggest that there may be a slight decrease of size with decreasing strength of spatial association between landslides (from 'inside' to 'partial' to 'touching' landslides).

The larger average size of follow-up landslides may be explained by conditions that have changed after occurrence of an earlier landslide, leading to increased landslide susceptibility. These changed conditions may relate to changes in surface morphology (Schuster and Highland 2003), sediment properties (Chen 2009), the vegetation (Singh et al. 2014), slope angle and land use (van Westen et al. 2006).

The shape of follow-up landslides differs from the shape of other landslides (Table 2). Landslides that are spatially associated with earlier landslides are less round (on average) than those that are not ($p < 0.01$). The difference in roundness can be explained by two mechanisms. First, weaker materials in an earlier landslide may move faster, and therefore further than other materials. This would then be reflected in less round shape. Second, especially for landslides of the 'touching' class, an earlier landslide may be higher in the landscape, deflecting landslides downslope, which leads to them being more elongated.

Both spatially associated landslides ('inside', 'partial' and 'touching' sub-inventories) and spatially un-associated landslides (outside sub-inventory) follow a power-law scaling in the right tail of inverse gamma distribution for large and medium landslides with an exponential rollover in the left tail of distribution. However, the exponent of the inverse power-law (α) and rollover (λ) differs between spatially associated landslides (even between sub-inventories) and spatially un-associated landslides (Fig. 9a–d). The exponent of inverse power-law (α) on average in spatially associated sub-inventories (i.e. 'inside', 'partial' and 'touching') is greater than un-associated sub-inventory ('outside'). For landslides that are spatially associated with earlier landslides, the exponent of power-law (α) increases when the strength of the spatial association (from 'inside' to 'touching') decreases (Table 2). The exponent of the inverse power-law (α) for 'touching' landslides ($\alpha = 1.78$) seems consistent with the most reported range of values in the literature ($1.5 < \alpha < 2.5$) (Borgomeo et al. 2014; Guzzetti et al. 2002; Malamud et al. 2004; Van Den Eeckhaut et al. 2007). The other landslides (i.e. 'inside', 'partial' and 'outside') exhibited lower values for exponent of the inverse power-law (α) (i.e. $1 < \alpha < 1.5$) (Table 2). The small exponent of the power-law in the sub-inventories suggests that larger landslides are contributing to each sub-inventory (Borgomeo et al. 2014; Van Den Eeckhaut et al. 2007).

The rollover (λ), which represents the size of the most frequent landslide, is larger for spatially associated landslides than for non-spatially associated landslides ('outside', Table 2). Interestingly, the rollover for the 'touching' landslides is three times larger than for the 'outside' landslides and larger than for the 'inside' and the 'partial' landslides. Analogously to our explanation for lower roundness, this may be explained by a boundary effect that can be caused by steeper slopes on the sides of an earlier landslide. The lateral boundary of a landslide is an intrinsically more disturbed part of a landslide, where the mechanical properties are weaker and infiltration is commonly larger. These factors contribute to instability, and hence to larger landslides.

Implications for landslide susceptibility assessment

The traditional division between landslide susceptibility and hazard is that the former describes spatial differences in landslide probability, and the latter describes the likelihood of landslide occurrence in time and space (Guzzetti et al. 1999; Varnes 1984). Our results suggest that a time-related internal control—the time that passed since an earlier landslide occurred—also plays a role in determining how susceptible a location is to landsliding. Apparently, susceptibility is not time-invariant and contains a temporal element that traditionally was seen as part of the definition of hazard. Therefore, spatial and temporal probabilities of landsliding are not independent (Guzzetti et al. 2005). This means that the susceptibility of a location changes over time, as the effects of an earlier landslide slowly disappear. This can be expressed as:

$$\text{susceptibility}_{s,t} = f\left(\text{conditioning attributes}_s, \text{previous landslides}_{s,t}\right) \quad (7)$$

where the susceptibility for landsliding of a certain location *s* at time *t* is not only a function of conditioning attributes such as slope or lithology, but also of the time passed since an earlier landslide in the same location or close by. In our dataset, the function describing the effect of earlier landslides on susceptibility is positive and decreases over time (e.g. Fig. 6). An earlier landslide appears to make a location more likely to experience landsliding again over a period of about 10 years, after which susceptibility appears to return to its previous value (Fig. 10). We explored possible mechanisms for this effect above.

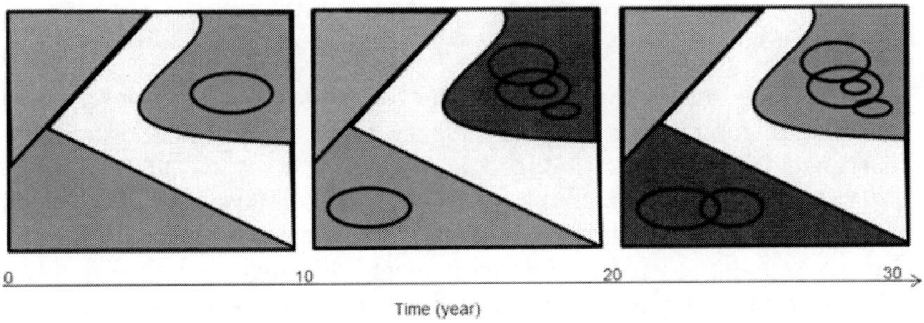

Figure 10. A landslide susceptibility map bases on slope units (coloured polygons) with the occurrence of landslides over time (ellipses with black outline). In the susceptibility map, *green* means low risk, *yellow* means moderate risk, *light red* means high risk and *dark read* means very high risk. The implication of our work is seen in the clustering of landslides after a first landslide happens within a time scale of about 10 years. This is reflected in a temporarily higher susceptibility in the affected slopes. The two temporarily highly susceptible slopes (*dark red* polygons) experience different landslide dynamics as a result of different landslide histories

The effect of the type of spatial association with (or the distance to) an earlier landslide is complex (Table 2). In places, a closer spatial association of a landslide with an earlier landslide leads to properties closer to spatially un-associated landslides (e.g. for the rollover), whereas in other places it leads to properties that are more different from un-associated landslides (e.g. the mean area) (Table 2). This complexity should be further researched since it may lead to a deeper understanding of the processes causing follow-up landsliding.

Other geomorphic settings could lead to different functions describing the effect of earlier landslides. For instance, exhaustion of sediment after landsliding, which occurs in locations where only little soil overlies bedrock (e.g. the steep basalt slopes of the Drakensberg in South-Africa (Singh et al. 2008)), would lead to a decreasing of landslide susceptibility instead of an increase. Given that weathering is required for new soil to form, it would also probably be a longer term effect than the 10 years that we observed (Fig. 6). The exhaustion of soft material apparently plays no important role in our study site—fitting with its geology of predominantly soft, easily weatherable rocks. A complicating factor is that when the duration of the effect of landslides on landslide susceptibility increases much beyond a decade, other slow changes will also start affecting susceptibility, such as land use changes (e.g. deforestation, afforestation and changes in agricultural practises) and climate changes. This makes it more difficult to observe the effect that landslides themselves have on susceptibility.

The decreasing temporal effect of earlier landslides on susceptibility is to some extent comparable with the legacy effect of earthquakes on the occurrence of earthquake-induced landslides (Lin et al. 2007; Marc et al. 2015; Parker et al. 2015). For instance, Marc et al. (2015) demonstrated that landslide rates were significantly elevated within 0.7–4.5 years after four earthquakes with magnitude higher than 6 Richter. After this, rates dropped gradually back to pre-earthquake levels.

Implications for mapping and monitoring of landslide populations

Most landslide inventories are not multi-temporal, although they do often describe landslides that occurred over a longer period or in response to several events. From such inventories, it is not possible to estimate the time-dependent effect of landslides on landslide susceptibility. In settings where that effect is strong, this may lead to susceptibility assessments that are biased. In statistical terms, this would be because the landslide observations used to estimate a susceptibility model or map are not independent. In geomorphic terms, if the landslide-landslide effect is positive and decreases over time (Fig. 6), resulting susceptibility maps would probably be biased to a smaller range of conditioning attributes than would be otherwise the case. Some places that are susceptible to landsliding would then possibly be considered safe. In case of a negative landslide-landslide effect, they would probably be biased to a larger range of conditioning attributes. In this case, some places that are safe would possibly be considered susceptible. How substantial these biases should be the subject of further study. Clearly, this calls for more multi-temporal landslide inventories.

CONCLUSIONS

In our study area, landslides follow landslides—and they do so more often than expected based on a random control. This landslide path dependency is strongest over a timescale of about 10 years, after which the effect of earlier landslides on follow-up landslides appears to decrease. This legacy effect has a considerable consequence on the relative area that is affected by landslides. Follow-up landslides differ from non-follow-up landsides in terms of size and shape: they are typically larger and less round. Also, the exponent of the power-law (α) and the value of rollover in spatially associated landslides on average are greater than in spatially un-associated landslides. From these findings, we conclude that landslide susceptibility in our study area should be considered as a dynamic measure that reflects path dependency and self-organization in landslides. Susceptibility changes because the effect of previous landslides on susceptibility, which in our case is positive, decreases over a period of about 10 years. Although our research concentrated on the Collazone region in Italy, we expect these results to be relevant for other landslide prone areas as well.

ACKNOWLEDGMENT

JS acknowledges financial support from the Ministry of Sciences, Research and Technology of Iran. The comments and suggestions of two anonymous reviewers have greatly helped us in preparing this manuscript. We gratefully acknowledge this help.

REFERENCES

1. Aleotti P, Chowdhury R (1999) Landslide hazard assessment: summary review and new perspectives. Bull Eng Geol Environ 58:21–44.
2. Ardizzone F, Cardinali M, Carrara A, Guzzetti F, Reichenbach P (2002) Impact of mapping errors on the reliability of landslide hazard maps. Nat Hazards Earth Syst Sci 2:3–14.
3. Ardizzone F et al. (2013) Very-high resolution stereoscopic satellite images for landslide mapping. Landslide Sci Pract 1:95–101.
4. Atkinson PM, Massari R (1998) Generalised linear modelling of susceptibility to landsliding in the central Apennines, Italy. Comput Geosci 24:373–385.
5. Barredo JI, Benavides A, Hervás J, Van Westen CJ (2000) Comparing heuristic landslide hazard assessment techniques using GIS in the Tirajana basin, gran Canaria Island, Spain. Int J Appl Earth Obs Geoinf 2000:9–23

6. Borgomeo E, Hebditch KV, Whittaker AC, Lonergan L (2014) Characterising the spatial distribution, frequency and geomorphic controls on landslide occurrence, Molise, Italy. Geomorphology 226:140–161
7. Brenning A (2005) Spatial prediction models for landslide hazards: review, comparison and evaluation. Nat Hazards Earth Syst Sci 5:853–862
8. Brunsden D (1993) Mass movement; the research frontier and beyond: a geomorphological approach. Geomorphology 7:85–128.
9. Cardinali M, Galli M, Guzzetti F, Ardizzone F, Reichenbach P, Bartoccini P (2006) Rainfall induced landslides in December 2004 in South-Western Umbria, Central Italy: types, extent, damage and risk assessment. Nat Hazards Earth Syst Sci 6:237–260
10. Carrara A, Cardinali M, Guzzetti F (1992) Uncertainty in assessing landslide hazard and risk. ITC J 2:172–183
11. Carrara A, Guzzetti F, Cardinali M, Reichenbach P (1999) Use of GIS Technology in the Prediction and Monitoring of landslide hazard. Nat Hazards 20:117–135.
12. Chen C-Y (2009) Sedimentary impacts from landslides in the Tachia River basin, Taiwan. Geomorphology 105:355–365. doi
13. Chiang S (2015) The analysis of size and shape of rainfall-induced landslides. Paper presented at the EGU General Assembly 2015, Vienna
14. Crozier MJ (1986) Landslides: causes, Consequences & Environment. Croom Helm Pub, London
15. Fiorucci F, Ardizzone F, Rossi M, Torri D (2015) The use of stereoscopic satellite images to map rills and ephemeral gullies. Remote Sens 7:14151–14178
16. Galli M, Ardizzone F, Cardinali M, Guzzetti F, Reichenbach P (2008) Comparing landslide inventory maps. Geomorphology 94:268–289.
17. Godt JW, Baum RL, Savage WZ, Salciarini D, Schulz WH, Harp EL (2008) Transient deterministic shallow landslide modeling: requirements for susceptibility and hazard assessments in a GIS framework. Eng Geol 102:214–226.
18. Guzzetti F, Carrara A, Cardinali M, Reichenbach P (1999) Landslide hazard evaluation: a review of current techniques and their application in a multi-scale study, Central Italy. Geomorphology 31:181–216.
19. Guzzetti F, Malamud BD, Turcotte DL, Reichenbach P (2002) Power-law correlations of landslide areas in Central Italy. Earth Planet Sci Lett 195:169–183.
20. Guzzetti F, Reichenbach P, Cardinali M, Galli M, Ardizzone F (2005) Probabilistic landslide hazard assessment at the basin scale. Geomorphology 72:272–299.
21. Guzzetti F, Reichenbach P, Ardizzone F, Cardinali M, Galli M (2006) Estimating the quality of landslide susceptibility models. Geomorphology 81:166–184.
22. Guzzetti F, Ardizzone F, Cardinali M, Galli M, Reichenbach P, Rossi M (2008) Distribution of landslides in the upper Tiber River basin, Central Italy. Geomorphology 96:105–122.
23. Guzzetti F, Ardizzone F, Cardinali M, Rossi M, Valigi D (2009) Landslide volumes and landslide mobilization rates in Umbria, Central Italy. Earth Planet Sci Lett 279:222–229.
24. Guzzetti F, Mondini AC, Cardinali M, Fiorucci F, Santangelo M, Chang K-T (2012) Landslide inventory maps: new tools for an old problem. Earth Sci Rev 112:42–66.
25. Hergarten S (2003) Landslides, sandpiles, and self-organized criticality. Nat Hazards Earth Syst Sci 3:505–514
26. Igwe O, Fukuoka H (2015) The effect of water-saturation on the stability of problematic slopes at the Iva Valley area, Southeast Nigeria. Arab J Geosci 8:3223–3233.
27. Lan HX, Zhou CH, Wang LJ, Zhang HY, Li RH (2004) Landslide hazard spatial analysis and prediction using GIS in the Xiaojiang watershed, Yunnan, China. Eng Geol 76:109–128.
28. Lin W-T, Chou W-C, Lin C-Y, Huang P-H, Tsai J-S (2007) Study of landslides caused by the 1999 Chi-Chi earthquake, Taiwan, with multitemporal SPOT images. Can J Remote Sens 33:289–302.

29. Malamud BD, Turcotte DL, Guzzetti F, Reichenbach P (2004) Landslide inventories and their statistical properties. Earth Surf Process Landf 29:687–711.
30. Mann HB, Whitney DR (1947) On a test of whether one of two random variables is stochastically larger than the other. Ann Math Stat 18:50–60
31. Marc O, Hovius N, Meunier P, Uchida T, Hayashi S (2015) Transient changes of landslide rates after earthquakes. Geology.
32. Maruca SL, Jacquez GM (2002) Area-based tests for association between spatial patterns. J Geogr Syst 4:69–83.
33. Parker R, Hancox G, Petley D, Massey C, Densmore A, Rosser N (2015) Spatial distributions of earthquake-induced landslides and hillslope preconditioning in northwest South Island, New Zealand. Earth Surf Dyn Discuss 3:1–52
34. Phillips JD (2006) Evolutionary geomorphology: thresholds and nonlinearity in landform response to environmental change Hydrol earth. Syst Sci 10:731–742.
35. Remondo J, González-Díez A, Díaz de Terán JR, Cendrero A (2003) Landslide susceptibility models utilising spatial data analysis techniques. A case study from the lower Deba Valley, Guipúzcoa (Spain). Nat Hazards 30:267–279.
36. Rossi M, Malamud BD (2014) D. 5.3. Prototype SW for determination of landslide statistics from inventory maps. CNR, KCL
37. Ruff M, Czurda K (2008) Landslide susceptibility analysis with a heuristic approach in the eastern alps (Vorarlberg, Austria). Geomorphology 94:314–324.
38. Santangelo M, Marchesini I, Bucci F, Cardinali M, Fiorucci F, Guzzetti F (2015) An approach to reduce mapping errors in the production of landslide inventory maps. Nat Hazards Earth Syst Sci 15:2111–2126.
39. Schuster R, Highland L (2003) Impact of landslides and innovative landslide-mitigation measures on the natural environment. Paper presented at the International Conference on Slope Engineering, Hong Kong, China
40. Singh RG, Botha GA, Richards NP, McCarthy TS (2008) Holocene landslides in KwaZulu-Natal, South Africa south African. J Geol 111:39–52.
41. Singh KK, Mehrotra A, Pal K (2014) Landslide detection from satellite images using spectral indices and digital elevation model. Disaster Adv 7:25–32
42. Stark CP, Hovius N (2001) The characterization of landslide size distributions. Geophys Res Lett 28:1091–1094.
43. Student (1908) The probable error of a mean. Biometrika 6:1–25.
44. Taylor FE, Malamud BD, Witt A (2015) What Shape is a Landslide? Statistical Patterns in Landslide Length to Width Ratio. Paper presented at the EGU General Assembly 2015, vienna
45. Temme AJAM, Keiler M, Karssenberg D, Lang A (2015) Complexity and non-linearity in earth surface processes—concepts, methods and applications. Earth Surf Process Landf 40:1270–1274.
46. Tien Bui D, Tuan T, Klempe H, Pradhan B, Revhaug I (2016) Spatial prediction models for shallow landslide hazards: a comparative assessment of the efficacy of support vector machines, artificial neural networks, kernel logistic regression, and logistic model tree. Landslides 13(2):361–378.
47. Turcotte DL, Malamud BD, Guzzetti F, Reichenbach P (2002) Self-organization, the cascade model, and natural hazards. Proc Natl Acad Sci 99:2530–2537.
48. Turcotte DL, Malamud BD, Guzzetti F, Reichenbach P (2006) A general landslide distribution applied to a small inventory in Todi, Italy Geological Society, London, Special Publications 261:105–111
49. Van Den Eeckhaut M, Poesen J, Govers G, Verstraeten G, Demoulin A (2007) Characteristics of the size distribution of recent and historical landslides in a populated hilly region. Earth Planet Sci Lett 256:588–603.
50. Van Westen CJ, Terlien MTJ (1996) An approach towards deterministic landslide hazard analysis in GIS. A case study from Manizales (Colombia). Earth Surf Process Landf 21:853–868.

51. Van Westen CJ, Rengers N, Terlien MTJ, Soeters R (1997) Prediction of the occurrence of slope instability phenomena through GIS-based hazard zonation. Int J Earth Sci 86:404–414
52. Van Westen CJ, Rengers N, Soeters R (2003) Use of geomorphological information in indirect landslide susceptibility assessment. Nat Hazards 30:399–419.
53. van Westen CJ, van Asch TWJ, Soeters R (2006) Landslide hazard and risk zonation—why is it still so difficult? Bull Eng Geol Environ 65:167–184.
54. van Westen CJ, Castellanos E, Kuriakose SL (2008) Spatial data for landslide susceptibility, hazard, and vulnerability assessment: an overview. Eng Geol 102:112–131.
55. Varnes DJ (1984) Landslide hazard zonation: a review of principles and practice. UNESCO Press, Paris, 63 pp
56. Wood JL, Harrison S, Reinhardt L (2015) Landslide inventories for climate impacts research in the European alps. Geomorphology 228:398–408.
57. Yeon YK, Han JG, Ryu KH (2010) Landslide susceptibility mapping in Injae, Korea, using a decision tree. Eng Geol 116:274–283.

CHAPTER 5

Assessment of Time–Spatial Changes of Shear Wave Velocities of Flysch Formation Prone to Mass Movements by Seismic Interferometry with the Use of Ambient Noise

Paulina Harba I Zenon Pilecki

The Mineral and Energy Economy Research Institute of the Polish Academy of SciencesKrakowPoland

ABSTRACT

This paper presents the results of monitoring flysch formation shear wave velocities at varying conditions of water content by seismic interferometry. Studies were conducted on a section of the active Just-Tegoborze landslide in the southern region of Poland. The landslide is intersected by the state road no. 75, which has a high traffic density. Measurements of the seismic interferometry were made on the basis of the registration of a local high-frequency noise generated mostly by passing heavy vehicles. Measurements were carried out in three series on January, March and July in 2015. Seismic noise vibration was registered using Güralp CMG-6TD broadband seismometers. One of the seismic profiles was located along a geological engineering cross-section prepared from research boreholes data. One of the main results of the work was the sections of changes of shear wave velocity. The interpreted seismic boundaries made it possible to correlate them with the lithological ones. The trough-like shape of the bedrock boundary may be an effect of weathering processes developed in the zone consisting of weak hieroglyphic beds. It is also supposed to be one of the slip surfaces. The seismic sections of shear wave velocity allowed to analyse velocity variations between three measurement series and to compare them with expected water content in colluvium deposits. The presented results of the study confirmed the utility of seismic interferometry based on ambient noise registration in evaluating the behaviour of landslides in the Carpathian flysch formation.

Keywords: Landslide Seismic interferometry Ambient noise Carpathian flysch

INTRODUCTION

The recognition of landslide behaviour in various geological conditions is crucial in landslide hazard evaluation (Aleotti and Chowdhury 1999; Anbarasu et al. 2010; Guzzetti et al. 1999; Dai et al. 2002; Barla et al. 2013; Bonnard and Corominas 2005; Cascini et al. 2005; Fukuoka et al. 2005; Mansour et al. 2011; Sassa et al. 2004, Song and Kim 2014). In this sense, geophysical methods may be helpful in monitoring landslides and, consequently, in assessing the landslide hazard (Bichler et al. 2004; Bièvre et al. 2015; Brückl et al. 2013; Bogoslovsky and Ogilvy 1977; Frasheri et al. 1998; Hack 2000; Méric et al. 2005; Jongmans and Garambois 2007; Jongmans et al. 2009; Lundström et al. 2009; Hibert et al. 2011; Reynolds 2011, Salas-Romero et al. 2015; Singh et al. 2013; Travelleti et al. 2013; Zarroca et al. 2014). In recent years, a new seismic method based on ambient noise has been intensively developed, known as seismic interferometry, which may be demonstrated to be useful in monitoring mass movements (Jongmans and Garambois 2007; Méric et al. 2007; Jongmans et al. 2009; Renalier et al. 2010a, 2010b; Mainsant et al. 2012b; Del Gaudio et al. 2013; Pilz et al. 2013).

Seismic interferometry is a method enabling the use of local seismic noise for analysis of changes of elastic properties within the geological medium. It is originated from the phenomenon of interference of seismic signals (Curtis et al. 2006; Schuster 2009). The principles and applications of seismic interferometry were broadly explained by Schuster (2009) and Wapenaar et al. (2010a, 2010b).

Seismic interferometry consists of retrieving the impulse response of the medium, the so-called Green's function, between a pair of receivers using the operations of cross-correlation, deconvolution or cross-coherence of seismic signals registered at the receivers. The method assumes a Rayleigh surface wave, which is dispersive, i.e., its velocity strictly depends on the frequency. Rayleigh waves are described by dispersion curves, which demonstrate the dependence of a phase velocity on wave frequency. Shear wave (S-wave) velocities have the dominant influence on the dispersion curve in a high-frequency range (>5 Hz) (Xia et al. 1999). As a result of inversion of a Rayleigh wave's phase velocity, an image of the changes in S-wave velocities is obtained. The S-wave velocity is sensitive to changes in properties of the ground medium, especially to changes in its rigidity, which could be also the consequence of changes in water content in the medium.

Seismic interferometry has increasing geo-engineering applications. Larose et al. (2015) presented recent developments in ambient seismic noise methods to monitor the variations in the mechanical properties of the subsurface and to identify new sources of seismic signals that occur outside the solid Earth. The authors demonstrated the use of seismic noise in marking the thermal variations in the subsoil, in buildings or in rock columns, the temporal and spatial evolution of a water table and the changes of the rigidity of the soil constituting a slope prone to mass movements.

Seismic interferometry studies analysing changes in the S-wave velocity preceding major landslide movements were broadly undertaken by Mainsant et al. (2012b). These researchers concluded that the observed changes in the S-wave velocity decreased two or even threefold before mass movement. The drop in S-wave velocities was correlated with the measurements of displacement performed on the terrain surface. Moreover, Mainsant et al. (2012a) studied the S-wave velocity in terms of rheological changes in clay-rich geological formations, which are responsible for many landslides. These researchers conducted rheometrical laboratory experiments to investigate the solid-fluid transition, which affects derangement of the particle network and is associated with a loss in rigidity. The results indicated that the S-wave velocity significantly varies with rheological changes in the clay, supposedly caused by the disorganization of the particle network. According to the outcomes of their investigations, the authors concluded that measuring the S-wave velocity could be applicable to monitoring rheological changes in landslides in clayey deposits and that the time-dependent variation in this parameter could be a beneficial indicator for earthflows. The authors continued laboratory experiments (Mainsant et al. 2015) in order to examine the physical origin of the S-wave velocity decreases which occur prior to the onset of mass movements. Additional rheometrical studies indicated that it is probably caused by a complex transient rheological response of the clay. These researchers supposed that such response is connected to the existence of the so-called "delayed fluidization" phases in which the shear modulus quickly decreases while the apparent viscosity varies more slowly and requires more time to reach its fluid level. The results are comparable with a real in-field situation of a landslide movement.

Renalier et al. (2010a) applied ambient seismic measurements to determine the deepest slip surface of a landslide in clayey deposits, which had not been detected by active seismic methods. This surface was found by obtaining the dispersion curves at lower frequencies than 3 Hz. During a 30-month period of seismic noise observations, they detected a slight but significant decrease in the surface wave velocities. Moreover, authors observed both lateral and vertical variations of the S-wave velocity which have been related to landslide activity and the location of slip surfaces. Additionally, Renalier et al. (2010b) successfully applied an ambient noise cross-correlation to obtain a 3D model of the S-wave velocity of a landslide affecting clayey deposits. They computed cross-correlations between the vertical components of recording stations which allowed them to retrieve the Rayleigh wave Green's functions and estimate their group velocity dispersion curves in the 1.7–5 Hz frequency range. The dispersion curves were then inverted with a neighbourhood algorithm which resulted in the 3D model of the S-wave velocity of the studied landslide.

The results of seismic studies described in this paper are a continuation of the work by Pilecki and Harba (2015) who presented the usefulness of the seismic interferometry method in mapping the structure and properties of the near-surface geological medium in the area of the active Just-Tegoborze landslide (southern Poland). Their preliminary results are based on the measurements carried out along single seismic profile in two measurement series. In this paper, we present the results developed with one additional series. Moreover, the results were extended with data from transverse seismic profiles. We also describe the geological conditions of the Just-Tegoborze landslide and the measurement methodology, as well as the methodology of seismic data processing and interpretation. The changes of elastic properties of geological medium are characterized by the changes in the S-wave velocity. The recognized seismic boundaries are correlated to lithological ones shown on the geological engineering cross-section.

GEOLOGICAL AND GEOMORPHOLOGICAL SETTING

Geographically, the Just-Tegoborze landslide is located in Swidnik and Tegoborze villages, in Nowy Sacz county in the southern part of Poland. It is situated on a slope of Jodlowiec Wielki Mountain and St. Just Mountain, slightly below St. Just pass of the Beskid Wyspowy range (Fig. 1).

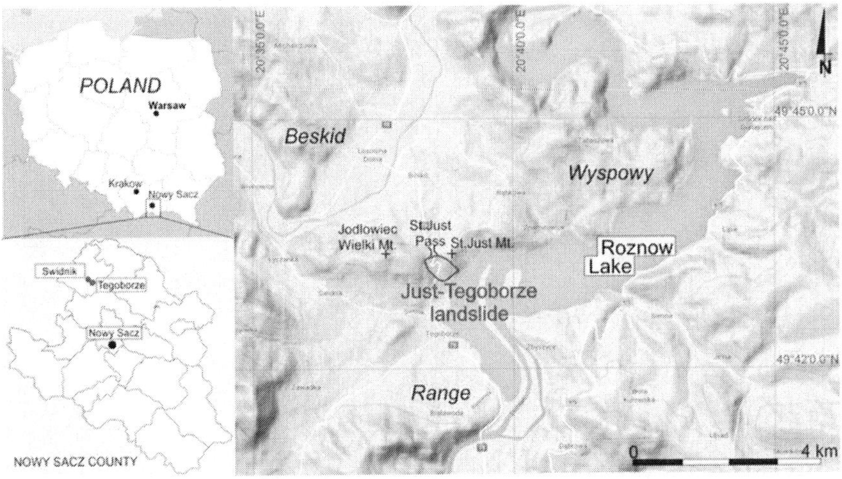

Figure 1. Geographic location map of the study area of the Just-Tegoborze landslide (on the basis of Google Maps)

From the geological point of view, the Just-Tegoborze landslide lies in Magura Nappe in the Outer (Flysch) Carpathians (Fig. 2). On the study area, Magura Nappe is represented by hieroglyphic beds, Magura beds and Sub-Magura beds. Hieroglyphic beds are mainly built of thin-bedded sandstones and shales. They are interbedded with medium and thick-bedded sandstones in the upper part, whereas shales constitute a majority in the lower part. Magura beds consist of muscovitic sandstones of 0.7 to 2 m thickness. Sub-Magura beds are built of sandstones and shales. Flysch formation, according to data from the 3I/P borehole located on the study area (Fig. 3), dips at angles from 60° to 84° toward the North.

Quaternary colluvial deposits cover discordant flysch bedrock on the study area. Strongly cohesive clayey colluvium has a thickness of 10 to 12 m (Fig. 4). It is very heterogenous and consists of clay, humic clay, silty clay, sandy clay, silts, clayey debris, clasts of sandstones, mudstones and shales. Soft clayey colluvium with relocated fragments of soft shales, locally interbedded with hard shales, mudstones and sandstones forms a layer of

3 to 5 m thickness below the near-surface clayey colluvium (Czudec 2012, unpublished data).

Several failure surfaces of the landslide were distinguished as a result of archival geological works (Chowaniec and Wojcik 2012). Failure surfaces were investigated on the study area within colluvial deposits at approximately 10 and 12 m depth and between colluvial deposits and the flysch bedrock at approximately 14 to 17 m depth (Fig. 4).

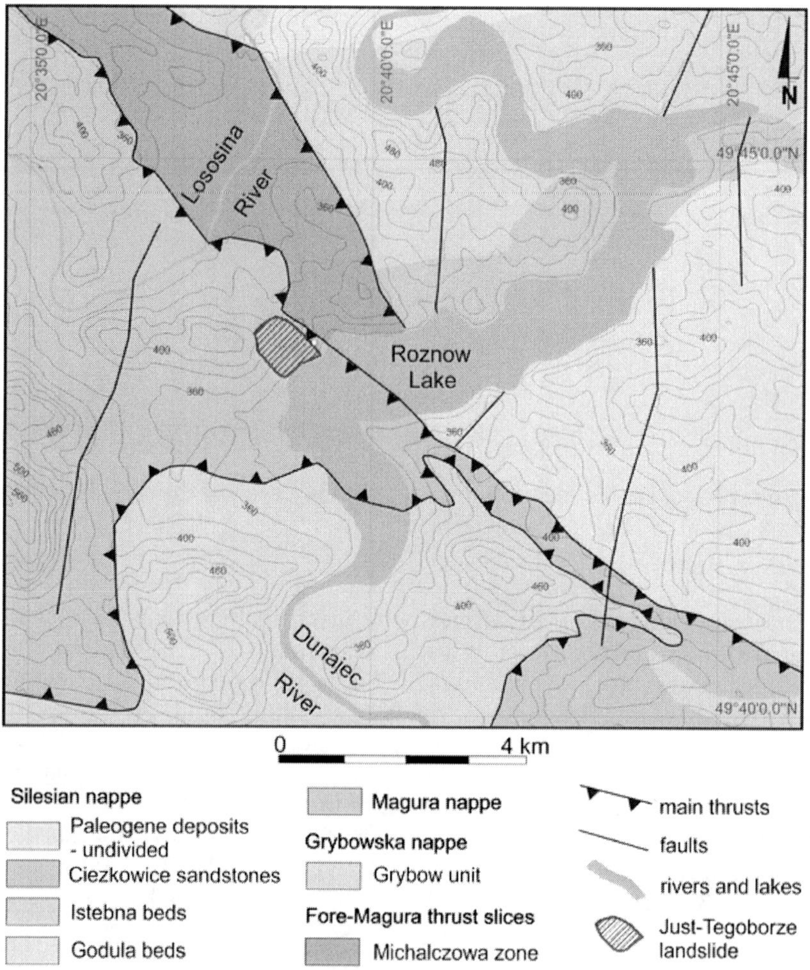

Figure 2. Geological map of the surrounding area of the Just-Tegoborze landslide with elevation contour lines marked (based on Cieszkowski and Waskowska 2010)

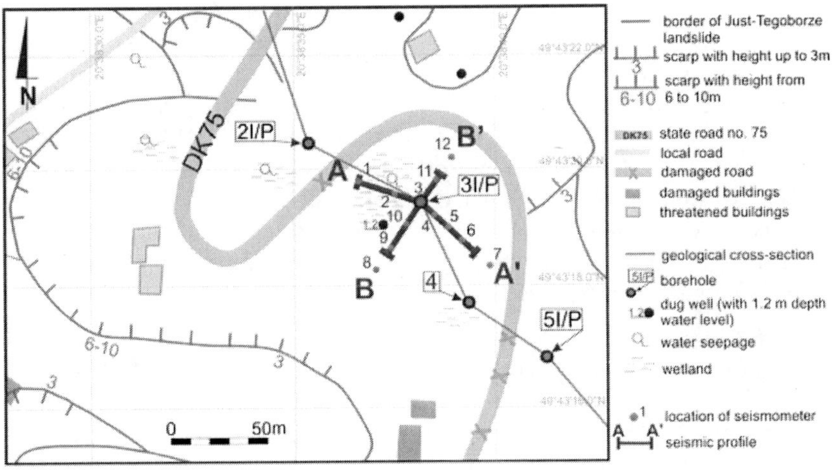

Figure 3. A part of the Just-Tegoborze landslide sketch on a base topographic map, including projection of a geological engineering cross-section with boreholes and seismic profiles (based on Chowaniec and Wojcik 2012)

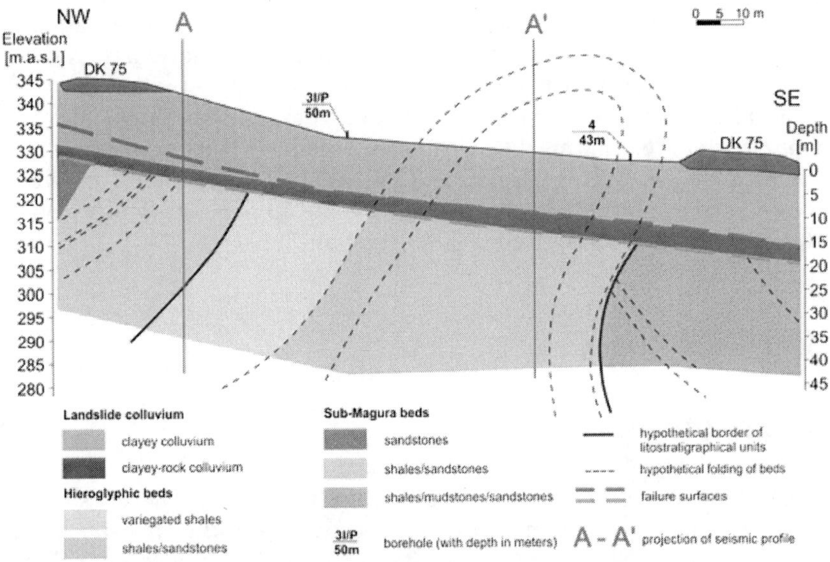

Figure . 4. A part of the geological engineering cross-section of the study area with the A-A′ projected seismic profile (based on Pilecki and Harba 2015)

The Just-Tegoborze landslide developed on the contact between shale and shale-sandstone complexes of Sub-Magura and hieroglyphic beds, and Magura sandstones. Shale and shale-siltstone complexes are less resistant and more susceptible to weathering than Magura sandstones. Magura sandstones have uniaxial compressive strength (UCS) from 19 to 91 MPa, whereas shales from hieroglyphic beds have UCS from 0.25 to 2.9 MPa (Czudec 2012, unpublished data). Other factors, which have an influence on the development of the Just-Tegoborze landslide, include the occurrence of weak variegated shales prone to swelling and complicated tectonics between Magura, Sub-Magura and hieroglyphic beds.

The Just-Tegoborze landslide has southeastern exposure and descends to the Roznow Lake (Fig. 1). The azimuth of its movement is 130°, and its average slope angle is 11.8° (Chowaniec and Wojcik 2012). From a geomorphological point of view, it consists of a number of mass movement areas, however, with no clear boundaries between them. Its slopes are of convex-concave type, and the terrain relief is strongly diversified. Slopes are mostly cut by small coomb valleys, dells and gullies. There are numerous scarps and landslide niches on the upper part of the landslide. On the study area, levels of water filtration were examined on 0.7, 2.0, 4.7 and 5.8 m. Water seepages are observed on the terrain surface close to the 3I/P (Fig. 3). Water level in a dug well, which is located on the study area, is 1.2 m deep, however, it reaches terrain surface during intense rainfalls and spring thaw. It leads to the conclusion that there are weakly permeable deposits which limit the infiltration of water deep into the geological medium. The analysis of the core from 3I/P borehole indicates the presence of low permeable, weakly and medium fractured shales and variegated shales at the depth of 13 m. Above them, there are very fractured shales and previously described colluvial deposits which have a water content up to 35% (Czudec 2012, unpublished data).

METHODOLOGY

Seismic acquisition

A local high-frequency seismic noise was used in the study with seismic interferometry, generated mainly by intense vehicle traffic on a state road no. 75 present in the immediate vicinity. Data acquisition was carried out by 12 Güralp CMG-6TD broadband seismometers working autonomously. They were installed along two profiles with lengths of 75 and 95 m, which crossed at the 3I/P borehole (Fig. 3). Distances between seismometers were from 10 to 20 m. Data recording lasted for 60 min with 10 ms sampling. Signals were recorded in the frequency band of 0.03–100 Hz.

Figure 5a shows an example of a seismogram of the seismic noise recorded, which is characterized by the strongest seismic energy within a frequency band between 5 and 22 Hz (Fig. 5b). A similar frequency band of vehicle traffic, used as a source for near-surface passive imaging, was studied by Kuzma et al. (2009).

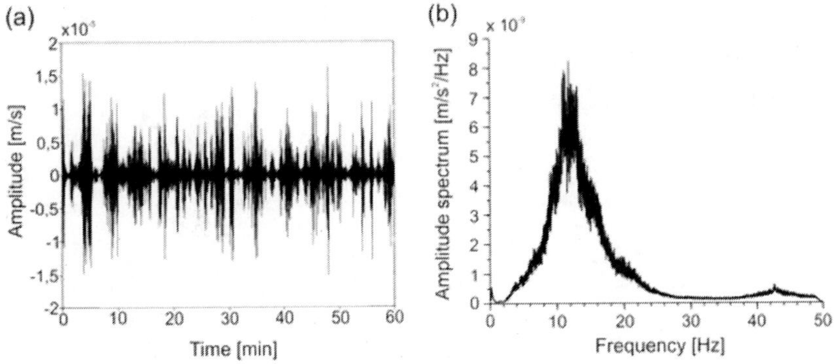

Figure 5. An example of a seismogram (**A**) and an amplitude spectrum (**B**) of seismic noise recorded in a period of 60 min at the Just-Tegoborze landslide

Measurements were carried out in three series in 2015: January 23rd (a dry period), March 9th (a period of an increased water content after the spring thaw) and July 9th (a period of an intense water infiltration during heavy rainfalls). Figure 6 shows a daily average precipitation and temperature measured at nearby meteorological stations from July 2014 to July 2015. In each seismic measurement series, the seismometers were installed at the same place. The recognition depth range of the geological medium was up to 20 m.

Processing and interpretation

The data processing and interpretation scheme was similar to those of Bensen et al. (2007) and Cheng et al. (2015).

The measured amplitude records of vertical velocity of ground vibrations on the Just-Tegoborze landslide were first band-pass filtered in the 7 to 35 Hz frequency band for smoothing amplitude spectra. Next, 1-bit normalization and cross-correlation were applied. Cross-correlation functions were computed between all available receiver pairs. As a result, series of empirical Green's functions were obtained and sorted afterwards into the so-called common virtual source gathers for every receiver. Each virtual source was considered as a bidirectional shot with two common virtual source gathers as the causal and acausal signal parts (Fig. 7).

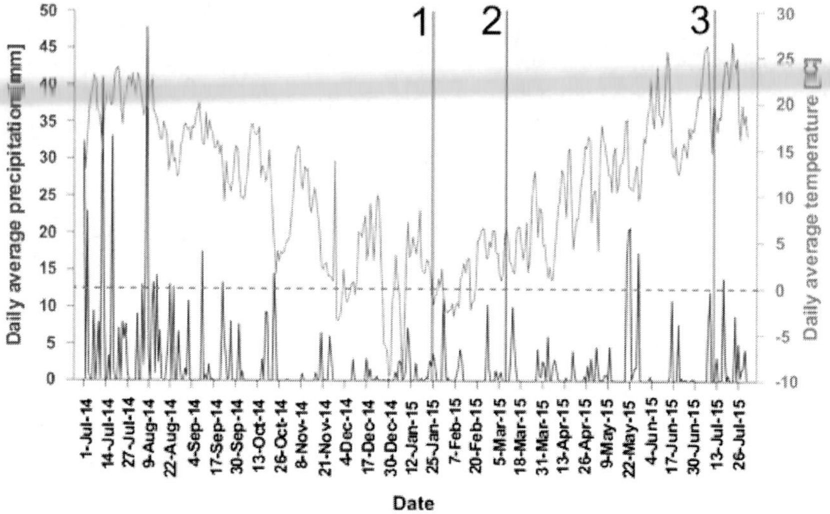

Figure 6. Daily average precipitation and daily average temperature data measured at the two meteorological stations in the vicinity of the Just-Tegoborze landslide from 1st July 2014 to 31st July 2015. Black vertical lines correspond to measurement series: 1, 2 and 3 respectively

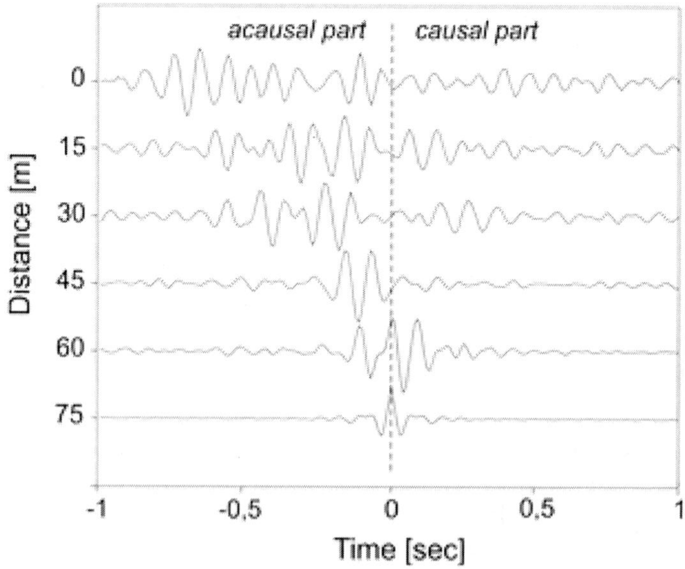

Figure 7. An example of a common virtual shot gather of the virtual source located at the 75 m point of the B-B′ seismic profile with acausal and causal parts indicated

The frequency–wavenumber (f–k) transform was subsequently computed to separate surface wave energy from those of body waves. Muting was applied (Fig. 8a) in order to improve the signal to noise ratio. Finally, the dispersion curve of the Rayleigh wave was extracted (Fig. 8b).

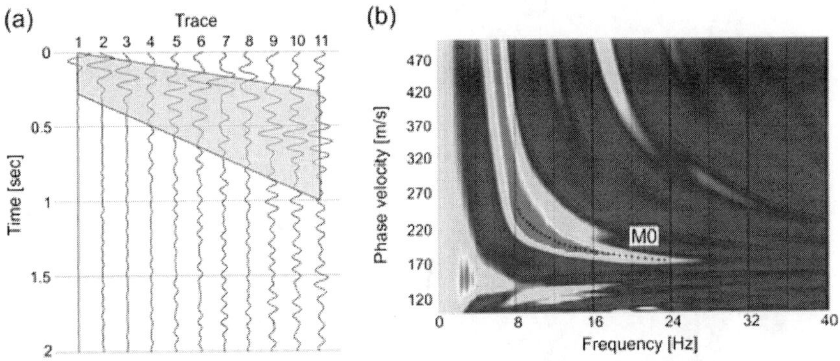

Fig. 8 An example of a seismic noise cross-correlogram with a muting window (**A**) and a dispersion image with a picked fundamental mode (*M0*) (**B**)

Data interpretation was based on the first picking fundamental modes of the Rayleigh wave from the dispersion curves. They were inverted afterwards to obtain 1D shear wave velocity models (Fig. 9). Inversion was conducted by means of a genetic algorithm. Finally, a 2D shear wave velocity section was constructed on the basis of the 1D information. Data processing and interpretation were carried out using own algorithms written in Matlab and supported by Geogiga Seismic Pro software.

RESULTS AND ANALYSIS

The seismic sections developed for the three measurement series are presented in Fig. 10 for longitudinal sections and in Fig. 11 for section transverse to the landslide axis.

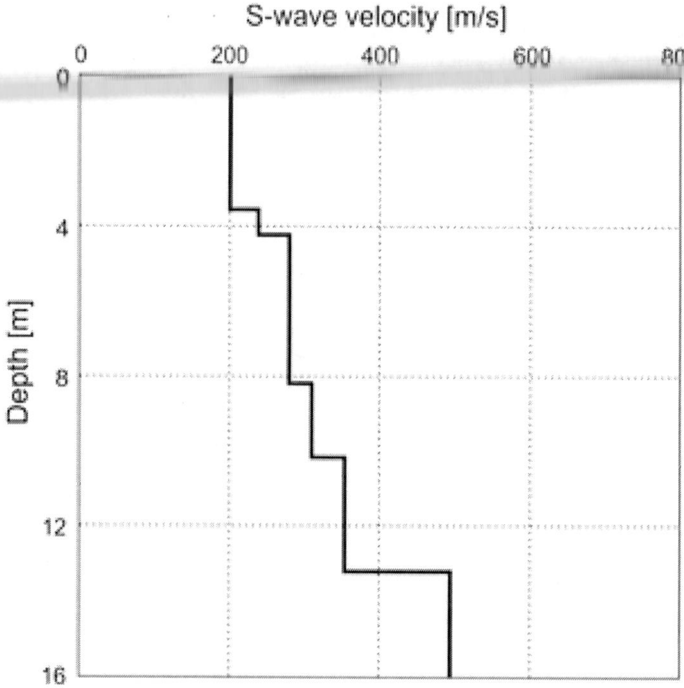

Figure 9. An example of a 1D S-wave velocity profile inverted from the dispersion curve

The 1st measurement series was carried out in the winter period in conditions which may be assumed to be dry (Fig. 6). Evident in this section are three layers with varied S-wave velocities (Fig. 10a). In the near-surface layer, with the lowest velocities ranging from 220 to 270 m/s, there is also a local minimum value in the central part. This layer is most likely connected with the increased water content in the medium, as demonstrated by the observed water table when executing borehole 3I/P and the visible dampness of the terrain surface while taking the measurements. The first layer is characterized by a thickness amounting to approximately 5 m along the entire length of the profile. The next, middle layer is characterized by the S-wave velocity ranging from 270 to 340 m/s. Its thickness is variable and primarily is approximately 8 to 10 m, locally decreasing to 5 m in the 70 to 75 m of the seismic profile (Fig. 10a). Thickness of the second layer is the largest in the central part of the profile. The deepest-located layer is characterized by the S-wave velocity values in the range of 340 to 400 m/s. The roof of this layer occurs at the depth of 10 to 15 m.

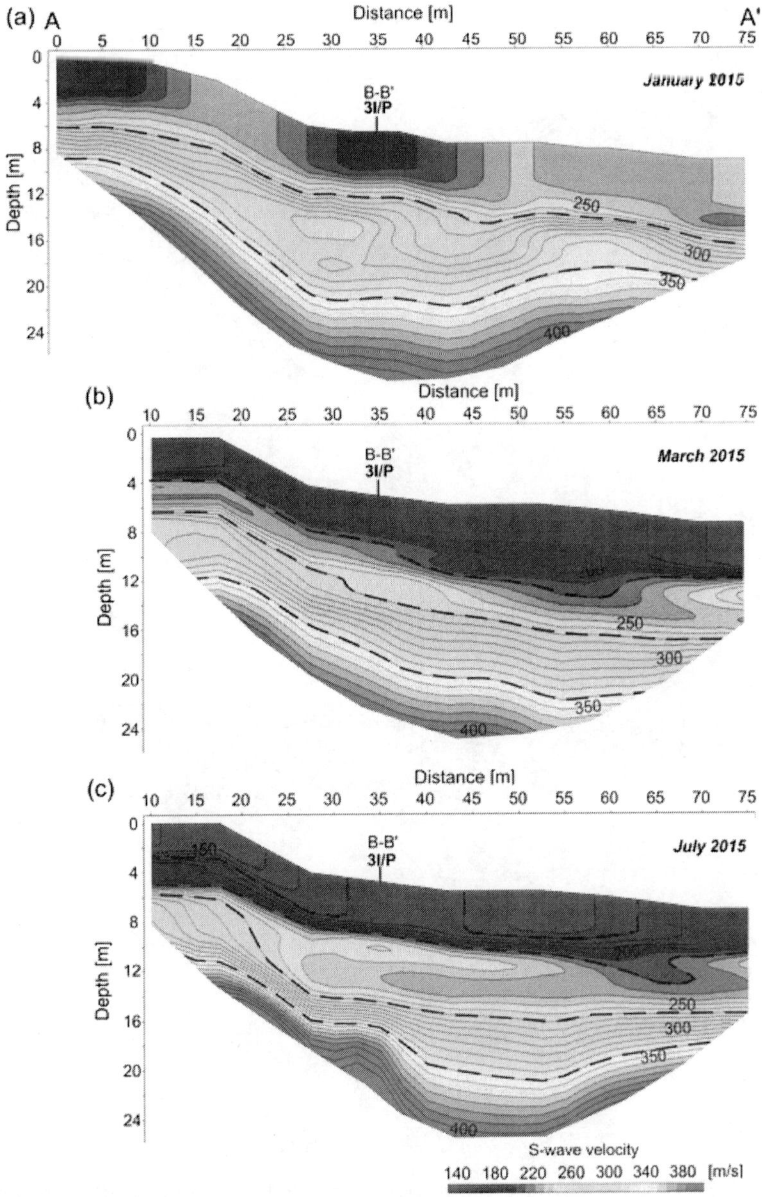

Figure 10. S-wave velocity section along the A-A′ seismic profile measured on 23rd January 2015 (**A**), 9th March 2015 (**B**) and 9th July 2015 (**C**)

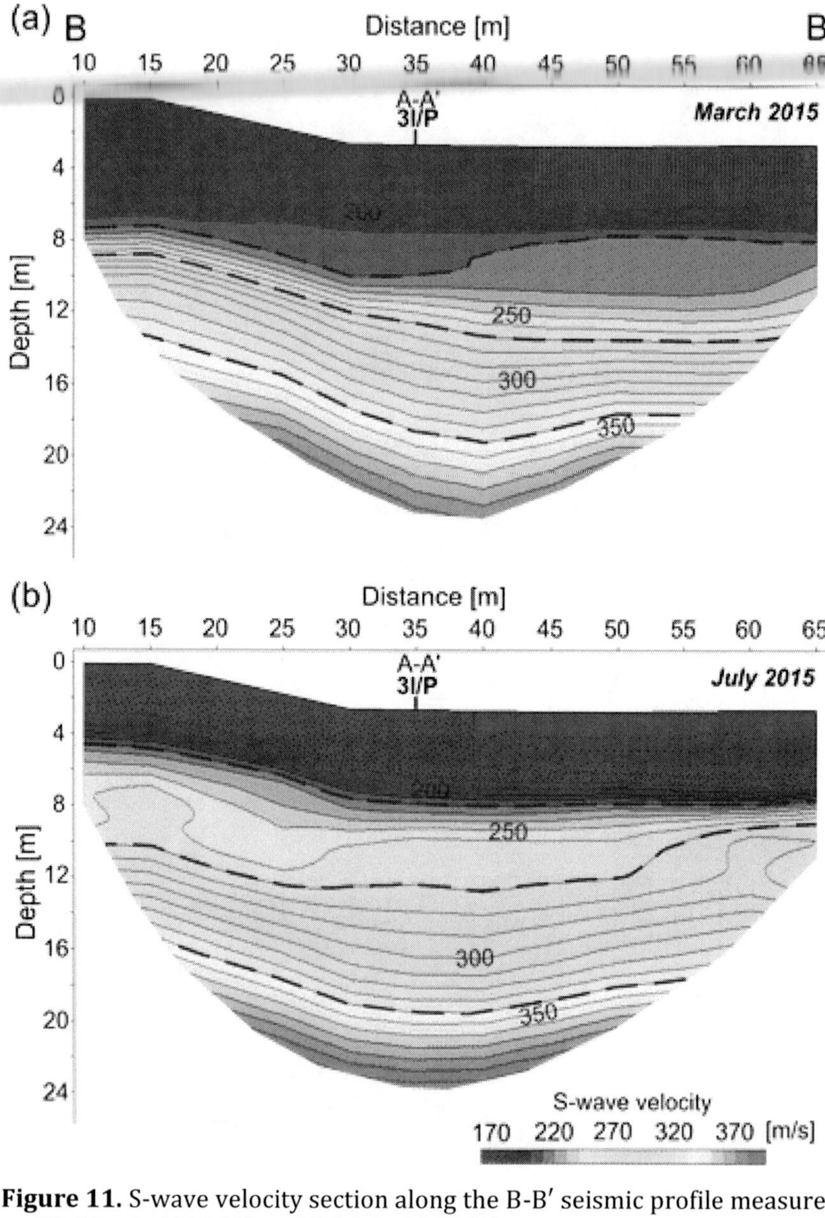

Figure 11. S-wave velocity section along the B-B′ seismic profile measured on 9th March 2015 (**A**) and 9th July 2015 (**B**)

The trough-like shape of the seismic boundary between the second and third layers occurs in the zone of the weaker rock mass of the hieroglyphic beds made of clayey variegated shales and shales with instances of minor sandstone interbedding. A harder sandstone-shale rock mass of Sub-Magura beds lies in the vicinity (Fig. 4). The roof of the third layer likely constitutes the surface of a less weathered bedrock.

The 2nd measurement series was carried out in the spring period with a melting snow cover. The near-surface ground might have not yet been fully thawed (Fig. 6). The longitudinal (Fig. 10b) and transverse (Fig. 11a) sections are characterized by a significant decrease in the S-wave velocity in the two surface layers of the medium. In order to emphasize the differences, layer boundaries were marked in the longitudinal section according to contour lines interpreted in Fig. 10a. The zone of weakened elastic properties stretched toward the drop in the terrain surface of the area. The velocity value of the seismic boundary of the bedrock was very slightly changed. It should be noted that the depth of this boundary was maintained.

The 3rd measurement series was conducted in the summer period directly after storm rainfalls (Fig. 6, Fig. 10c, Fig. 11b). Increased infiltration into the medium caused a greater weakening of elastic properties, especially in the first layer, compared to the 2nd series. The velocity value of the seismic boundary of the bedrock again only slightly changed. The depth of this boundary was also maintained. A similar effect may be observed in the transverse sections (Fig. 11a, b).

In relation to the geological engineering cross-section of the research area (Fig. 4), one may notice that the first two near-surface seismic layers correspond to clayey colluvium formations. The seismic boundary occurring at a depth of approximately 10 to 15 m may be correlated to the roof of the less weathered rock mass of the shale and shale/sandstone series. In the seismic image, the bedrock (third layer) did not undergo significant changes as a result of infiltration, which confirms lesser disturbance by landslide processes in comparison with colluvium layers. Presumably, the bedrock roof boundary is one of the most important failure surfaces of the landslide. The trough-like shape of this seismic boundary in the central part of the seismic section may be connected with the greater range of weathering in the zone of weaker hieroglyphic beds.

CONCLUSIONS

This study has demonstrated that it is possible to monitor the changes of shear wave velocities of a geological compound medium, as it is monitored in the Just-Tegoborze landslide by the seismic interferometry method using a local high-frequency ambient noise. The results clearly show changes of the S-wave velocity which are most likely caused by various conditions of infiltration and water content in the geological medium. Generally, an increasing water content in the medium results in the decrease of the S-wave velocity value and also the medium's rigidity.

The interpreted seismic boundaries may be correlated to lithological boundaries shown in a geological engineering cross-section. In the seismic image, the course of the bedrock boundary is more complex in comparison with the geological engineering cross-section. The trough-like shape of this seismic boundary in the central part of the seismic section may be related to the decrease of rigidity of weak hieroglyphic beds compared to the stronger surrounding rock mass, however, it requires further detailed studies. The investigated bedrock boundary is presumably one of the failure surfaces.

The complicated course of the bedrock boundary may also have an influence on the water flow paths in the colluvium layer affected by mass movements. In consequence, it has an influence on the stability of the state road structure.

The presented results of seismic interferometry may be helpful in evaluating the behaviour of the Just-Tegoborze landslide. These findings can be also used in designing protections against mass movements.

ACKNOWLEDGEMENTS

The article was prepared as a result of the LOFRES project No. PBS1/A2/13/2013 performed within the 1st call of the Applied Research Programme co-financed by the Polish National Centre for Research and Development. We thank the editor and anonymous reviewers for valuable comments and discussions.

REFERENCES

1. Aleotti P, Chowdhury R (1999) Landslide hazard assessment: summary review and new perspectives. Bull Eng Geol Environ 58:21–44
2. Anbarasu K, Sengupta A, Gupta S, Sharma SP (2010) Mechanism of activation of the Lanta Khola landslide in Sikkim Himalayas. Landslides 7:135–147
3. Barla G, Antolini F, Barla M (2013) Slope stabilization in difficult conditions: the case study of a debris slide in NW Italian Alps. Landslides 10:343–355

4. Bensen GD, Ritzwoller MH, Barmin MP, Levshin AL, Lin F, Moschetti MP, Shapiro NM, Yang Y (2007) Processing seismic ambient noise data to obtain reliable broad-band surface wave dispersion measurements. Geophys J Int 169:1239–1260
5. Bichler A, Bobrowsky P, Best M, Douma M, Hunter J, Calvert T, Burns R (2004) Three-dimensional mapping of a landslide using a multi-geophysical approach: the Quesnel Forks landslide. Landslides 1:29–40
6. Bièvre G, Jongmans D, Goutaland D, Pathier E, Zumbo V (2015) Geophysical characterization of the lithological control on the kinematic pattern in a large clayey landslide (Avignonet, French Alps). Landslides. doi:10.1007/s10346-015-0579-0
7. Bogoslovsky VA, Ogilvy AA (1977) Geophysical methods for the investigation of landslides. Geophysics 42(3):562–571
8. Bonnard C, Corominas J (2005) Landslide hazard management practices in the world. Landslides 2:245–246
9. Brückl E, Brunner FK, Lang E, Mertl S, Müller M, Stary U (2013) The Gradenbach Observatory—monitoring deep-seated gravitational slope deformation by geodetic, hydrological, and seismological methods. Landslides 10:815–829
10. Cascini L, Bonnard C, Corominas J, Jibson R, Montero-Olarte J (2005) Landslide hazard and risk zoning for urban planning and development. In: Hungr O, Fell R, Couture R, Eberhardt E (eds) Landslide risk management, proceeding of the international conference on landslide risk management, Vancouver, Canada. A.A. Balkema Publishers, Taylor & Francis Group, London, pp. 199–235
11. Cheng F, Xia J, Xu Y, Xu Z, Pan Y (2015) A new passive seismic method based on seismic interferometry and multichannel analysis of surface waves. J Appl Geophys 117:126–135
12. Chowaniec J, Wojcik A (eds) (2012) Landslides in Malopolskie voivodeship, A guidebook (in Polish). Carthographic Publishing House Compass, Krakow, pp. 89–91
13. Cieszkowski M, Waskowska A (2010) Castles from the Roznow and Czchow lakes area as geotouristic lapidarium and role of the Istebna Sandstones in their construction (in polish). Geotourism 2(21):3–18
14. Curtis A, Gerstoft P, Sato H, Snieder R, Wapenaar K (2006) Seismic interferometry – turning noise into signal. Lead Edge 25:1082–1092
15. Czudec G (2012) Geological-engineering documentation for developing landslide stabilization no. MPL0051, on the state road no. 75, 51 + 900 do 52 + 700 km in Tegoborze-Just, Lososina Dolna commune, Nowy Sacz county, Malopolskie voivodeship. Geotech Sp. z o.o. Archive no. 1266 (in Polish) – unpublished data
16. Dai FC, Lee CF, Ngai YY (2002) Landslide risk assessment and management: an overview. Eng Geol 64:65–87
17. Del Gaudio V, Wasowski J, Muscillo S (2013) New developments in ambient noise analysis to characterise the seismic response of landslide-prone slopes. Nat Hazards Earth Syst Sci 13:2075–2087
18. Guzzetti F, Carrara A, Cardinali M, Reichenbach P (1999) Landslide hazard evaluation: a review of current techniques and their application in a multi-scale study, Central Italy. Geomorphology, vol 31:181–216
19. Frasheri A, Kapllani L, Dhima F (1998) Geophysical landslide investigation and prediction in the hydrotechnical works. Journal of the Balkan Geophysical Society 1(3):38–43
20. Fukuoka H, Sassa K, Wang G, Wang F, Wang Y, Tian Y (2005) Landslide risk assessment and disaster management in the Imperial Resort Palace of Lishan, Xian, China. In: Sassa K, Fukuoka H, Wang G, Wang F (eds) Landslides, Risk Analysis and Sustainable Disaster Management. Springer, Berlin Heidelberg, pp. 81–89
21. Hack R (2000) Geophysics for slope stability. Surv Geophys 21:423–448
22. Hibert C, Grandjean G, Bitri A, Travelletti J, Malet J-P (2011) Characterizing landslides through geophysical data fusion: example of the La Valette landslide (France). Eng Geol 128:23–29

23. Jongmans D, Bièvre G, Renalier F, Schwartz S, Beaurez N, Orengo Y (2009) Geophysical investigation of a large landslide in glaciolacustrine clays in the Trièves area (French Alps). Eng Geol 109:45–56
24. Jongmans D, Garambois S (2007) Geophysical investigation of landslides: a review. Bulletin de la Société Géologique de France 178(2):101–112
25. Kuzma HA, Fernández-Martínez JL, Zhao Y, Dunson C, Zhai M, Mangriotis M, Rector JW (2009) Vehicle traffic as a source for near-surface passive seismic imaging. Symposium on the Application of Geophysics to Engineering and Environmental Problems 2009:609–615
26. Larose E, Carrière S, Voisin C, Bottelin P, Baillet L, Guéguen P, Walter F, Jongmans D, Guiller B, Garambois S, Gimbert F, Massey C (2015) Environmental seismology: what can we learn on Earth processes with ambient noise? J Appl Geophys 116:62–74
27. Lundström K, Larsson R, Dahlin T (2009) Mapping of quick clay formations using geotechnical and geophysical methods. Landslides 6:1–15
28. Mainsant G, Chambon G, Jongmans D, Larose E, Baillet L (2015) Shear-wave-velocity drop prior to clayey mass movement in laboratory flume experiments. Eng Geol 192:26–32
29. Mainsant G, Jongmans D, Chambon G, Larose E, Baillet L (2012a) Shear-wave velocity as an indicator for rheological changes in clay materials: lessons from laboratory experiments. Geophys Res Lett 39:L19301
30. Mainsant G, Larose E, Brönnimann C, Jongmans D, Michoud C, Jaboyedoff M (2012b) Ambient seismic noise monitoring of a clay landslide: toward failure prediction. J Geophys Res 117:F01030
31. Mansour MF, Morgenstern NR, Martin CD (2011) Expected damage from displacement of slow-moving slides. Landslides 8:117–131
32. Méric O, Garambois S, Jongmans D, Wathelet M, Chatelain JL, Vengeon JM (2005) Application of geophysical methods for the investigation of the large gravitational mass movement of Séchilienne, France. Can Geotech J 42:1105–1115
33. Méric O, Garambois S, Malet J-P, Cadet H, Gueguen P, Jongmans D (2007) Seismic noise-based methods for soft-rock landslide characterization. Bulletin de la Société Géologique de France 178(2):137–148
34. Pilecki Z, Harba P (2015) Structure and properties of a landslide investigated with seismic interferometry using high-frequency seismic noise—preliminary results (in polish). Bulletin of the Mineral and Energy Economy Research Institute of the Polish Academy of Sciences 89:63–76
35. Pilz M, Parolai S, Bindi D, Saponaro A, Abdybachaev U (2013) Combining seismic noise techniques for landslide characterization. Pure Appl Geophys 171(8):1729–1745
36. Renalier F, Bièvre G, Jongmans D, Campillo M, Bard P-Y (2010a) Clayey landslide investigations using active and passive Vs measurements. In: Miller RD, Bradford JH, Holliger K (eds) Advances in near-surface seismology and ground-penetrating radar, Geophys. dev. Ser, vol 15. Society of Exploration Geophysicists, Tulsaa, pp. 397–414
37. Renalier F, Jongmans D, Campillo M, Bard P-Y (2010b) Shear wave velocity imaging of the Avignonet landslide (France) using ambient noise cross correlation. J Geophys Res 115:F03032
38. Reynolds JM (2011) An introduction to applied and environmental geophysics, 2nd edn. Wiley-Blackwell Publishing, Chichester
39. Salas-Romero S, Malehmir A, Snowball I, Lougheed BC, Hellqvist M (2015) Identifying landslide preconditions in Swedish quick clays – insights from integration of surface geophysical, core sample- and downhole property measurements, DOI:10.1007/s10346-015-0633-y
40. Sassa K, Wang G, Fukuoka H, Wang F, Ochiai T, Sugiyama M, Sekiguchi T (2004) Landslide risk evaluation and hazard zoning for rapid and long-travel landslides in urban development areas. Landslides 1:221–235
41. Schuster GT (2009) Seismic interferometry. Cambridge University Press, New York

42. Singh RP, Dubey CS, Singh SK, Shukla DP, Mishra BK, Tajbakhsh M, Ningthoujam PS, Sharma M, Singh N (2013) A new slope mass rating in mountainous terrain, Jammu and Kashmir Himalayas: application of geophysical technique in slope stability studies. Landslides 10:255–265
43. Song Y-S, Kim K-S (2014) Geotechnical properties of landslide sites in Korea with differing geology. In: Sassa K, Canuti P, Yin Y (eds) Landslide science for a safer Geoenvironment, Volume 2: Methods of Landslide Studies. Springer International Publishing Switzerland, Cham, pp. 87–92
44. Travelleti J, Malet J-P, Samyn K, Grandjean G, Jaboyedoff M (2013) Control of landslide retrogression by discontinuities: evidence by the integration of airborne- and ground-based geophysical information. Landslides 10:10–54
45. Wapenaar K, Draganov D, Snieder R, Campman X, Verdel A (2010a) Tutorial on seismic interferometry: part 1—basic principles and applications. Geophysics 75(5):75A195–75A209
46. Wapenaar K, Slob E, Snieder R, Curtis A (2010b) Tutorial on seismic interferometry: part 2—underlying theory and new advances. Geophysics 75(5):75A211–75A227
47. Xia J, Miller RD, Park CB (1999) Estimation of near-surface shear-wave velocity by inversion of Rayleigh waves. Geophysics 64(3):691–700
48. Zarroca M, Linares R, Roqué C, Rosell J, Gutiérrez F (2014) Integrated geophysical and morphostratigraphic approach to investigate a coseismic (?) translational slide responsible for the destruction of the Montclus village (Spanish Pyrenees). Landslides 11:655–671

CHAPTER 6

Assessment of Data Availability for the Development of Landslide Fatality Curves

Tonje Grahn [1 2], HenrikJaldell[3]

[1.]*Centre for Climate and SafetyKarlstad UniversityKarlstadSweden*

[2.]*Department of Environmental and Life SciencesKarlstad UniversityKarlstadSweden*

[3.]*Department of EconomicsKarlstad UniversityKarlstadSweden*

ABSTRACT

Quick clay landslides are a special feature of Norwegian and Swedish geologies. Vibrations or small initial landslides can cause a quick clay layer to collapse and liquefy, resulting in rapid landslides with little or no time for evacuation, making them a real threat to human life. Research concentrating on damages due to landslides is scarce, and analyses of loss of human lives caused by quick clay landslides in the scientific literature are, to our knowledge, non-existing. Fatality quantification can complement landslide risk assessments and serves as guidance for policy choices when evaluating efficient risk-reducing measures. The objectives of this study were to assess and analyze available damage information in an existing data set of 66 historical landslide events that occurred in Norway and Sweden between 1848 and 2009, and access its applicability for quantifying loss of human life caused by quick clay landslides. Fatality curves were estimated as functions of the number of exposed persons per landslide. Monte Carlo simulations were used to account for the uncertainties in the number of people actually exposed. The results of the study imply that the quick clay fatality curves are non linear, indicating that the probability of losing lives increases exponentially when the number of exposed persons increases. Potential factors affecting human susceptibility to landslides (e.g., landslide-, area-, or individual-specific characteristics) could not be satisfyingly quantified based on available historical records. Future research should concentrate on quantifying susceptibility factors that can further explain human vulnerability to quick clay landslides.

Keywords: Landslide Quick clay Loss of life Landslide fatalities Data availability Landslide damage

INTRODUCTION

In a wide context, natural hazards impose a real threat to human life all over the world. From 1994 to 2013, the average death toll was greater than 99,700 per year (Cred 2015). Statistics from The Center for Research on the Epidemiology of Disasters (CRED) show that 17% of all fatalities from natural hazards worldwide were caused by landslides (Lacasse et al. 2010). In contrast to other natural hazards (e.g., floods or earthquakes), which affect large areas, landslides are generally limited in area but can occur with a high frequency in a specific region (JRC 2011). Landslides can be caused by both natural and anthropogenic actions, and landslide disasters are increasing worldwide (Andersson-Sköld et al. 2014; Guzzetti 2016; Ho et al. 2013; Nadim et al. 2006; SOU 2007). The reasons are suspected to be increased susceptibility of surface soil to instability because of deforestation and overexploration of natural resources, greater vulnerability of exposed population as a result of urbanization and uncontrolled land use, and climate change resulting in a greater potential for weather extremes (Lacasse et al. 2010; Nadim et al. 2006). Pereira et al. (2014) have found that most landslides occur during the wettest months, reflecting the importance of rainfall-triggering mechanisms. Landslides can cause serious damage to buildings and infrastructure, to the environment, and to human life (Devoli et al. 2007; Jiménez-Perálvarez et al. 2009; Lacasse et al. 2010). Further, landslides in populated areas can create health consequences that exceed the local capacity to respond, increasing both mortality and morbidity from disasters (Keim 2008). These types of disaster risks therefore need to be taken into account in present and future societal planning at local, regional, and national levels.

Quantifying the consequences can assist the risk management process by guiding policy interventions to choose the most effective mitigation project. This evokes a large demand from policy makers on local, regional, and national levels for quantitative damage estimates as guidance and support so that efficient decisions can be made in the use of public and private resources for risk-reducing measures. Quantification of loss of life or injuries are however problematic because information often is incomplete.

The main scope of landslide hazard assessments is to provide quantitative expertise on future slope failures to planners, decision-makers, civil defense authorities, insurance companies, developers, and individual landowners (Guzzetti et al. 2005a). Within the field of risk assessment, methods for estimating hazard probabilities are relatively well-established, but when determining risks related to natural hazards, many studies neglect investigating the damage potential or are limited to the assessment of immobile objects such as buildings (Jonkman et al. 2010). Landslide risk (R) is usually described as a function of probability and the consequences of a landslide (e.g., Andersson-Sköld et al. 2014)

$$R = P \times C_i$$

where P is the probability of a hazard, and $\boldsymbol{C}$ is a vector C_i of all potential consequences. Fatalities (C_f) can then be expressed as a function of human susceptibility (S_f) to landslides when exposed and the number of exposed humans (E).

$$C_f = S_f \times E$$

The largest proportion of landslide risk analyses or landslide susceptibility analyses has had a technical approach focusing on the first part of the function, the hazard probability (P), by performing and developing methods for mapping landslide risk, stability analysis, and mechanisms triggering landslides (e.g., Guzzetti et al. 2005a, 2006, 2012; Jiménez-Perálvarez et al. 2009; Melchiorre and Tryggvason 2015; Poli and Sterlacchini 2007; Roslee and Jamaluddin 2012; Salas-Romero et al. 2015; Van Den Eeckhaut et al. 2012). Potential landslide consequences (C_i) have not attained the same scientific interest despite the obvious importance it has on the estimated risk product (R). Further, estimations have been found to be more sensitive to uncertainties in damage susceptibility (S) than uncertainty in hazard exposure (P) (Jongman et al. 2012).

Few published studies address quantitative landslide analysis of consequences (C_f) to human life. Among the scientifically published literature, an extensive portion focuses on Italian landslide scenarios. Several studies base their research on variations of the same historical landslide catalogs (e.g., Guzzetti et al. 2003, 2006; Salvati et al. 2010).

Analyses of experienced risk factors affecting fatality rates can, however, serve as input to risk assessments and disaster response by increasing the understanding of vulnerable groups and ease assistance in the direct aftermath of a disaster (Agrawal et al. 2013). Previous studies have shown that historical information from landslides can be valuable when quantifying landslide threat to human lives (Cascini et al. 2008).

This study analyses consequences of quick clay slides by focusing on exposure (E), and human susceptibility (S_f) to loss of life.

$$R_f = P \times (S_f \times E)$$

The objective of this study is to assess available damage data from previous quick clay landslides in order to identify and quantify factors affecting human susceptibility to landslides. Fatality functions are derived and compared to fatality rates previously applied in risk assessment for policy purposes. We will use data from the Swedish Geotechnical Institute (SGI 2011, 2012). To our knowledge, the Göta River investigation is the only investigation thus far that included quantification of the loss of human life from landslides in the Nordic countries. However, the fatality rate is burdened with large uncertainties due to uncertainties in the number of exposed humans (E). It is therefore interesting to perform a more in-depth analysis of the estimations of loss of life using this data set.

LANDSLIDE CHARACTERISTICS AND METHODS FOR QUANTIFICATION

Experienced fatalities in European landslides

Fatality numbers and fatality rates vary between countries and regions. Landslides differ in their characteristics, such as temporal perspectives and warning opportunities. Since European landslides often are small, isolated events, this leads to an underestimation of European landslides reported to global databases (Van Den Eeckhaut et al. 2012). For example, EM-DAT reports landslide events where more than 10 people were killed, 100 or more were affected, or international assistance or a state of emergency were called for (Van Den Eeckhaut et al. 2012). Based on the Northern Portugal Landslide Database (NPLD), 136 people lost their lives in 436 landslides between 1900 and 2010 in Portugal (Pereira et al. 2014). This yields an average fatality rate of 0.3. In Switzerland, Hilker et al. (2009) found by assessing The Swiss Federal Research Institutes (WSL) landslide database, that fatalities averaged 3 per year during the time period 1972–2007. Three hundred forty-four landslides occurring in the Umbria region in Italy between the years 1941 and 1997 caused 12 fatalities, yielding an average fatality rate of 0.03 (Guzzetti et al. 2006). A national Italian investigation on landslide risk in the twentieth century revealed that at least 7799 casualties were reported, 5831 lives were lost, 108 missing, and 1860 injured (Guzzetti et al. 2003). Extending the time period to the years 1279–2002 and including landslides causing death, missing people, injuries, and homelessness, landslides caused a total of 10,111 fatalities yielding and average fatality rate of 10.1 (Guzzetti et al. 2005b). This data set includes the disastrous landslide at the Vajont dam in 1963 that generated an enormous wave, killing almost 2000 people. Guzzetti (2000) found that the national landslide mortality rate in Italy decreased between 1950 and 1999. Compared to other areas, the fatality frequency was found to be lower in the Alp region, Canada, and Hong Kong, but higher in Japan and China (Guzzetti 2000). Giannecchini and D'Amato Avanzi (2012) found that, in the Versilia River Basin, landslides and flood together caused between 1 and 13 fatalities per event.

Considering the Nordic countries, approximately 2000 people have been killed by all kinds of landslides in Norway (Jaedicke et al. 2009) during the last 150 years. No such compilation has been performed for Sweden, but since 1950, at least 10 people have lost their lives, and 160 were injured as a result of landslides (MSB 2016).

Factors affecting susceptibility (S_f)

Damage caused by landslides mostly affects private homes, road networks, and other infrastructure (Guzzetti et al. 2006). This is where humans are most likely to get exposed to landslides. Fast-moving landslides, including rock falls, rockslides, rock avalanches, and debris flows, have historically caused the largest number of landslides deaths (Guzzetti 2000). The threats that landslides induce on human life are often a main concern. Humans are an essential part of the damage potential because landslides are associated with high rates of traumatic injury and mortality caused by trauma and suffocation (Keiler et al. 2005; Keim 2008). Research quantifying risks to human lives when exposed to landslides do exist, but is not extensive. Agrawal et al. (2013) examined landslides in Uganda and focused on risk factors affecting the extent of impact to human life and health. Despite the differences between Uganda and the Nordic countries in terms of factors such as socioeconomic aspects, the risk factors identified in that study are of interest. The following risk factors were assessed: number of exposed individuals, sex, age, type of injury, severity of injury, type of medical assistance, and where the injury occurred (e.g., at home, inside, outside, or in a vehicle). The study also showed that in

Uganda, most people are injured or killed outside and not at their place of residence. This diverges from the data describing Norwegian quick clay landslides, where loss of life occurred most frequently in the person's place of residence (Jaedicke et al. 2009). Zhang and Zhang (2014) examined factors involved with human flight behavior related to rapid landslides along roads in China. They showed that the susceptibility of humans when exposed inside buildings depends on the characteristics of the landslide and the technical resistance of the buildings, but also that individual behavior and personal attributes (i.e., age, sex, disability, running speed, response, education, and prior experience with landslides) are of large importance for successfully fleeing a landslide. People with reduced health statuses are also more vulnerable when exposed to disasters (Keim 2008). Personal attributes being an important factor for human susceptibility are confirmed by Viscusi (2006) claiming that the magnitude of loss of human life is influenced by factors such as individuals' exposure to hazards and their levels of self-protective behavior. Lacasse et al. (2010) further points out that day of week, time of day, and functioning warning systems are risk factors that should be taken into account. Furthermore, healthy people are less likely to suffer disaster-related mortality and are therefore more disaster resilient (Keim 2008).

Susceptibility to flood exposure is also relevant to consider since landslides near rivers, lakes, and coasts can be flooded when large amounts of soil and clay slide into the water, creating large waves rapidly flooding the surrounding areas with risk to human life, livestock, and other asset values.

Methods for quantifying fatalities (C_f)

General methods for estimating consequences such as loss of life have not been standardized to the same extent as estimations of hazard probabilities (Jonkman et al. 2010). Jonkman et al. (2010) presented a general approach for estimating loss of life for "low probability-large consequence" accidents and disasters. This general approach includes an assessment of physical effects associated with an event, determination of the number of exposed persons, and determination of the mortality among the exposed population.

Previous scientifically published quantification on fatalities are, as far as we know, calculated frequency rates for average number of fatalities per event, year, month, day, intensity, area of occurrence or expressed as the number of deaths per 100,000 inhabitants for a given population over a predefined time period (Cascini et al. 2008; Giannecchini and D'Amato Avanzi 2012; Guzzetti et al. 2005b; Hilker et al. 2009; Pereira et al. 2014; Salvati et al. 2010). Methodologically, the studies differ concerning which observation inclusion criteria that are used when deriving a fatality (or mortality) rate. Guzzetti et al. (2005b) combined the number of missing people with the number of deaths when estimating fatalities. Salvati et al. (2010) also included injured humans as well as those killed and declared missing. Giannecchini and D'Amato Avanzi (2012) did not distinguish between flood and landslide fatalities. The studies also differ in the length of time periods used for estimating a landslide fatality rate. Guzzetti et al. (2005b) includes information from landslides going back to 1279, and Salvati et al. (2010) as far back as year 68. How well the oldest landslides reflect present human susceptibility to landslides is questionable.

Salvati et al. (2010) also performed a more detailed risk evaluation for Italy for the time period 1950–2008. They compared magnitude of events between regions and time periods by modeling distributions of flood and landslide events with causalities taking into account the frequency and intensity of events. They performed the analysis using a Zipf distribution which is a discrete distribution often used for modeling rare events for a finite population size. The analysis performed by Salvati et al. (2010) requires an abundant amount of data. This is available in Italy since Italy has access to more data, due to that the country has suffered substantial losses to landslides, has a strong tradition of historical research, and that an extensive part of the landslide damage literature stems from Italian research groups.

To increase the knowledge of the risk of landslides in Sweden, the Swedish Geotechnical Institute (SGI), at the Swedish government's initiative, performed comprehensive risk analyses of landslide risks in the Göta River valley. The main focus was on hazard events (i.e., stability analysis), but a large proportion of the analyses was also dedicated to identifying, quantifying, and, as far as possible, monetizing potential future damage from landslides (Göransson et al. 2014; SGI 2012). SGI developed a method for quantifying risk of losing lives and add this type of consequence to other types of consequences (i.e., building, infrastructure, and industrial damage, etc.). A prerequisite for applying the method is to quantify human susceptibility when exposed to landslides (C_f). C_f was calculated as the relative frequency of fatalities and exposed population (evacuated population not included) using historical records (SGI 2011).

Characteristics of quick clay landslide

Quick clay landslides occur in Norwegian and Swedish geologies where large land areas with quick clay deposits are seen (Jaedicke et al. 2009; SGI 2012; Salas-Romero et al. 2015). Quick clay can develop both in marine and fresh water sediments, but most Scandinavian quick clay developed in glaciomarine sediments deposited during deglaciation 14,000 to

10,000 years ago (Salas-Romero et al. 2015). Quick clay is a soil with high water content and weak binding between the particles (Andersson-Sköld et al. 2014). Vibrations or small initial landslides can cause a quick clay layer to collapse and liquefy, resulting in rapid landslides (Andersson-Sköld et al. 2011). This can be caused by natural processes (e.g., precipitation and/or erosion) or by human processes (e.g., overloading and excavation) (SGI 2012). Quick clay landslides occur without warning signs leaving little or no time for warning and evacuation (Melchiorre and Tryggvason 2015; SGI 2012). Signs of creep have, however, on some occasions been documented prior to landslides (Melchiorre and Tryggvason 2015). Quick clay landslides occur in Norway and Sweden on a regular basis and become hazardous events when they occur in developed and populated areas. Large quick clay landslides in populated areas are not frequent in Scandinavia, but areas are continuously being mapped because of the concern about potentially large and unacceptable consequences. One example is the most landslide prone area in Sweden, the highly developed and populated Göta River valley, with a number of landslides every year (Andersson-Sköld et al. 2014). However, because of the low frequency of large quick clay landslides, area-specific data available for damage analyses and damage prognoses are scarce.

DATA AND METHODS

Sample data set

In this study, we analyze the data set compiled by the SGI for use in landslide risk assessment of the Göta River valley. The data set has in the present study been subject to content analysis in order to identify susceptibility factors (S_f) and other quantitative damage information identified as relevant by previous research (see "Landslide characteristics and methods for quantification" section) and to analyze the uncertainty of the derived susceptibility factor. The data set contains 66 landslides collected from two data sources. Among them, 55 quick clay landslides that occurred between 1848 and 2009 were extracted from the Norwegian landslide database, Skrednett. The data quality related to documentation of historical events is strongly influenced by the personal engagement of local observers and observational routines. (Jaedicke et al. 2009). Despite the uncertainties, the database is seen as a unique source for statistical analysis, including risk analysis (Jaedicke et al. 2009). The remaining 11 landslides were extracted from the Swedish natural hazards information system and occurred in Sweden during a shorter time period, 1950 to 2006 (MSB 2016). The landslide frequencies differ between Norway and

Sweden. While the frequency of landslides in Norway is approximately the same as for Austria and Italy, it has been lower in Sweden which has the same frequency as Switzerland (Andersson-Sköld et al. 2013).

We have been looking at the characteristics describing exposure and vulnerability in the specific landslide events in the data set that could have been affecting human susceptibility in these events, and evaluating the potential for quantification of the characteristics.

Exposure and fatalities

Fatalities per landslide ranged from 0 to 116. Overall, 167 people lost their lives in the 66 landslides, yielding an average landslide fatality rate of 2.5 (Table 1). Humans were exposed in only 52 to 62% of the landslides, and people died in 27% of them, meaning that many events resulted in zero people exposed and zero fatalities. Exposed persons ranged from 0 to 375 per landslide. Because of uncertainties in the data set concerning the actual number of people exposed per landslide, the number of exposed humans is represented by an interval for some landslides. The number of people exposed overall, summing up all the landslides in the sample is between 734 and 1594. Where human exposure was burdened with uncertainty and therefore given as an interval, the SGI gave a *best guess* and used this as a point estimator together with other certain observations to derive an average fatality rate (C_f). Combining the sum of these *best guesses* for number of people exposed (1035) with the total 167 fatalities produced an average fatality rate of 0.161 and was applied in the governmentally initiated risk assessment representing human susceptibility of people potentially exposed to future landslides in the Göta River Valley.

The data shows that humans were exposed in 34 events, and for 13 events, it described non-fatal injuries, from minor to severe and irreversible. For two events, humans did not suffer physical damage, but in one, a victim was not able to escape because of a physical disability. Summarizing the fatal landslides, they have the following distribution: In 7 landslides, one died; in 2 landslides, three died; in 4 landslides, four died (Fig. 1); there were one landslide each with six, seven, eight, and nine fatalities. Concerning number of fatalities, one landslide event stands out. This is the Verdal landslide in Norway in May 1893. It occurred at nighttime in Verdal in Northern Tröndelag killing 116 persons of the 250 persons exposed to the landslide.

Table 1. Overview of the data set by country, county, year, month, and time of occurrence, number of exposed humans, number of fatalities, injuries, cause of landslide, warning signs, prior slide in the area, and if a landslide triggered a flood damaging wave. Empty spaces in the columns means that the information is unknown and not documented in the databases

Number	Country	County	Year	Month	Time of day	Exposed humans	Fatalities
1	Norway	Trondheim	1848	September	03.00	3	1
2	Norway	Nord-Tröndelag	1859	September	Daytime	2	0
3	Norway	Sör-Tröndelag	1871	December	Daytime	4	0
4	Norway	Nord-Tröndelag	1874	April	Daytime	1	0
5	Norway	Akershus	1883	November	Nighttime	10	6
6	Norway	Nord-Tröndelag	1893	May	01.30	250	116
7	Norway	Troms	1898	December	05.00	12	7
8	Norway	Nord-Tröndelag	1900	May	Daytime	4	3
9	Norway	Nordland	1902	November	Daytime	12	4
10	Norway	Nord-Tröndelag	1909	July	Daytime	0	0
11	Norway	Sör-Tröndelag	1920	September	07.00	1	0
12	Norway	Nord-Tröndelag	1921	Springtime		1	0
13	Norway	Östfold	1925	April		0	0
14	Norway	Nordland	1925	July		0	0
15	Norway	Nordland	1925	April		2.5	0
16	Norway	Buskerud	1927	Mars		0	0
17	Norway	Nord-Tröndelag	1932	Mars		3	0
18	Norway	Buskerud	1935	January		6	4
19	Norway	Vestfold	1937	January	10.30	2	0
20	Norway	Telemark	1943	July		50	0
21	Norway	Akershus	1950	October		0	0
22	Norway	Akershus	1953	December	Nighttime	0	0
23	Norway	Nord-Tröndelag	1959	April	08.50	2	1
24	Norway	Nord-Tröndelag	1962	September	Nighttime	1–48	1
25	Norway	Sör-Tröndelag	1965	June		1	0
26	Norway	Nord-Tröndelag	1965	April		0–5	0
27	Norway	Nordland	1965	January		0–3	0
28	Norway	Östfold	1967	October	11.36	4–45	4
29	Norway	Sör-Tröndelag	1973	December		1	0
30	Norway	Östfold	1974	December	16.30	0–5	0
31	Norway	Telemark	1976			3	1
32	Norway	Sör-Tröndelag	1978	April	Daytime	1–40	1
33	Norway	Östfold	1980	August		0	0
34	Norway	Östfold	1980			0	0
35	Norway	Sör-Tröndelag	1982	March		0	0
36	Norway	Nord-Tröndelag	1982	October	Morning	0	0
37	Norway	Nordland	1984	July		0	0
38	Norway	Troms	1984	May	11.00	0	0
39	Norway	Sör-Tröndelag	1985	October	11.30	7	0
40	Norway	Östfold	1986			0	0

41	Norway	Sör-Tröndelag	1987	December		1–3	0
42	Norway	Östfold	1988	Springtime		0	0
43	Norway	Sör-Tröndelag	1989	November		1–3	1
44	Norway	Hedemark	1994	October	20.00	0	0
45	Norway	Nordland	1996	June	Nighttime	7	4
46	Norway	Hedemark	1999			0	0
47	Norway	Akershus	2000	November		0	0
48	Norway	Buskerud	2000	Autumn		1–13	0
49	Norway	Troms	2001	June		0	0
50	Norway	Sör-Tröndelag	2002	April	04.00	0–5	0
51	Norway	Nord-Tröndelag	2002	April	Nighttime	1	0
52	Norway	Nord-Tröndelag	2007	March	Daytime	0	0
53	Norway	Nord-Tröndelag	2007	May		0–5	0
54	Norway	Östfold	2008	April		0	0
55	Norway	Nord-Tröndelag	2009	March	11.50	7–100	0
56	Sweden	Västra Götaland	1950	September	08.00	90–375	1
57	Sweden	Västra Götaland	1953	April	14.00	0–6	0
58	Sweden	Västra Götaland	1957	June	11.25	6–310	3
59	Sweden	Värmland	1969	April	04.00	0	0
60	Sweden	Stockholm	1972	October	Nigttime	0	0
61	Sweden	Västra Götaland	1977	November	16.05	200	9
62	Sweden	Ångermanland	1987	November	Nighttime	0	0
63	Sweden	Västra Götaland	1996	April	18.40	0	0
64	Sweden	Södermanland	1997	May	00.59	10	0
65	Sweden	Örebo	2006	December	Morning	0	0
66	Sweden	Västra Götaland	2006	December	19.00	28	0

Number	Injuries	Cause of slide	Warning signs	Prior slides in area	Damaging wave
1				Yes	
2					
3			Yes, saw the slide coming		Yes
4	1				Yes
5		Rainy autumn	Initial slide 2 weeks earlier	Yes	Yes
6	Many injuries		Small slides occurring in the area during springtime, the river turned gray despite no rain	Yes	Yes
7			Several slides occurring in the area prior to the main slide	Yes	
8		Construction of railway, possibly caused by dynamite			
9	8	Construction of railroad			
10	0				
11	Light injuries	Construction			No

12					
13	0	Occurred after rainy period	Small initial slides led to evacuation of people and animals	Yes	
14	0			Yes	
15					
16	0				
17		Possibly the lowering of water levels in the river	People started running when land started moving		
18					
19					
20	Minor injuries	Bombing (WW2)			
21	0				
22	0		People used to landslides. Land movement detected the day before led to evacuation	Every year	
23	0	Erosion in river			
24					
25					Yes
26					
27					Yes
28		Rainy period	Land movement had been observed, and a crack in a window had also been observed		No
29					
30				Yes	
31					
32		Human element			Yes
33	0	Human element			
34	0				
35	0			Yes	
36	0			Yes	
37	0				
38	0				Yes
39	7 injured	Intensive rain might have triggered the slide			
40	0			Yes	
41	1	Heavy rain			
42	0	Probably water undermining the ground			
43	0	Probably triggered by heavy rain		Yes	

44	0			
45		Possibly road construction		
46	0			Yes
47	0			
48				
49	0			Yes
50	0	Land filling		
51				
52	0	Possibly triggered by road construction		
53		Erosion from the river Reina		
54	0	Potentially triggered by heavy precipitation		
55	Minor damages	Partly triggered road construction		
56	90	Pilework had been done a few days prior to the slide and the vibrations might have triggered the slide	Cracks where observed prior to the slide	
57	0		Slide	Yes
58	3	Erosion of the river	Cracks where observed prior to the slide and led to evacuation of the area	
59	0	Erosion of the river	A sinking in the road had been observed the day prior to the slide	
60	0	Depositing of the soil on top of slope in combination with construction work		
61	60			
62	0	Probably erosion of the river		
63	0	Erosion in the river in the combination with low water levels		
64	5 minor injuries	Combination of erosion, development/exploitment, and rain	Minor slides had been observed in the area the week prior to slide	Yes
65	0	Construction and rain	Sinking and gliding of soil in the area had been observed	
66	3 injured	Construction and rain		

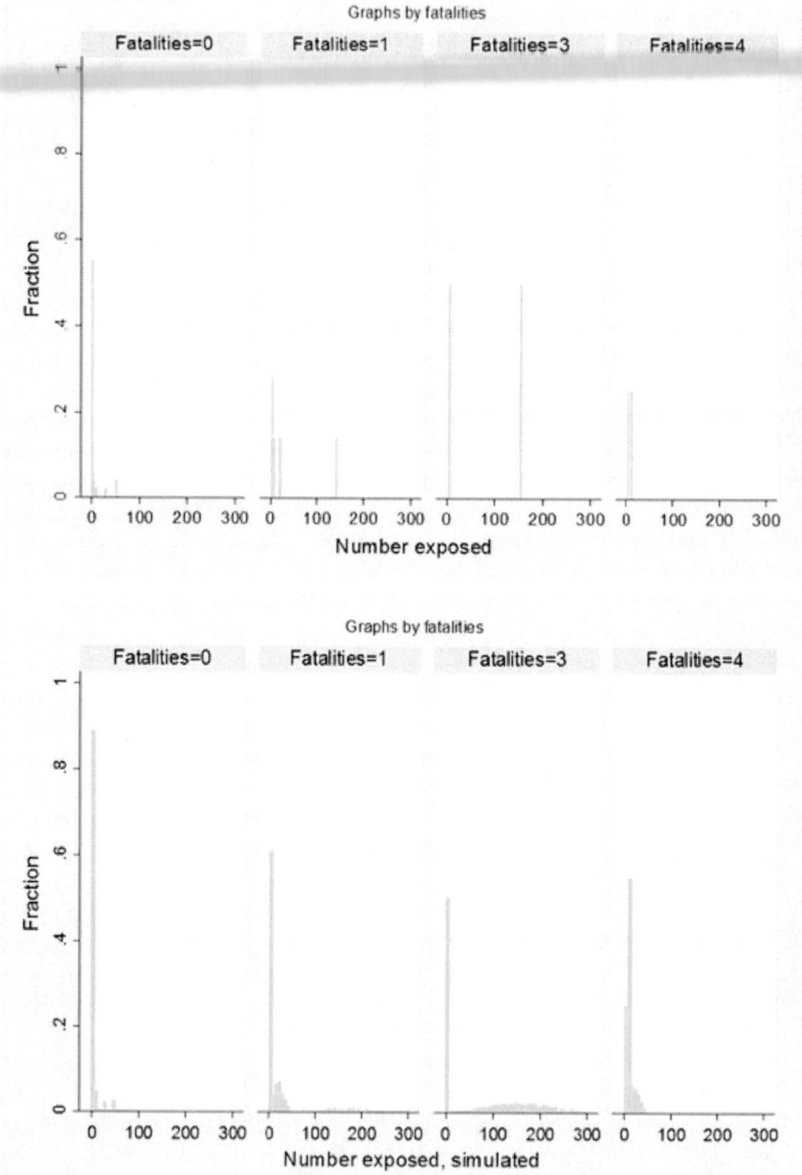

Figure1. Histograms of SGI data (*upper*) and simulated (*lower*) number of people exposed for 1, 3, and 4 fatalities

The data set also documents eight landslide events near lakes and rivers causing flood exposure to nearby settlements resulting in a range of poorly documented tangible damages.

Warning signs and previous experience

For 29 events, some kind of warning signal (e.g., gliding, creep, cracks, sinking of the ground, or small initial landslides) had indicated that a landslide was underway (Table 1). Anecdotal descriptions could be found, such as the detection of a crack in a window observed prior to the landslide that demolished the house. The crack was in the aftermath of the slide interpreted as signs of soil instability affecting the structure of the house. For two events, residents in the area saw the landslide coming and were able to escape to safety. Further, 16 events occurred in areas previously exposed to landslides.

Time of occurrence

Other characteristics interesting to quantify is time of occurrence. For all events, the year of occurrence was recorded but the month was recorded for only 59 events, and in 3 others, only the season was recorded. For 31 events, the accurate time of day was recorded, but for 17, it was recorded only as day or night. For the remaining slides (42%), the time of day is unknown. Of those slides for which the time of day was recorded, 13 slides occurred during the night. These landslides exposed a total of eight humans and caused the loss of six lives (Table 1).

Statistical methods

While the data presented in "Sample data set" section can serve as a basis for identification of risk factors, it is of utmost importance that the amount of data is sufficient relative to the number of parameters in a model (Jonkman et al. 2010). The amount of available data determines the detail level in the model used for estimation. Country, year, and number of people exposed were the only potential risk factors in the sample set adequately documented to serve as input data to the statistical analysis estimating the loss of life functions. Further, the 26 landslides where humans definitely were not exposed were dropped from further consideration in the statistical analysis. Two statistical approaches were applied to the remaining data set. (1) Regression analysis was used to explore the relationship between fatalities and number of exposed humans, country of occurrence and year of occurrence, under the hypothesis that number of persons exposed per landslide, country of occurrence and year of occurrence affect the number of fatalities per landslide. (2) Monte Carlo simulation was performed to take the uncertainty in the relative frequency of landslide fatality rates as a function of the number of exposed humans per landslide into account.

Regression analysis

Two different types of regression models are applied to the data set. One linear model using ordinary least squares (OLS) and one non-linear count data regression analysis applying two different distributions. The models have all been estimated using one dependent variable (the number of fatalities) and three independent variables (country [Norway or Sweden], year, and number of people exposed based on SGI's *best guess* values). Because the Verdal landslide is a statistical outlier, the models were estimated both with and without it. The estimated function for the linear model is given by the following:

$$\text{Fatalities} = \beta_0 + \beta_1 \times \text{Number Exposed} + \beta_2 \times \text{Norway} + \beta_3 \times \text{Year}.$$

Count data models were applied because of the properties of the fatality variable. The values for fatalities are discrete, have relatively low numbers, and include zeros. The tested count data models were the Poisson model and the negative binomial model. The Poisson model assumes that the expected value and variance are equal. A negative binomial model takes into account that the variance is greater than the expected value (overdispersion). Overdispersion is estimated by a parameter alpha (see, e.g., Greene 2008). The count data models were compared using various test statistics using likelihood values, the Vuong test, and comparing the actual to the predicted probabilities following the procedures suggested by Long and Freese (2014). The count data model that best fit the data was the negative binomial regression model.[1]

Monte Carlo simulation

There are, however, two major drawbacks with using the ordinary least square model and the negative binomial model to the data set in this study. The first is that we have only 40 observations, which makes statistical inference difficult. The second is that there are uncertainties about number of people exposed, in some cases, a single number represents those exposed, but for others, a range is recorded.

By making Monte Carlo simulations with the data, it is possible to account for the uncertainties in the exposure variable. The Monte Carlo simulation was performed by simulating 1000 trials from each landslide.

To do so, we made assumptions about the statistical distributions of the uncertainty in the exposure variable based on information documented in the landslide databases. Four different distributions were applied. (1) If a single number represented those exposed, we assumed the value was certain, and it was used (no distribution). (2) If a particular value could be assumed to be more probable, but an interval surrounding this value also was given, we applied a triangular distribution covering the interval given but using the most probable value as the mode. (3) If the most probable value was zero, but there was uncertainty, we applied a right triangle distribution (skewed distribution). (4) If only an interval was recorded, but no probable value, we applied a rectangular distribution. For 25 landslides, no distribution was applied. A triangular distribution was applied for 11 events, a rectangular distribution for three, and a skewed distribution was applied for one event where the probable value was assumed to be zero, but with uncertainty. This landslide event was not included in the basic model regression analysis but is included in the Monte Carlo simulation since there is an uncertainty interval for number of exposed human in this observation. The simulated data set thus included 40,000 simulated observations.

The distributions of the original SGI data are compared to the simulated data for one, three, and four fatalities in the histograms in Fig. 1. The exposed numbers for more than four fatalities are all considered certain and therefore have no intervals. It can be noted that the distribution for one fatality is more spread out for the simulated data than for the original data of two, three, and, especially, 140 people exposed. The distribution for three fatalities is also more spread out for 155 people exposed.

RESULTS OF STATISTICAL ANALYSIS

Table 2 shows the results from the regressions using original SGI data (not simulated), where four models are estimated: the OLS models (model 1) and the negative binomial model (model 2) with and without the Verdal landslide. Results show a statistically significant positive relationship between the number of people exposed and fatalities meaning that the number of fatalities increases when more people were exposed.[2] This relationship is less strong when the outlier (the Verdal landslide) is excluded. The variable for country is not statistically significant in three of the four models. This indicates that there are no significant differences in fatalities for Norwegian and Swedish landslides, and that we can merge the data from the two countries. The year estimate is negative, but not statistically significant, meaning that we cannot conclude that the number of fatalities during the analyzed time period has decreased.

Table 2. Parameter estimates basic model. Standard errors in parenthesis

	Model 1a (incl. Verdal) OLS	Model 1b (excl. Verdal) OLS	Model 2a (incl. Verdal) Negative binomial	Model 2b (excl. Verdal) Negative binomial
Intercept	[illegible]	[illegible]	[illegible]	[illegible]
β_0	(85.295)	(14.579)	(12.235)	(13.141)
Number of people exposed	0.283**	0.031**	0.019**	0.035*
β_1	(0.038)	(0.010)	(0.006)	(0.021)
Norway	22.709**	1.281	1.855	4.265
β_2	(6.071)	(1.211)	(1.129)	(3.419)
Year	−0.038	−0.011	−0.010*	−0.011*
β_3	(0.043)	(0.007)	(0.006)	(0.006)
Alpha			1.416**	1.581**
			(0.701)	(0.766)
R-square	0.64	0.29		
Log-likelihood	−148.8	−77.7	−59.3	−52.8
Number of observations	39	38	39	38

*$p < 0.10$, **$p < 0.01$

Table 3 shows the results of the regressions using data from the Monte Carlo simulations. Because 40,000 observations are included, all parameter estimates are statistically significant. Because neither a country effect nor a trend effect was found in the first estimations (Table 2), they were not taken into account. The Verdal landslide very much affects the results. In the OLS model (model 3), the risk of fatality increases linearly by 0.18 for each additional person exposed if the Verdal landslide is included; it increases only 0.02 if it is excluded. The count data models (model 4) are non-linear as can be seen from the predicted values shown graphically in Figs. 2 and 3. In Table 3 and Fig.3, the results of a model only using data for up to 20 people exposed are also shown to investigate the potential that the landslide fatality data consist of separate distributions depending on the number of persons exposed. The coefficient for number of people exposed for this model is higher than for the other models (except for 3a), indicating that the risk of fatality is higher in the sample set with 0 to 20 people exposed.

Table 3. Monte Carlo simulation. Parameter estimates basic model. Standard errors in parentheses

	Model 3a (incl. Verdal) OLS	Model 3b (excl. Verdal) OLS	Model 4a (incl. Verdal) Negative binomial	Model 4b (excl. Verdal) Negative binomial	Model 4c (max 20 no. exposed) Negative binomial
Main part					
Intercept	−0.801***	0.932***	−0.112***	−0.022**	−1.079***
	(0.078)	(0.011)	(0.010)	(0.010)	(0.018)
Number of people exposed (simulated)	0.178***	0.017***	0.016***	0.008**	0.166***
	(0.001)	(0.001)	(0.0001)	(0.0001)	(0.002)
Alpha			2.575***	2.457***	1.877***
			(0.031)	(0.033)	(0.029)
R-square	0.37	0.15			
Log-likelihood	−163,296	−82,822	−64,130	−56,884	−42,263
Number of simulated observations	40,000	39,000	40,000	39,000	33,000

***$p < 0.001$

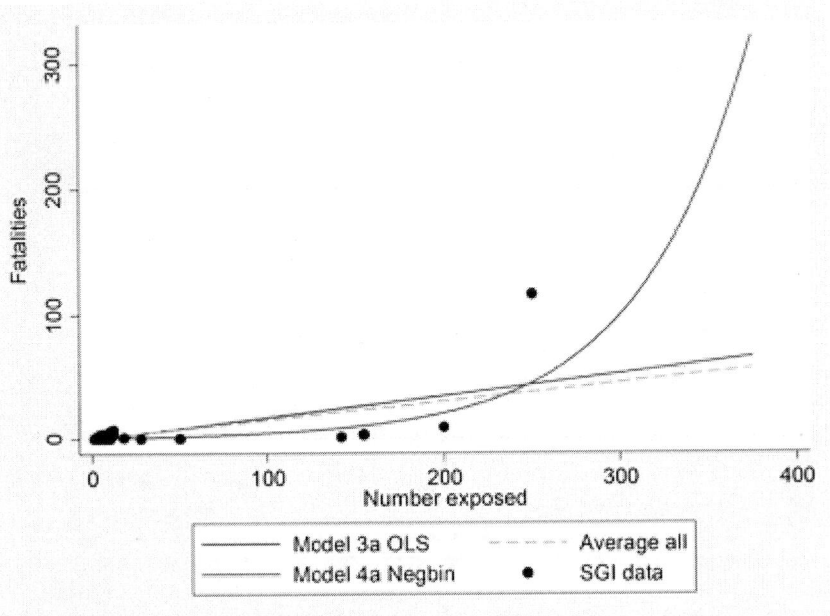

Figure 2. Observed and predicted values. The SGI data values are the most probable number of people exposed according to the Swedish Geological Institute (SGI). The average is the average loss of life rate based on SGI's observed values of exposed persons. The predicted values (models 3a and 4a) are calculated from the estimated models using the Monte Carlo simulated data, which considers the uncertainty in number of exposed persons

Figure 2 shows observed data, the average statistic (i.e., 0.16), and the predicted values using all observations (including the Verdal landslide) for the OLS and the negative binomial models. It illustrates that the OLS model yields a result very similar to the average. The count data model on the other hand predicts a non-linear relationship between numbers of people exposed and expected fatalities. When few people are exposed or when around 250 people are exposed, the linear and non-linear models predict the same number of fatalities. However, between these values, the linear models overestimate the number of fatalities. Note that the SGI data values shown in Fig. 2 are the numbers SGI assumed to be the most probable numbers of people exposed and not necessarily actual number of exposed humans. The simulated observations used in the estimation of the models are not shown. However, the curves are drawn from the predicted values using the models with simulated data.

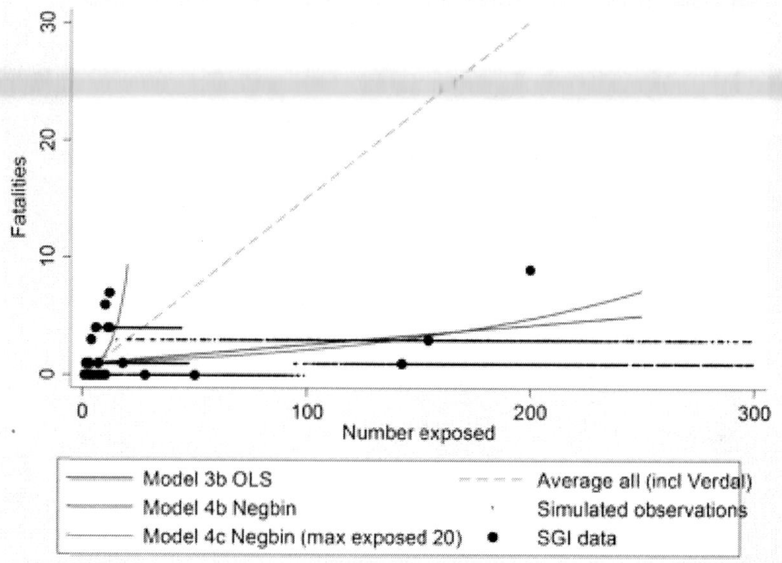

Figure 3. Observed and predicted values excluding the Verdal slide. The SGI data values are the most probable number of people exposed according to the Swedish Geological Institute (SGI). The average is the average loss of life rate based on SGI's observed values of exposed persons. The predicted values (model 3b, 4b, and 4c) are calculated from the estimated models using the Monte Carlo simulated data, which considers the uncertainty in number of exposed persons. The spread in uncertainty is also shown by the *horizontal dots* called "simulated observations"

Figure 3 shows observed data, the average statistic, and the predicted values excluding the outlier, the Verdal landslide, for three models applied to the Monte Carlo simulated data. Figure 3 also shows the spread of the simulated observations. However, note that it only shows the spread, not the density of the spread (cf., Fig. 1). The negative binomial model and the OLS model are similar, which means that the count data model predicts relationships that are nearly linear up to 200 exposed people. By removing the outlier, fewer fatalities are predicted. However, the graph of the negative binomial regression model only considering up to 20 people exposed shows that the risk of fatality is increasing more, compared to the models including all observations. This model predicts even more fatalities than the average value used by SGI. Figure 3 therefore shows that it may be reasonable to use different fatality curves depending on the number of people exposed.

DISCUSSION

In this study, quick clay fatality curves were derived as the relative frequency between exposure and lives lost. The results imply that loss of life increases exponentially with the size of the exposed population, but also that there might be different subsets of loss of life distributions with individual slope gradients depending on the size of the exposed population. This should be taken into account by policymakers when deciding on risk-reducing policy measures. Previously, in Sweden, a mean fatality index has been applied to risk assessments (SGI 2011). Our results imply that this approach overestimates the number of fatalities when the number of exposed individuals is between 20 and 250, underestimates it when the number of exposed individuals is high (over 250), but also underestimates it for under 20 people exposed. Probably diversified loss of life curves accounting for the size of the exposed population may be more appropriate to apply.

In our data set, individuals were exposed at their place of residence, indoors or outdoors, while working, or as a visitor in the area, and we recognize that more factors beyond exposure are likely to affect human susceptibility to quick clay landslides. Several potential factors were identified using peer-reviewed literature on landslide hazards. A functioning warning system was one such factor (Lacasse et al. 2010). Quick clay landslides, however, usually come without warning signs and therefore leave little time for emergency warning and evacuation. Despite this fact, signs had on some occasions prior to a landslide been noticed and also documented in the data set we analyzed. This was signs of gliding, creep, cracks, sinking of the ground, or small initial landslides. These signs cannot be equated to having a warning system, but could be interpreted as warning signs, if they were recognized as such. Prior experience to landslides is acknowledged as affecting human susceptibility (Zhang and Zhang 2014), and it is likely to believe that prior experience or specific knowledge would be necessary to acknowledge the "warning signs" mentioned above as a forecast for oncoming landslides. In our data set, however, only 16 slides occurred in areas previously exposed limiting the number of observations were we can assume that inhabitants had previous experience. Scandinavians ability to detect landslide threats are also affected by quick clay landslides occurring on very gentle slopes which are therefore presumed by inhabitants to be stable (Melchiorre and Tryggvason 2015). There is also reason to believe that time of day might have impacted humans preparedness and therefore affected their fatality risk. If slides occurred when inhabitants were asleep rather than in daylight, people were not to the same extent able to detect an oncoming landslide. The time of occurrence was, however, poorly documented in the data set and the uncertainty about time accuracy limited the feasibility of analyzing how this had affected human susceptibility in our cases.

Several more factors were identified as potentially affecting human susceptibility to landslides ("Landslide characteristics and methods for quantification" section) but that was too poorly documented to be quantified with the purpose of statistically measuring their effect. One reason is that damaging landslides in populated areas are low frequency events, and therefore, available data from landslides in specific areas applicable to damage analyses and damage prognoses are scarce. The quality and detail level for historical landslides also varies and can depend heavily upon the observer (Guzzetti et al. 1999; Guzzetti 2000; Guzzetti et al. 2006; Jaedicke et al. 2009; Salvati et al. 2010).

Extensive research has been performed using Italian landslide data. There are, however, some distinct differences between Italian data sets and the data set used in our study. The number of observations in the Italian data sets by far outnumbers our observations, but they do also cover a much longer time period which can raise questions concerning the applicability to present socioeconomic conditions. Further, different types of landslides are merged as opposed to the inclusion of only quick clay landslides in our study. Some Italian data sets also compile missing and injured people into a category of fatalities causing uncertainty as to the exact number of lives lost per landslide. One similarity between the data sets is that the actual number of landslides where people lost their lives is relatively few, but even in this measure, dissimilarities are distinct. We found that 27% of the 66 landslides from our data set resulted in deadly outcomes, whereas 6% was found in the Italian data by Giannecchini and D'Amato Avanzi (2012). The difference might be attributed to differences in warning and evacuation potentials between the types of landslides included. The special characteristic of quick clay landslides, that are not represented in previously derived fatality estimations, implicate that their consequences might need to be estimated separate from landslides with other characteristics, e.g., slower oncoming landslides where the land movements can be monitored and where warnings and evacuations can be exercised. Further, even the more well-documented Italian databases are burdened with uncertainties, e.g., uncertainties are found in poor data quality regarding the date of occurrence, number of casualties (injured, deaths, and missing people), and difficulties in differentiating landslide events from flooding events (Guzzetti 2000; Salvati et al. 2010).

Scandinavia has in modern times been relatively spared of exposure to natural hazards with large numbers of casualties. Historically, however, there have been events, e.g,. the Verdal landslide, that were devastating to communities and other events that could have been disastrous if they had occurred closer to human settlements, at another time of day, or at the present time with the present population. It is important to emphasize the effect the Verdal landslide had on the estimated functions in our study, and reflecting on whether this observation is an outlier or a low frequency observation that would recur repeatedly if the sample was to be expanded. Looking outside the sampling made by SGI, there have been incidents just as disastrous as the Verdal landslide both in Norway and Sweden. For example, in 1918 in Norrköping, Sweden, a clay landslide caused a railroad accident, killing 41. In 1726 in Verdal, Norway, a quick clay landslide killed 8. In 1648, in the Göta River valley, between 85 and 127 people were killed. These landslides are low frequency events, but they do occur and should therefore be represented when deriving loss of life estimates for application in ex-ante analysis so that the potential of experiencing more disastrous outcomes in future landslides is considered. With growing populations and urbanization, we are developing land areas more prone to quick clay landslides and thereby potentially exposing more people to landslide risks.

CONCLUSION

This study focused on susceptibility factors for exposed elements at risk, in this case human life. The amount of information available from historical information varied between the observations in the sample. For some quantitative information (e.g., time of day, warning signs, building characteristics, cause of death, physical ability, and age of the exposed population), the informational value of the data set was very limited. We found that potential risk factors such as country, year, and number of people exposed were the only risk factors in the sample set adequately documented to serve as input data to the statistical analysis estimating loss of life functions and the derived loss of life estimates were functions of the number of exposed persons. We do recognize that this risk factor only explain a limited part of human susceptibility when exposed to quick clay landslides slides, but it also contribute with important quantitative information to ex-ante analysis with the purpose of guiding policy decision on risk-reducing efforts. Records of historical events are one of the few sources at hand for evaluating the actual fatality risk, and despite uncertainties, they can contribute valuable information when assessing and communicating landslide risk. Concerning availability and quality of quantifiable damage and loss information in the Nordic historical records, with respect to statistical inference, the data sample set was limited. One known way of circumventing this barrier can be to transfer values derived in other studies, derived in other countries with quick clay deposits (e.g., Canada and Russia), or for other types of landslides. Based on our results, however, fatality rates obtained for other types of landslides can probably not be applied to quick clay landslide risk assessments. Future research should concentrate on quantifying risk factors that can further explain human vulnerability to quick clay landslides. The obtained results of this study also emphasize that future research regarding quick clay landslides is in need of, besides historical data, also other types of data. Such data can be obtained by performing detailed studies, e.g., surveys, investigation, in situ and laboratory tests, monitoring, analysis, and modeling, on representative cases of landslides in quick clay in order to generalize this type of studies.

FOOTNOTES

1 More complex zero-inflated count data models were also tested. The predicted values from them were quite similar to the negative binomial model, and they are therefore not presented.

2 The null hypotheses are that the coefficients are zero.

ACKNOWLEDGEMENT

The study is financially supported by Länsförsäkringar AB, Sweden. We also like to thank Lars Nyberg, Magnus Johansson, Kurt Petersen, and four anonymous reviewers for constructive comments.

Disclaimer

The paper has not been previously published and is not under consideration elsewhere.

REFERENCES

1. Agrawal S, Gopalakrishnan T, Gorokhovich Y, Doocy S (2013) Risk factors for injuries in landslide and flood-affected populations in Uganda. Prehospital and Disaster Medicine 28(4):314–321.
2. Andersson-Sköld Y, Falemo S, Suer P, Grahn T (2011) Landslide risk and climate change—economic assessment of consequences in the Gota river valley. Proceedings of the 15th European Conference on Soil Mechanics and Geotechnical Engineering. Geotechnics of hard soils - weak rocks. Part 1 to 3
3. Andersson-Sköld Y, Bergman R, Johansson M, Persson E, Nyberg L (2013) Landslide risk management—a brief overview and example from Sweden of current situation and climate change. International Journal of Disaster Risk Reduction 3(1):44–61.

4. Andersson-Sköld Y, Falemo S, Trembaly M (2014) Development of methodology for quantitative landslide risk assessment—example Göta river valley. Nat Sci 6(3):130–143.
5. Cascini L, Ferlisi S, Vitolo E (2008) Individual and societal risk owing to landslides in the Campania region (Southern Italy). Georisk 2(3):125–140.
6. Cred (2015) The human costs of natural disasters 2015. A global perspective. Centre for research on the epidemiology of disasters. http://emdat.be /human _cost_natdis (2016–09-20)
7. Devoli G, Morales A, Høeg K (2007) Historical landslides in Nicaragua-collection and analysis of data. Landslides 4(1):5–18.
8. Giannecchini R, D'Amato Avanzi G (2012) Historical research as a tool in estimating hydrogeological hazard in a typical small alpine-like area: the example of the Versilia River basin (Apuan Alps, Italy). Phys Chem Earth 49:32–43.
9. Göransson G, Norrman J, Larson M, Alén C, Rosén L (2014) A methodology for estimating risks associated with landslides of contaminated soil into rivers. Sci Total Environ 472:481–495.
10. Greene W (2008) Econometric analysis, Sixth edn. Pearson, Prentice-Hall, Upper Saddle River
11. Guzzetti F (2000) Landslide fatalities and the evaluation of landslide risk in Italy. Eng Geol 58(2):89–107
12. Guzzetti F (2016) Forecasting natural hazards, performance of scientists, ethics, and the need for transparency. Toxicol Environ Chem 98(9):1043–1059.
13. Guzzetti F, Carrara A, Cardinali M, Reichenbach P (1999) Landslide hazard evaluation: a review of current techniques and their application in a multi-scale study, Central Italy. Geomorphology 31(1–4):181–216
14. Guzzetti F, Reichenbach P, Cardinali M, Ardizzone F, Galli M (2003) The impact of landslides in the Umbria region, Central Italy. Natural Hazards and Earth System Science 3(5):469–486
15. Guzzetti F, Reichenbach P, Cardinali M, Galli M, Ardizzone F (2005a) Probabilistic landslide hazard assessment at the basin scale. Geomorphology 72(1–4):272–299.
16. Guzzetti F, Stark CP, Salvati P (2005b) Evaluation of flood and landslide risk to the population of Italy. Environ Manag 36(1):15–36
17. Guzzetti F, Galli M, Reichenbach P, Ardizzone F, Cardinali M (2006) Landslide hazard assessment in the Collazzone area, Umbria, Central Italy. Natural Hazards and Earth System Science 6(1):115–131
18. Guzzetti F, Mondini AC, Cardinali M, Fiorucci F, Santangelo M, Chang K (2012) Landslide inventory maps: new tools for an old problem. Earth Sci Rev 112(1–2):42–66.
19. Hilker N, Badoux A, Hegg C (2009) The swiss flood and landslide damage database 1972-2007. Natural Hazards and Earth System Science 9(3):913–925
20. Ho KKS, Chao PA, Lau TMF, de Silva S (2013) Investigation of the 20 August 2005 fatal landslide at Fu Yung Shan Tsuen, Hong Kong. Landslides 10(3):285–297.
21. Jaedicke C, Lied K, Kronholm K (2009) Integrated database for rapid mass movements in Norway. Natural Hazards and Earth System Science 9(2):469–479
22. Jiménez-Perálvarez JD, Irigaray C, El Hamdouni R, Chacón J (2009) Building models for automatic landslide-susceptibility analysis, mapping and validation in ArcGIS. Nat Hazards 50:571–590.
23. Jongman B, Kreibich H, Apel H, Barredo JI, Bates PD, Feyen L, Gericke A, Neal J, Aerts JCJH, Ward PJ (2012) Comparative flood damage assessment: towards a European approach. Natural hazard and earth system sciences 12(3733–3752):2012.
24. Jonkman SN, Lentz A, Vrijling JK (2010) A general approach for the estimation of loss of life due to natural and technological disasters. Reliab Eng Syst Saf 95(11):1123–1133.
25. JRC (2011) Safeland. Living with landslide risk in Europe: Assessment, effects of global change, and risk management strategies. 7th Framewoek Program, Cooperation Theme 6 Environment, sub-activity 6.1.3 Natural hazards

26. Keiler M, Zischg A, Fuchs S, Hama M, Stötter J (2005) Avalanche related damage potential- changes of persons and mobile values since the mid-twentieth century, case study Galtür. Natural Hazards and Earth System Science 5(1):49–58
27. Keim ME (2008) Building human resilience. The role of public health preparedness and response as an adaptation to climate change. Am J Prev Med 35(5):508–516.
28. Lacasse S, Nadim F, Kalsnes B (2010) Living with landslide risk. Geotechnical Engineering Journal of the SEAGS & AGSSEA Vol. 41 No.4 December 2010 ISSN 0046–5828
29. Long JS, Freese J (2014) Regression models for categorical dependent variables using Stata, Third edn. Stata Press, College Station, TX
30. Melchiorre C, Tryggvason A (2015) Application of a fast and efficient algorithm to assess landslide-prone areas in sensitive clays in Sweden. Nat Hazards Earth Syst Sci 15(12):2703–2713.
31. MSB (2016) Swedish Civil Contigency Offic, Swedish Natural Hazards Information System. http://ndb.msb.se
32. Nadim F, Kjekstad O, Peduzzi P, Herold C, Jaedicke C (2006) Global landslide and avalanche hotspots. Landslides 3(2):159–173.
33. Pereira S, Zêzere JL, Quaresma ID, Bateira C (2014) Landslide incidence in the North of Portugal: analysis of a historical landslide database based on press releases and technical reports. Geomorphology 214:514–525.
34. Poli S, Sterlacchini S (2007) Landslide representation strategies in susceptibility studies using weights-of-evidence modeling technique. Nat Resour Res 16(2):121–134
35. Roslee R, Jamaluddin TA (2012) Landslide hazard vulnerability (LHV): review of literature and a proposed new approach in landslide risk management for Malaysia. Bulletin of the Geological Society of Malaysia 58:75–88
36. Salas-Romero S, Malehmir A, Snowball I, Lougheed BC, Hellqvist M (2015) Erratum to: identifying landslide preconditions in Swedish quick clays—insights from integration of surface geophysical, core sample- and downhole property measurements. Landslides 1.
37. Salvati P, Bianchi C, Rossi M, Guzzetti F (2010) Societal landslide and flood risk in Italy. Natural Hazards and Earth System Science 10(3):465–483
38. SGI (2011) Kartläggning, exponering, sårbarhet och värdering av liv. Metodik konsekvensbedömning, GÄU- delrapport 15. Swedish Geotechnical Institute. Linköping
39. SGI (2012) Landslide risks in the Göta River Valley in a Changing Climate. Final report, part 1- Societal consequences. Swedish Geotechnical Institute. Linköping.
40. SOU (2007) Sweden facing climate change—threats and opportunities. Final report from the Swedish Commission on Climate and Vulnerability. Swedish Government Official Reports (SOU) 2007:60, Ministry of the Environment and Energy. Stockholm
41. Van Den Eeckhaut M, Hervás J, Jaedicke C, Malet J, Montanarella L, Nadim F (2012) Statistical modelling of Europe-wide landslide susceptibility using limited landslide inventory data. Landslides 9(3):357–369
42. Viscusi WK (2006) Natural disaster risks: an introduction. J Risk Uncertain 33(1–2):5–11.
43. Zhang S, Zhang LM (2014) Human vulnerability to quick shallow landslides along road: fleeing process and modeling. Landslides 11(6):1115–1129.

CHAPTER 7

Stability and Run-out Analysis of Earthquake-induced Landslides

Yingbin Zhang[1]

[1] Department of Geotechnical Engineering, School of Civil Engineering, Southwest Jiaotong University, Chengdu, China

1. INTRODUCTION

A large number of landslides can be caused by a strong earthquake and they have been the source of significant damage and loss of people and property. Therefore, it is very important to predict the stability of slope and the movement behaviors of a potential landslide under an earthquake loading, i.e., stability and run-out analysis (Figure 1).

Earthquake-induced landslides have been the source of significant damage and loss of people and property. One of the most serious event is the 1970 Peru earthquake. This event caused a huge rock avalanche that killed almost 54,000 people and buried two cities [143]. Another example is, in the 1920 Haiyuan earthquake, a large number of landslides caused widespread damage to infrastructure and buildings and killed at least 100,000 people, almost half of the total earthquake deaths [82].

Therefore, it is very important to predict the earthquake-induced landslides and to take countermeasures for potential landslides.

Main topics of earthquake-induced landslides are the following:

1. Investigation of recent and historical earthquake-induced landslides and their impacts so as to produce inventories of historical earthquake-induced landslides

2. Prediction of potential earthquake-induced landslides, including (i) failure mechanism and stability analysis of seismic slopes, (ii) movement mechanism and behaviours of earthquake-induced landslides, and (iii) instrumentation and monitoring technologies for potential earthquake-induced landslides or post-earthquake landslides.

3. Preventive countermeasures for earthquake-induced landslides, including (i) Stabilization and disaster mitigation of earthquake-related landslides, (ii) risk assessment and management of earthquake-related landslides, and (iii) hazard map and early warning system for earthquake-related landslides

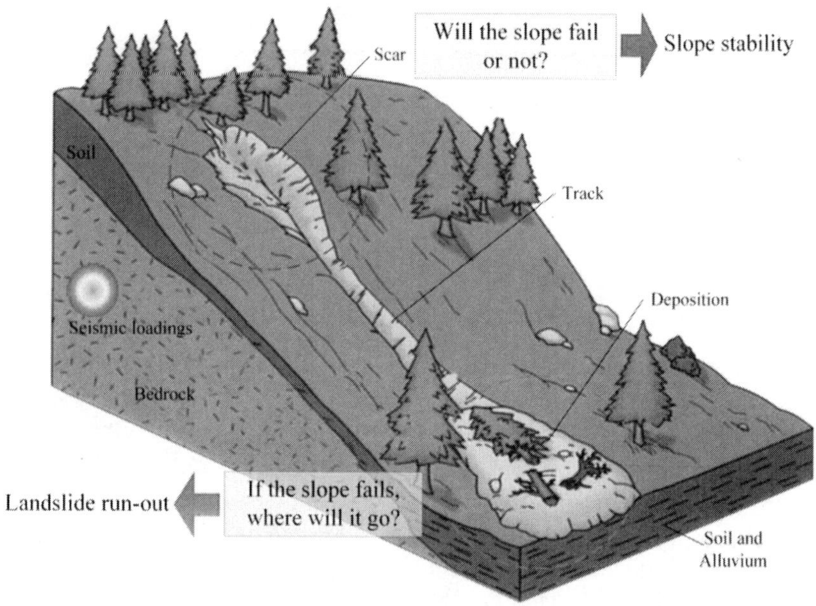

FIGURE 1. Stability and run-out analysis of earthquake-induced landslides

This chapter focuses on the prediction of potential earthquake-induced landslides. The prediction of potential landslide can be carried out using detailed geotechnical investigations and stability calculations. (i) Failure mechanism and stability analysis of seismic slopes, i.e. seismic slope stability analysis and (ii) movement mechanism and behaviours of earthquake-induced landslides, i.e. landside run-out analysis are outlined firstly, and then the merits and demerits of each method are clarified in this chapter.

2. SEISMIC SLOPE STABILITY ANALYSIS

So far, methods developed to analyze the stability of earthquake slopes can be divided into three types: (1) pseudo-static methods, (2) dynamic sliding block methods, and (3) stress-strain methods. These three types of methods can be applied in different cases due to each of them has merit and demerit [73].

2.1. Pseudo-Static Methods

[166] first presented the pseudo-static method, which is a simple method for evaluating of seismic stability of a slope. This type of method can be used to man-made or natural slopes based on either analytical method or numerical method. The earthquake force, acting on the an element or whole of the slope, is writed by a horizontal force and/or a vertical volum force equal to the gravitation force multiple a coefficient *k*, called the pseudo-static coefficient as shown in Figure 2 and Equation (1).

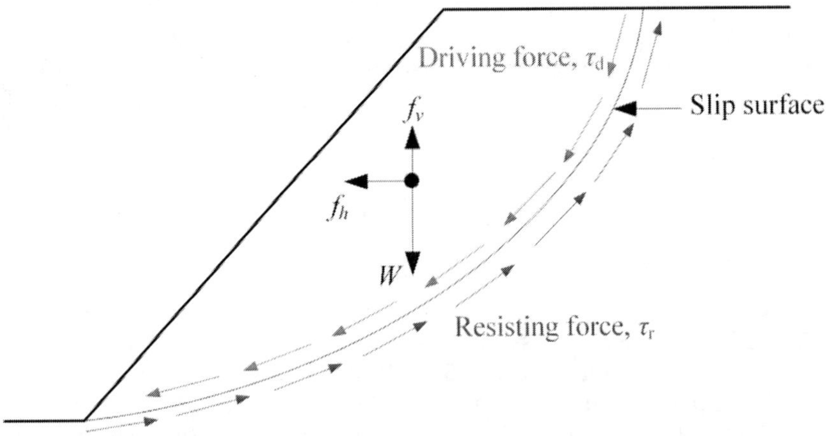

FIGURE 2. Forces acting on a slope in pseudo-static slope stability analysis

Thus, *k* times the gravitational acceleration *g*, i.e. *a*=*kg* forms the assumed seismic acceleration *a*. The assumed pseudo-static forces acting on a potential sliding mass of weight *W* will be

$$f_h = \frac{a_h W}{g} = k_h W \qquad (1)$$

where *ah* and *av* are horizontal and vertical pseudo-static accelerations, respectively, *kh* and *kv* are horizontal and vertical pseudo-static coefficients, respectively. The factor of safety(FOS) is represented as the ratio of the resisting force to the driving force, Equation (2).

$$f_v = \frac{a_v W}{g} = k_v W \tag{2}$$

From Equation (1), the pseudo-static force is determined by the seismic coefficient. The key problem for the pseudo-static procedure is how to select an appropriate seismic coefficient under an acceptable FOS. There have been studies for determining the most appropriate pseudo-static coefficient by a matter of experience and judgment.

[166] classical paper made the original suggestion to use of *kh*=0.1 for severe earthquakes, *kh*=0.2 for violent and/or destructive earthquakes, and of *kh*=0.5 for catastrophic earthquakes.

[103] presented a minimum pseudo-static FOS of 1.5 based on a slope material strength reduction factor (SRF) of 0.8 and the following acceleration values associated with two different earthquake magnitudes *M*. The same values of seismic coefficients for magnitude 6.5 and 8.25 earthquakes are recommended by [154], but with an acceptable FOS of 1.15.

$$\begin{aligned} a &= 0.1g \text{ for } M = 6.5 \text{ implying } k = 0.1 \\ a &= 0.15g \text{ for } M = 8.25 \text{ implying } k = 0.15 \end{aligned} \tag{3}$$

[137] also presented the pseudo-static coefficient related to earthquake magnitude. In detail, for an 8.25, 7.5, 7.0 and 6.5 magnitude earthquakes, if the seismic coefficients equal to 1/2, 1/3, 1/4 and 1/5 of the PGA, respectively, the computed FOSs are larger than 1.0, the accumulated displacements of slope are likely to be acceptably small.

In the report published by the International Commission of Large Dams (ICOLD), [154] shows a list of the minimum FOS value and horizontal seismic coefficients for 14 large dams worldwide, in which the minimum FOSs range from 1.0 to 1.5 and the horizontal earthquake coefficients range from 0.1 to 0.15. The Corps of Engineers Manual recommended a earthquake coefficient of 0.1 or 0.15 for areas where major and great earthquake threats are estimated, respectively, and a FOS of no larger than 1.0 for all magnitude earthquakes.

Some references related the earthquake coefficient value to the peak ground acceleration (PGA) [10, 67, 108]. [108] related a pseudo-static coefficient of 1/3 to 1/2 of the PGA at the top of a double-side slope (a dam in the source reference), whereas [67] related a pseudo-static coefficient of 1/2 of the PGA of bedrock (PGA_{rock}) with a FOS of no larger than 1.0 and a SRF of 20%. And, [10] recommended the pseudo-static coefficient of 0.6 or 0.75 times of the PGA of bedrock (0.6 or $0.75PGA_{rock}$). It should be noted that the value given by [10] is conservative because the original study is designed for solid-waste landfills, where the allowable deformation are relatively small. [89] pointed that although engineering judgment is required for all cases, the criteria of [67] should be appropriate for most slopes.

[91] suggested one-half of PGA to use in an area of low seismicity (peak acceleration <0.15g) for the stability of earth embankments. This can be obtained from the peak horizontal motion (mean) from Modified Mercalli Intensity (MMI), magnitude-distance attenuation and the probability of a 50-year, 90% nonexceedance. However, in an area of moderate to strong seismicity ($0.15g \leq PGA \leq 0.40g$), PGA is obtained from the peak horizontal motion, from MMI, magnitude-distance attenuation and probability of 250-year, 90% nonexceedance.

[76] suggested a minimum FOS of 1.0, also based on a slope material SRF of 0.8 and the following values of pseudo-static coefficient: a equals to 0.17PGA or 0.5PGA for the dynamic response analysis is to be performed for the slope or earthquake structure or not.

[163] developed an expression for the earthquake coefficient in terms of characters of ground motion and magnitude of earthquake based on the data of [10].

It is almost common that only the horizontal acceleration is considered in evaluating the stability and deformation of a slope because the horizontal acceleration is the principal de-stabilizing force that acts on earth structures as well as the principal source of damage observed in earthquakes [4].

From Figure 2, the horizontal force clearly increases the driving force and decreases the FOS. The vertical pseudo-static force generally has less influence on the FOS than the horizontal pseudo-static force does because the vertical pseudo-static reduces both the driving and resisting forces. Hence, the effects of vertical seismic loading are frequently omitted in pseudo-static analysis [89].

Several investigators performed some analyses and have shown that the inclination of seismic loading have a significant influence on the seismic stability of slope by coupling the vertical and horizontal components of seismic force [20, 100].

In summary, pseudo static method can be simply and directly used to identify the FOS and the critical seismic coefficient *kc*. In addition, performance of slope is closely related to permanent displacement, but the results of pseudo-static method are difficult to interpret the performance of slope after a seismic event because this method provides no information about permanent displacement. Because the pseudo-static analysis method provides only a rough assessment of seismic slope, it should be only used for the preliminary procedures. More accurate methods can be used to the followed process [73, 163, 170].

2.2. Dynamic Sliding Block Methods

Displacement-based dynamic slidng block method is another alternative approach to evaluate the seismic slope stability, as permanent displacement is a useful index of slope performance, especially for those man-made slopes constructed for special purposes such as dams, embankments et al. This method has been widely used in earthquake geotechnical engineering.

In 1965, [119] proposed the dynamic sliding block method for estmating the permanent displacement of embankment affected by a seismic loading. In this method, sliding would be induced once the seismic loading exceed the critical seismic force of a potential failure surface as shown in Figure 3. The sliding would be accumulated until the end of seismic loading. We can evaluate the accumulated permanent displacement to assess the seismic stability of a slope.

Newmark's method showes that the yield acceleration of a potential block is a function of the FOS and slope angle, as:

$$a_c = (\mathrm{FOS} - 1) g \sin \alpha \tag{4}$$

where *ac* is in terms of the gravity acceleration *g*; FOS is the static factor of safety; and α is the slope angle.

Since then, the method has been numerous extensions and applications. The section 2.2.1 and section 2.2.2 will give reviews for these two aspects, respectively. In addition, a regional scale application of the dynamic sliding block method is reviewed in section 2.2.3.

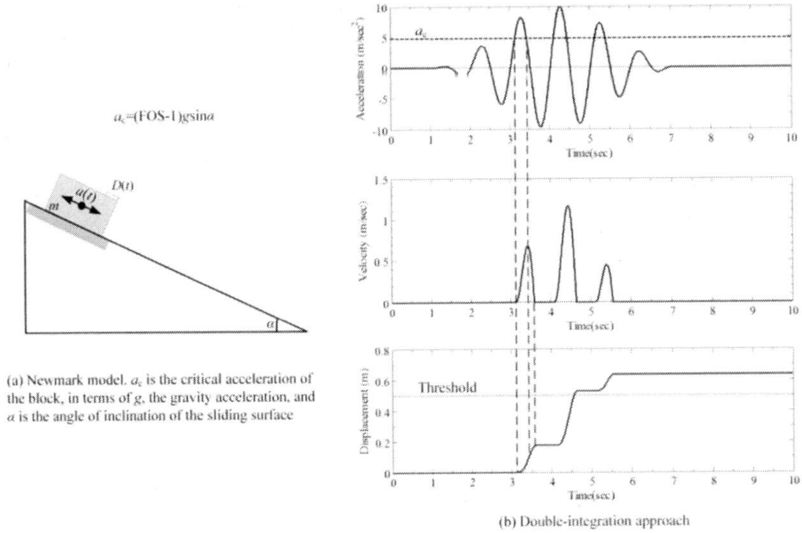

FIGURE 3. Illustration of the original Newmark's method

2.2.1. Extensions

More attention has been focused over the last decades on developing methods to more accurately analyze the seismic stability of a slope for dams, embankments or other important structures by modeling the dynamic response of a slope more rigorously.

After the first dynamic sliding rigid block method, [155 and 97] published more sophisticated methods to account for the un-rigid block. Similar studies also given by [103]. As the classification given by [73], methods for estimating the permanent displacement of a sliding system induced by earthquake loading can been grouped into: (1) rigid-block model [119], (2) decoupled model [10, 104], and (3) coupled model [11, 97, 139].

2.2.2. Applications

Since the rigid-block method was published in 1965 by Newmark, it has seen numerous applications, four of which are shown in Figure 4. The applications in recent years include (1) the seismic deformation analysis of earth dams and embankments [1, 22, 48, 49, 89, 90, 97, 103, 138, 144, 145, 150, 155, 179, 180]; (2) the displacements associated with landslides [34, 53, 70, 171]; (3) the seismic deformation of landfills with geosynthetic liners [10, 181]; (4) the seismic settlement of surface foundations [141]; and (5) the potential sliding of concrete gravity dams [32, 47, 95]. The extension of the analogue by [140] to gravity retaining walls has met worldwide acceptance, and has found its way into seismic codes of practice. Several other generalised applications have also appeared (e.g. [2, 3, 45, 99, 139, 162, 169]).

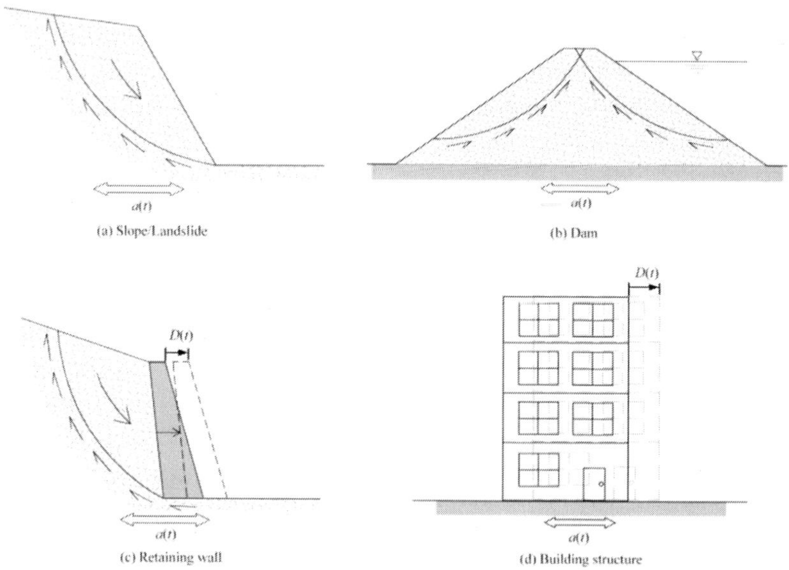

FIGURE 4. Applications of dynamic sliding block method in geotechnical engineering

2.2.3. Regional Scale Analysis

Except a single slope analysis, where the landslides are likely to occur and what kind of seismic conditions will cause it failure are two important topics in seismic hazard assessment, i.e.regional scale analysis [59].

For a regional scale analysis, slope stability analysis menthds will be not suitable [143, 168]. With the development of Geographic Information Systems (GIS) tools in recent years, regional scale analyses by the dynamic sliding block method have been proposed, in which ground shaking characteristic parameters, geotechnical material and topographic data are considered (e.g. [34, 71, 75, 106, 114, 151, 155]).

The Newmark analysis (which combines slope stability calculations with seismic ground-motion records) is widely used to evaluate the potential for landslides that could be triggered by earthquake shaking [70, 71, 72, 74, 113].

2.3. Stress-Strain Methods

With the developments of the simulation approach and computer technology in recent years, the stress-strain method is becoming increasingly used in seismic slope stability analysis. These methods can be grouped into continuous methods, e.g. finite element method (FEM) [21], finite difference method (FDM) [116], boundary element method (BEM) [12], and discontinuous methods, e.g. rigid block spring method (RBSM) [77; 80], discontinuous deformation analysis (DDA) [159, 160] and discrete element method (DEM) [31].

2.3.1. Continuous Methods

[21] developed and named FEM of engineering analysis, in which the studied system is meshed into small many elements. This method can be applied to estimate the slope stability including dynamic stability analysis.

Some applications of the continuous methods have been proposed, e.g. [89, 94, 153] and [156]. Recently, nonlinear in-elastic soil models have been developed and implemented in two-dimension (2-D) and three-dimension (3-D) models (e.g., [42, 50, 135, 164]). In addition, [93] and [183] studied the seismic slope stability by using FDM.

2.3.2. Discontinuous Methods

For the analysis of a potential failure mass consisting of multiple blocks as shown in Figure 5, the discontinuous methods are more applicable [120]. Some applications of RBSM and DEM can be found in some literature (e.g. [8, 52, 77, 79, 80, 85121, 127128, 129, 130, 131, 136, 182]).

DDA is also a discontinuous method developed for the modeling of the behaviors of multiple block systems. Since the novel formulation and the numerical code of DDA were presented, DDA draws more and more attention and many extensions and modifications to the original method have been proposed to overcome some limitations [19, 37, 38, 81, 87, 98] and make it more suitable, practical and efficient to seismic slope stability.

The DDA can be used both to static rock slope engineering (e.g. [17, 81, 102, 123, 176, 187]) and the seismic rock slope stability analysis [56, 57, 54].

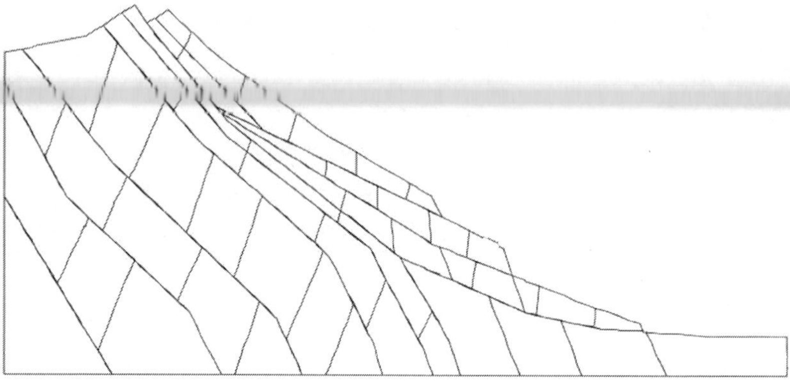

FIGURE 5. A jointed rock slope (modified from Bhasin and Kaynia, 2004)

In summary, stress-strain method represents a powerful alternative approach for seismic slope stability analysis which is accurate, versatile and requires fewer a priori assumptions, especially, regarding the shape of failure surface.

3. LANDSLIDE RUN-OUT ANALYSIS

It is important to estimate the movement behaviour of a potential landslide. For example, the movement distance is an important parameter in risk assessment and measure design. There are many run-out analysis methods, which can fall into four categories: (1) experimental methods, (2) empirical methods, (3) analytical methods, and (4) numerical simulation methods. The states of the art of these methods are reviewed in the following four subsections 3.1 - 3.4.

3.1. Experiment Methods

Physical modelling typically involves using scale models to capture the motion of landslides. Physical experiments are usually preferred to models because models require more assumptions than direct measurements. But for landslides, direct experiment is difficult, dangerous, expensive, and of limited utility. Based on laboratory experiments and filed investigation data, there are many different available models developed for calculating run-out zones.

Some full-scale direct experiments with artificial landslides have been completed [118, 122124, 125, 126] and others). However, since landslides are frequently heterogeneous and single event cannot be repeated carefully through adjusting only one factor, direct experiment is difficult, dangerous, expensive and of limited utility. And observing conditions are complicated by the danger of being in close proximity to a landslide and the difficulty of measuring a material with properties that change when observed in-situ or when isolated for measurement. But laboratory experiments are still the first qualitative and quantitative observations on the obtained results became fundamental for a better understanding of movement runout behaviour.

3.2. Empirical Methods

Several empirical methods for assessing landslide travel distance and velocity have been developed based on historical data and on the analysis of the relationship between parameters characterizing both the landslide, e.g. the volume of the landslide mass, and the path, e.g. local morphology, and the distance travelled by the failure mass [65]. Regression model-based methods and geomorphology-based methods are two kinds of common methods.

3.2.1. Regression Model-Based Methods

The regression model-based methods are developed on an apparent inverse relationship between landslide volume and angle of reach (also called as fahrböschund by [58]). Several linear regression equations have been proposed [25, 96, 153]. Introduced by [58], the angle of reach is the inclination of the line connecting the crest of the source with the toe of the deposit, as measured along the approximate streamline of motion. The angle of reach is considered an index of the efficiency of energy dissipation, and so is inversely related to mobility. Similar correlations between volume and other simple mobility indices have been proposed [33, 60, 142]. Given estimated source location, volume and path direction, these methods provide estimates of the distal limit of motion [111].

Improved empirical model notable performing regressions on subsets with varying scopes were presented by [13, 25, 69] and others.

Regression model-based models play a valuable role in landslide run-out analysis due to the regression model-based methods are simple. But the regression model-based methods are difficult to apply in practice with a high degree of certainty. For example, the correlation coefficients for some of regression models are 0.7-0.8, while a value of larger than 0.95 generally indicates a strong correlation. And it is difficult in this method to take account of influences of the ground condition, the micro-topography, the degree of saturation of the landslide mass and et al. For this point, geomorphology-based method is another alternative approach to predict the run-out of landslide.

3.2.2. Geomorphology-Based Methods

Field work and photo interpretation are the main sources of the geomorphological analysis for determining the travel distance of landslides [65]. The outer margin of the landslide deposits give an appraisal of the maximum distances that landslides have been able to reach during the present landscape (Figure 6). Several authors have provided these studies (e.g. [23, 24, 26, 88]).

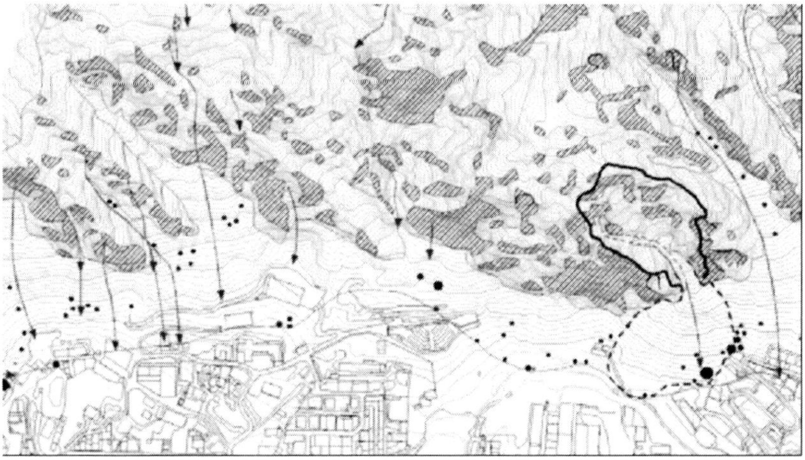

FIGURE 6. Boundary of the potential rockfall run-out area in Santa Coloma (Principality of Andorra), defined by the line that links the farthest fallen blocks observed in the field [**24**]. Arrows indicate historical rockfall paths and solid circles are large fallen boulders [**65**]

The geomorphological approach does not give any clue of the emplacement mechanism. Furthermore, the slope geometry and the circumstances responsible for past landslides might have changed. Therefore, results obtained in a given place cannot be easily exported to other localities.

In summary, empirical methods, both regression model-based methods and geomorphology-based methods, typically predict travel distances, while the deformation characteristics or the slide velocities of the landslide are not predicted. These models may be applied to establish initial hazard characteristics for preliminary run-out analysis, which may be later refined by other models.

3.3. Analytical Methods

In contrast to empirical methods, analytical methods are based on mechanics and involve the solution of motion equations [111]. The simplest analytical model is the classical sliding block model as shown in Figure 7, which is based on work-energy theory [6, 9, 43, Müller-Bernet in 58, 63, 83, 84, 132, 147]. Internal deformation and it's associated energy dissipation are neglected and the landslides is treated as a lumped mass. At any position along the path, the sum of the energies including the potential energy, kinetic energy and net energy loss equals the initial potential energy. This energy balance can be visualized using the concept of energy grade lines, as shown in Figure 9. The concept of energy grade lines is useful for visualizing the energy balance. v is the velocity of the block, g is the vertical acceleration due to gravity and $v^2/2g$ is known as the velocity head, which is the kinetic energy of the block normalized by the product of its mass and g. The same normalization of net energy loss is known as head loss. Note that the positions of the energy lines are referenced to the centre of mass of the block and that the true energy line and mean energy line do not necessarily coincide. Given the initial position of the center of mass and a suitable relationship to approximate the energy losses, the position and velocity of the block can be determined at any given time.

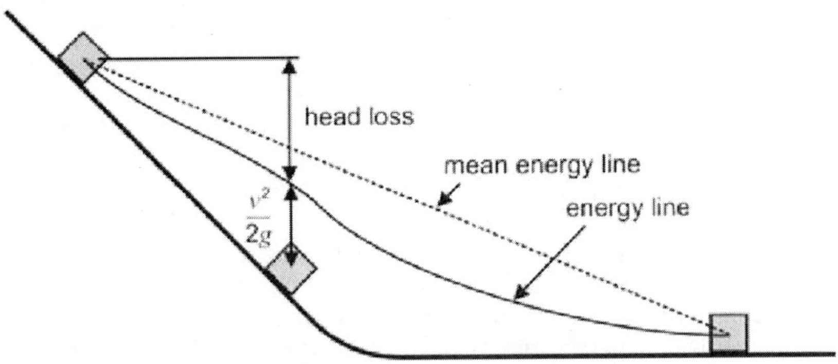

FIGURE 7. The classical sliding block model, based on work-energy theory [111]

Three-dimensional analysis for investigating runout of a slope were also proposed [36, 40, 51, 92 and 109]. These models require a high resolution Digital Elevation Model (DEM).

Generally speaking, the use of analytical methods is somewhat motivated by the limitations of purely empirical methods, as the unique geometry and materials involved in each case can be accounted for explicitly and a statistically-significant database of previous events is not necessarily required. The simplicity of a lumped mass allows analytical solutions, fast and effectively [66]. However, because the landslide is reduced to a single point, lumped mass models cannot provide the exact maximum runout distance, but only the displacement concerning the centre of mass [44, 62].

3.4. Numerical Simulation Methods

The single-block model should be only applied to the motion of the center of mass of a rigid body, but more complex continuum deformable mass or multi-block system is often appeared in practice. Some numerical simulation methods have been developed to account explicitly for deformation during motion.

3.4.1. Continuous Methods

When considering that the dimensions of a typical particle is much smaller than the depth and length of the debris, the debris mass is treated as continuum. According to depth averaged Saint Venant approach, the material is assumed to be incompressible and the mass and momentum equations are written in a depth-averaged form. Many numerical methods now exist to investigate the run-out process of landslide (e.g. [18, 27, 28, 29, 30, 35, 62, 111, 112, 133, 149, 161, 165]). These methods are usually based on continuum mechanics and assume that the avalanche thickness is very much smaller than its extent parallel to the bed, i.e. thin layer depth-averaged models. The primary differences are their representation of basal resistance force and the constitutive relations describing the mechanical behaviour of the considered material. These models can accurately take account of detailed topography effects, shown to be significant, with a reasonable computational time, making it possible to perform sensitivity studies of the parameters used in the model. They can provide effective properties that make it possible to roughly reproduce not only the deposit shape but also the dynamic as shown in [46) and 117] for examples. However, conventional continuum approaching models, which neglects the contact between rocks, makes it impossible to trace the position of individual rock during a landslide.

3.4.2. Discontinuous Methods

When the landslide mass consists of large fragments and boulders, the run-out mass is modelled as an assembly of blocks moving down a surface. Some authors take circular shape models in their run-out analysis to evaluate maximum runout and final deposit position of past or potential events (e.g., [134]). Although polygonal shapes have the disadvantages due to the complexity of the contact patterns and penalty in computational time, methods using non-circular shapes will be required for more real-world problems. It is more appropriate when problems are limited in finite blocks. Discontinuous numerical simulation methods are powerful tools in simulation of failure and run-out process of rock avalanche controlled by weakness surface. DEM [31] and DDA [159, 157] are two of the most commonly used methods.

Both DEM and DDA employ the equations of dynamic motion which are solved at finite points in time, in a series of time steps, but there are some subtle but significant differences in their formulations of the solution schemes and contact mechanics. In the solution schemes, equations of motion in DDA are derived using the principle of minimization of the total potential energy of the system, while the equations of motion as implemented in DEM are derived directly from the force balance equations, which still resultant unbalanced force after a time step and damping is necessarily used to dissipate energy. In the contact mechanics, the DDA used a penalty method in which the contact is assumed to be rigid. No overlapping or interpenetration of the blocks is allowed as the same as real physical cases, whereas soft contact approach is used in DEM. The soft contact approach requires laboratory or field measured joint stiffness, which may be difficult to obtain in many cases. Many comparisons of basic models (sliding, colliding and rolling models) between the DEM and DDA were carried out and show that the results from DDA are more close to the analytical values than that from DEM [188]. Compared to DEM, DDA has a simpler and more straightforward physical meaning [172].

Applications of DEM can be found in some literatures, such as [85, 128, 129, 131, 136, 182].

DDA can be used for estimating the affected area of an earthquake-induced landslide.

[55] first validated the applicability of DDA for the dynamic behaviour of block sliding on an slope. Based on the same inputs model of seismic loadings, [7576886105, 158, 167, 178] studied the dynamic response or/and stability analysis of tunnel, slope, dam, foundation or ancient masonry structure by using DDA. Alternatively, the seismic loadings also can be applied to the base block [146, 147], which is different from the original DDA. Later, [173, 174, 175, 177] applied DDA to simulate the kinematic behavior of sliding rock blocks in the Tsaoling landslide and the Chiu-fen-erh-shan landslide induced by the 1999 Chi-Chi earthquake. Recently, [184] applied newest DDA program to simulate the largest landslides induced by the 2008 Wenchuan earthquake.

4. COMPARISONS OF VARIOUS METHODS

The studies in the field of the earthquake-induced landslides are generally reviewed. Two parts of contents, i) seismic stability analysis and ii) run-out analysis are reviewed and compared. Some conclusions can be drawn:

1. Three categories methods can be used to analyse the seismic stability of a slope. Each of these types of methods has strengths and weaknesses and each can be appropriately applied in different situations. In detail, pseudo-static methods can simply and directly determine the FOS and the critical coefficient *kc* of a slope, while the widely used Newmark's methods and its extensions can determine the co-seismic deformation of a slope. And the Newmark's methods can be used to estimate where earthquake-induced landslides are likely to occur and what kind of shaking conditions will trigger them based on the GIS technology. More sophisticated analysis for real dynamic process of a seismic slope should be carried out by stress-strain methods, including both continuous methods and discontinuous methods.
2. Four kinds of methods can be used to analyse the run-out of a landslide. In detail, experiment method can provide the qualitative and quantitative observations on the obtained results although this method is difficult, dangerous, expensive and of limited utility. Empirical method can be directly used for assessing landslide travel distance and velocity based on historical data and on the analysis of the relationship between parameters characterizing both landslide and the path. Analytical method can be more directly used without the need of statistically-significant database of previous events. Numerical simulation method can be used to provide more information for the landslide composed by the complex continuum deformable mass or multi-blocks.

5. A CASE STUDY: THE DAGUANGBAO LANDSLIDE [185, 186]

5.1. Background

The Daguangbao landslide is located in the hanging wall only 6.5 km away from the Yingxiu-Beichuan fault. It is a typical bedding landslide.

Figure 8 gives a pre- and post-earthquake 3-D topographies, from which cross-section of the Daguangbao landslide can be obtained (Figure 9). The extent of the damage caused by the Daguangbao landslide is reflected in the following statistics [61]:

1. The affected area covered 7.3~10 km^2;
2. The accumulation body width is 2.2 km;
3. Estimated volume of collapsed rock mass is 750~840 million m^3;
4. The failure zone is more than 1 km;
5. The failure mass moved about 4.5 km;
6. Formed a 600m high landslide dam.

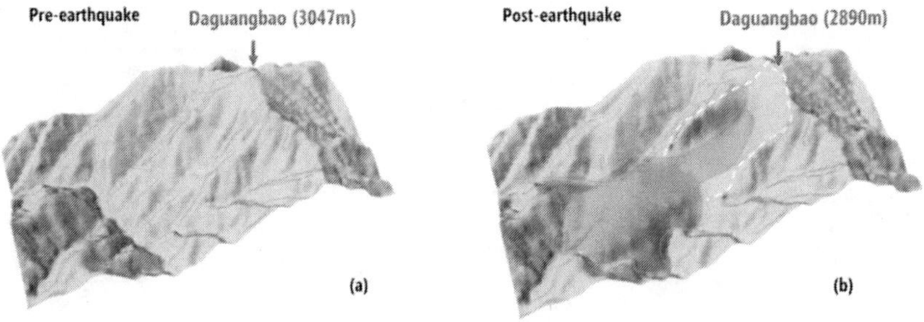

FIGURE 8. Pre- and post-earthquake 3-D topographies of the Daguangbao landslide

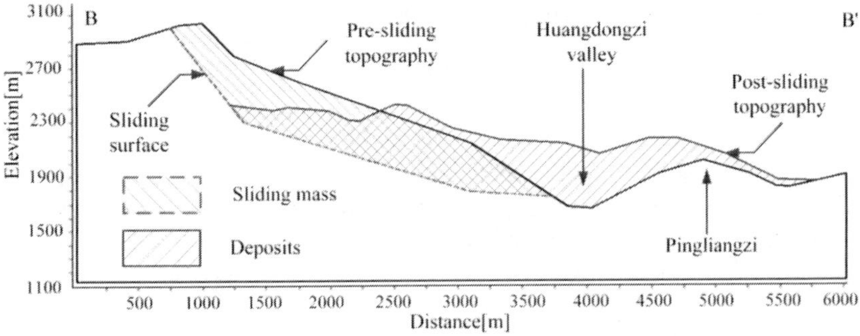

FIGURE 9. Cross-section of the Daguangbao landslide before and after the Wenchuan earthquake.4

5.2. Material Properties And Ground Motion

5.2.1. Material Properties

The Daguangbao landslide is so huge that the size effect must be considered. To account for this discrepancy, experience equations based on Hoek-Brown failure criterion, which size effect can be considered, is used to back calculate the material strength. Table 1 lists the material properties of the Daguangbao landslide.

TABLE 1. Material properties of the Daguangbao landslide in FLAC^{3D} and DDA

	Material 1	Material 2	Materia 3	
			FLAC^{3D}	DDA
Density (ρ): g/cm^3	2.5	2.6	2.6	260000
Unit weight of rock (γ) : kN/m^3	25	26	26	0
Elastic modulus (E) : Gpa	1.86	2.63	14.76	
Poisson's ratio (v)	0.2	0.2	0.1	
Friction angle of discontinuities (φ) : º	10.8	12.18	23.53	
Cohesion of discontinuities (c) : Mpa	1.276	1.576	4.052	
Tensile strength of discontinuities (σ_t) : kPa	12	32	556	

5.2.2. Ground Motion

The horizontal earthquake wave is the projection combination in the main sliding direction (N60ºE) use the MZQP acceleration records in E-W and N-S directions as Equation (5). The inputted vertical earthquake wave is the MZQP acceleration records in U-D direction.

$$a_H = a_{E-W} \cdot \sin 60^\circ + a_{N-S} \cdot \cos 60^\circ \tag{5}$$

Figure 10 shows the input combined acceleration records. Velocity and displacement time histories can be obtained by first and second integration from acceleration record. The duration of earthquake wave is 60s.

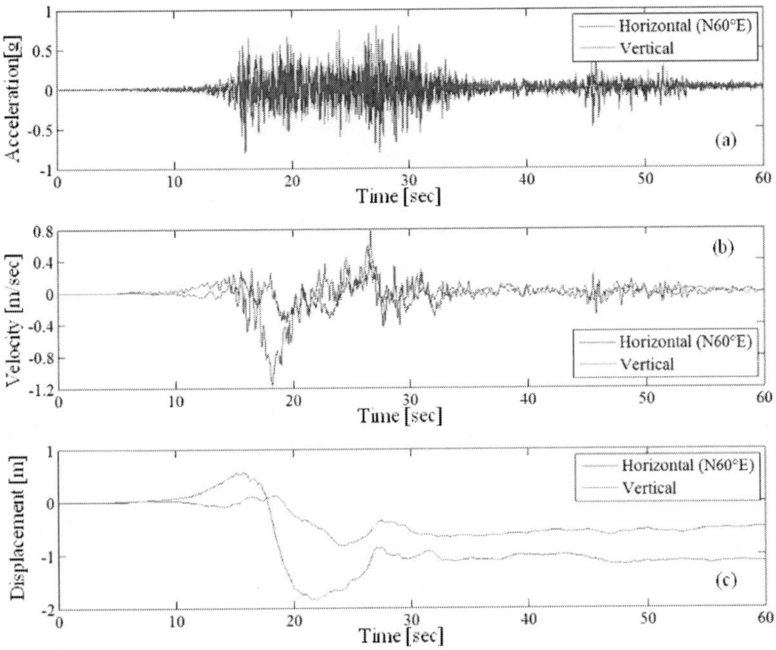

FIGURE 10. Input post-corrected horizontal and vertical ground records projected to N60ºE direction

5.3. Numerical Simulations-Run-Out Analysis

Seismic DDA can successfully simulate the movement of earthquake induced landslide. Two main features determine the Daguangbao landslide is a unique case, one is near-fault location (≈6.5 km) and the other one is huge scale ($\approx 800\times10^6$ m^3). The near-fault location determines the Daguangbao landslide must be shocked by the extreme ground motion from the strong Wenchuan earthquake. And the Daguangbao landslide located on the meizoseismal area where the vertical seismic component is very large. In addition, the landslide is so huge that the size effects must be considered. The friction coefficient measured in the laboratory is no longer suitable for stability and run-out analysis.

To these two features, the Daguangbao landslide is simulated by the newest seismic DDA code in which multi-direction seismic forces can be applied in the base block directly, and experience equations based on Hoek-Brown failure criterion is applied to back-calculate the material strength by trying to consider the size effect.

TABLE 2. Control parameters for DDA

Parameter	Value
Assumed maximum displacement ratio (g_2)	0.001
Total number of time steps	20,000
Time step (g_1)	0.005s
Contact spring stiffness (g_0)	5.0×10^5 kN/m
Factor of over-relaxation	1.3

5.3.1. Geometry Of Sliding Blocks

The main sliding direction of the Daguangbao landslide, N60ºE, is selected as analysis profile. The DDA model is depicted in Figure 11. In this simulation, based on the shape of failure surface and the character of slope topography, the whole slope is divided into three parts: base block, upper sliding mass, and lower sliding mass. Then two sliding masses are divided into the smaller discrete deformable blocks based on pre-existing discontinuities.

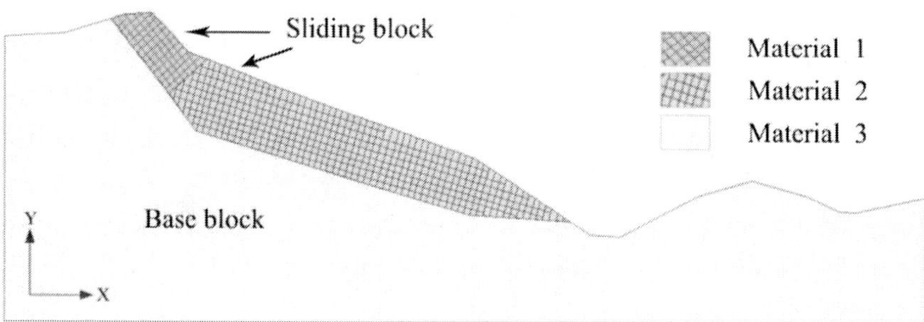

FIGURE 11. DDA model of the Daguangbao landslide

5.3.2. Results

Figure 15 shows the post-failure behavior of the Daguangbao landslide simulated by the seismic DDA code. Simulated results show that the sliding blocks climb over the Pingliangzi. After overlapping the final step of DDA calculation with the topographic cross-section at the Daguangbao landslide, the deposit pattern of the simulated Daguangbao landslide under horizontal-and-vertical situation coincides well with local topography.

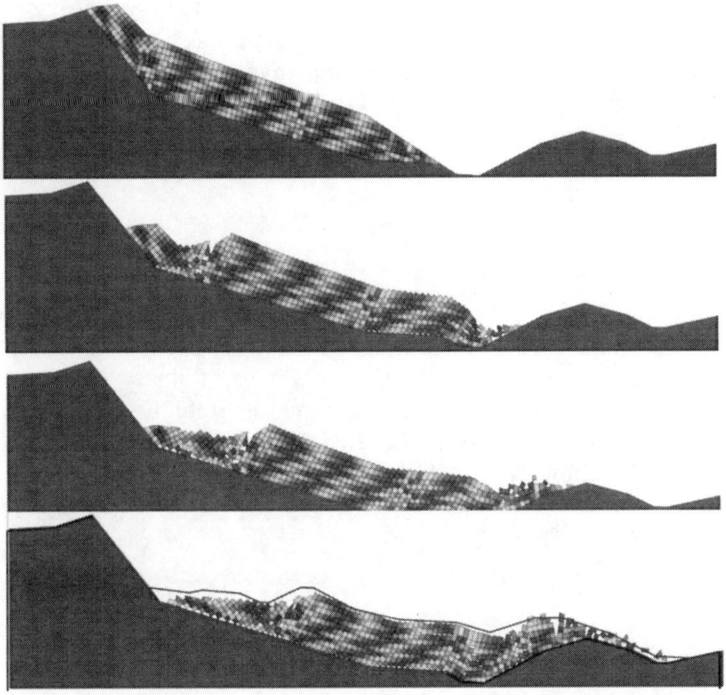

FIGURE 12. Simulation results of the Daguangbao landslide

6. CONCLUSIONS

Five cases are performed using finite difference program FLAC^{3D}, under the real seismic waves near the study site. The results show that the seismic conditions cause a significant reduction in factor of safety than static situation. It also found that the vertical seismic has a significant influence on tension failure of block, although it has an insignificant influence on change of the factor of safety. Another important conclusion is the effect of vertical seismic force on relative displacement of potential sliding mass is significant. In addition, large area of tension failure caused by the combined seismic forces at back edge of the slope applies the evidence of effect of vertical seismic force on failure mechanism of slope stability.

A comparison of simulation results from three situations, static, only-horizontal and horizontal-and-vertical, is carried out. Seismic force has a significant influence on the arrival distance, and shape of post-failure. Arrival distance from horizontal-and-vertical situation is larger than that from only-horizontal situation. In addition, the deposit pattern of the simulated Daguangbao landslide under horizontal-and-vertical situation coincides well with local topography. The vertical seismic force should be considered for landslide assessment and management, especially in the situation that the studied site located on the meizoseismal area during the earthquake.

7. ACKNOWLEDGEMENTS

The authors gratefully acknowledge the financial support provided by the National Natural Science Foundation of China (No. 51408511), the Scientific Research Foundation for the Returned Overseas Chinese Scholars, State Education Ministry and the opening fund of State Key Laboratory of Geohazard Prevention and Geoenvironment Protection (Chengdu University of Technology) (No. SKLGP2014K015).

REFERENCES

1. Ambraseys, N., and Sarma, S. 1967. The response of earth dams to strong earthquakes. Geotechnique, 17(3): 181-213.
2. Ambraseys, N., and Menu, J. 1988. Earthquake-induced ground displacements. Earthquake Engineering & Structural Dynamics, 16(7): 985-1006.
3. Ambraseys, N., and Srbulov, M. 1994. Attenuation of earthquake-induced ground displacements. Earthquake Engineering & Structural Dynamics, 23(5): 467-487.
4. Anderson, D., and Kavazanjian Jr, E. 1995. Performance of landfills under seismic loading. In Proc., 3rd Int. Conf. on Recent Advances in Geotechnical Earthquake Engineering and Soil Dynamics. Univ. of Missouri Rolla, MO, Vol.3, pp. 277-306.
5. Army, U.C.o.E. 1960. Stability of Earth and Rockfill Dams, EM 1110-2-1902.
6. Azzoni, A., La Barbera, G., and Zaninetti, A. 1995. Analysis and prediction of rockfalls using a mathematical model. In International Journal of Rock Mechanics and Mining Sciences & Geomechanics Abstracts. Elsevier, Vol.32, pp. 709-724.
7. Bakun-Mazor, D., Hatzor, Y., and Glaser, S. 2012. Dynamic sliding of tetrahedral wedge: The role of interface friction. International Journal for Numerical and Analytical Methods in Geomechanics, 36(3): 327-343.
8. Bhasin R, Kaynia AM. 2004. Static and dynamic simulation of a 700-m high rock slope in western Norway. Engineering Geology, 71(3-4): 213-226.
9. Bozzolo, D., and Pamini, R. 1986. Simulation of rock falls down a valley side. Acta Mechanica, 63(1-4): 113-130.
10. Bray, J.D., and Rathje, E.M. 1998. Earthquake-induced displacements of solid-waste landfills. Journal of Geotechnical and Geoenvironmental Engineering, 124(3): 242-253.
11. Bray, J.D., and Travasarou, T. 2007. Simplified procedure for estimating earthquake-induced deviatoric slope displacements. Journal of Geotechnical and Geoenvironmental Engineering, 133(4): 381-392.
12. Brebbia, C.A., and Wrobel, L. 1980. The boundary element method. Computer methods in fluids.(A 81-28303 11-34) London, Pentech Press, Ltd., 1980: 26-48.
13. Cannon, S.H. 1993. An empirical model for the volume-change behavior of debris flows. In Hydraulic Engineering '93, San Francisco.

14. CDCDMG:California Department of Conversation, Division of Mines and Geology, 1997. Guidelines for evaluating and mitigating seismic hazards in California. CDMG Special Publication.
15. Chang, K.-T., Lin, M.-L., Dong, J.-J., and Chien, C.-H. 2011. The Hungtsaiping landslides: from ancient to recent. Landslides, 9(2): 205-214.
16. Chen, L., and Zhao, W. 1979. Longling earthquake, 1976. Earthquake press, Beijing.
17. Chen, G., and Ohnishi, Y. 1999. Slope stability analysis using Discontinuous Deformation Analysis method. Rock Mecganics for Industry: 535-541.
18. Chen, H., and Lee, C. 2000. Numerical simulation of debris flows. Canadian Geotechnical Journal, 37(1): 146-160.
19. Cheng, Y.M. 1998. Advancements and improvement in discontinuous deformation analysis. Computers and Geotechnics, 22(2): 153-163.
20. Chopra, A.K. 1966. The importance of the vertical component of earthquake motions. Bulletin of the Seismological Society of America, 56(5): 1163-1175.
21. Clough, R.W. 1960. The finite element method in plane stress analysis. In 2nd Conference on Electronic Computation, Pittsburgh, PA.
22. Constantinou, M., and Gazetas, G. 1987. Probabilistic seismic sliding deformations of earth dams and slopes. In Probabilistic Mechanics and Structural Reliability (1984). ASCE, pp. 318-321.
23. Copons, R., and Vilaplana, J.M. 2008. Rockfall susceptibility zoning at a large scale: From geomorphological inventory to preliminary land use planning. Engineering Geology, 102(3): 142-151.
24. Copons, R., Vilaplana, J.M., Corominas, J., Altimir, J., and Amigó, J. 2004. Rockfall Risk Management in High-Density Urban Areas. The Andorran Experience. Landslide hazard and risk: 675-698.
25. Corominas, J. 1996. The angle of reach as a mobility index for small and large landslides. Canadian Geotechnical Journal, 33(2): 260-271.
26. Costa, J.E. 1984. Physical geomorphology of debris flows. In Developments and applications of geomorphology. Springer. pp. 268-317.
27. Crosta, G., and Agliardi, F. 2003a. A methodology for physically based rockfall hazard assessment. Natural Hazards and Earth System Science, 3(5): 407-422.
28. Crosta, G., and Frattini, P. 2003b. Distributed modelling of shallow landslides triggered by intense rainfall. Natural Hazards and Earth System Science, 3(1/2): 81-93.
29. Crosta, G.B., Frattini, P., and Fusi, N. 2007. Fragmentation in the Val Pola rock avalanche, Italian Alps. Journal of Geophysical Research: Earth Surface (2003–2012), 112(F1).
30. Crosta, G., Imposimato, S., Roddeman, D., Chiesa, S., and Moia, F. 2005. Small fast-moving flow-like landslides in volcanic deposits: the 2001 Las Colinas Landslide (El Salvador). Engineering Geology, 79(3): 185-214.
31. Cundall, P. 1971. A computer model for simulating progressive, large scale movements in blocky rock system. In In Symposium of International Society of Rock Mechanics, Nancy, France, pp. 11–18.
32. Danay, A., and Adeghe, L. 1993. Seismic-induced slip of concrete gravity dams. Journal of Structural Engineering, 119(1): 108-129.
33. Davies, T.R. 1982. Spreading of rock avalanche debris by mechanical fluidization. Rock Mechanics, 15(1): 9-24.
34. Del Gaudio, V., Pierri, P., and Wasowski, J. 2003. An approach to time-probabilistic evaluation of seismically induced landslide hazard. Bulletin of the Seismological Society of America, 93(2): 557-569.
35. Denlinger, R.P., and Iverson, R.M. 2001. Flow of variably fluidized granular masses across three-dimensional terrain: 2. Numerical predictions and experimental tests. Journal of Geophysical Research: Solid Earth (1978–2012), 106(B1): 553-566.
36. Descoeudres, F., and Zimmermann, T. 1987. Three-dimensional dynamic calculation of rockfalls. In 6th ISRM Congress.

37. Doolin, D.M. 2005. Unified displacement boundary constraint formulation for discontinuous deformation analysis (DDA). International Journal for Numerical and Analytical Methods in Geomechanics, 29(12): 1199-1207.
38. Doolin, D.M., and Sitar, N. 2004. Time Integration in Discontinuous Deformation Analysis. Journal of Engineering Mechanics, 130(3): 249-258.
39. Dorren, L.K. 2003. A review of rockfall mechanics and modelling approaches. Progress in Physical Geography, 27(1): 69-87.
40. Dorren, L., and Heuvelink, G.B. 2004. Effect of support size on the accuracy of a distributed rockfall model. International Journal of Geographical Information Science, 18(6): 595-609.
41. Dreyfus, D., Rathje, E.M., and Jibson, R.W. 2013. The Influence of Different Simplified Sliding-Block Models and Input Parameters on Regional Predictions of Seismic Landslides Triggered by the Northridge Earthquake. Engineering Geology.
42. Elgamal, A.-W.M., Scott, R.F., Succarieh, M.F., and Yan, L. 1990. La Villita dam response during five earthquakes including permanent deformation. Journal of Geotechnical Engineering, 116(10): 1443-1462.
43. Evans, S., and Hungr, O. 1993. The assessment of rockfall hazard at the base of talus slopes. Canadian Geotechnical Journal, 30(4): 620-636.
44. Evans, S., Hungr, O., and Enegren, E. 1994. The Avalanche Lake rock avalanche, Mackenzie mountains, northwest territories, Canada: description, dating, and dynamics. Canadian Geotechnical Journal, 31(5): 749-768.
45. Fardis, M.N. 2009. Seismic design, assessment and retrofitting of concrete buildings: based on EN-Eurocode 8. Springer.
46. Favreau, P., Mangeney, A., Lucas, A., Crosta, G., and Bouchut, F. 2010. Numerical modeling of landquakes. Geophysical Research Letters, 37(15): L15305.
47. Fenves, G., and Chopra, A.K. 1986. Simplified analysis for earthquake resistant design of concrete gravity dams. Earthquake Engineering Research Center, University of California.
48. Franklin, A.G., and Chang, F.K. 1977. Permanent displacements of earth embankments by Newmark sliding block analysis.
49. Gazetas, G., and Uddin, N. 1994. Permanent deformation on preexisting sliding surfaces in dams. Journal of Geotechnical Engineering, 120(11): 2041-2061.
50. Griffiths, D., and Prevost, J.H. 1988. Two-and three-dimensional dynamic finite element analyses of the Long Valley Dam. Geotechnique, 38(3): 367-388.
51. Guzzetti, F., Malamud, B.D., Turcotte, D.L., and Reichenbach, P. 2002. Power-law correlations of landslide areas in central Italy. Earth and Planetary Science Letters, 195(3): 169-183.
52. Hamajima, R., Kawai, T., Yamashita, K., and Kusabuka, M. 1985. Numerical analysis of cracked and jointed rock mass. In the 5th International Conference on Numerical Methods in Geomechanics, Nagoya, Japan, pp. 207-214.
53. Harp, E.L., and Jibson, R.W. 1995. Inventory of landslides triggered by the 1994 Northridge, California earthquake. US Geological Survey.
54. Hatzor, Y.H. 2003. Fully Dynamic Stability Analysis of Jointed Rock Slopes. In Proceedings of the 10th ISRM Congress, pp. 503-514.
55. Hatzor, Y.H., and Feintuch, A. 2001. The validity of dynamic block displacement prediction using DDA. International Journal of Rock Mechanics & Mining Sciences.
56. Hatzor, Y.H., Arzi, A.A., and Tsesarsky, M. 2002. Realistic dynamic analysis of jointed rock slopes using DDA. In 5th Int. Conf. on Analysis of Discontinuous Deformation - Stability of rock structures, Abingdon, Balkema, Rotterdam, The Netherlands, pp. 47–56.
57. Hatzor, Y., Arzi, A.A., Zaslavsky, Y., and Shapira, A. 2004. Dynamic stability analysis of jointed rock slopes using the DDA method: King Herod's Palace, Masada, Israel. International Journal of Rock Mechanics and Mining Sciences, 41(5): 813-832.
58. Heim, A. 1932. Bergsturz und Menschenleben, Zurich: Fretz and Wasmuth Verlag.

59. Hsieh, S.-Y., and Lee, C.-T. 2011. Empirical estimation of the Newmark displacement from the Arias intensity and critical acceleration. Engineering Geology, 122(1-2): 34-42.
60. Hsü, K.J. 1975. Catastrophic debris streams (sturzstroms) generated by rockfalls. Geological Society of America Bulletin, 86(1): 129-140.
61. Huang, R., Pei, X., Fan, X., Zhang, W., Li, S., and Li, B. 2012. The characteristics and failure mechanism of the largest landslide triggered by the Wenchuan earthquake, May 12, 2008, China. Landslides, 9(1): 131-142.
62. Hungr, O. 1995. A model for the runout analysis of rapid flow slides, debris flows, and avalanches. Canadian Geotechnical Journal, 32(4): 610-623.
63. Hungr, O., and Evans, S. 1988. Engineering evaluation of fragmental rockfall hazards. In Proceedings of the Fifth International Symposium on Landslides, Lausanne, AA Balkema, Rotterdam, Netherlands, pp. 685-690.
64. Hungr, O., and Evans, S. 2004. Entrainment of debris in rock avalanches: An analysis of a long run-out mechanism. Geological Society of America Bulletin, 116(9-10): 1240-1252.
65. Hungr, O., Corominas, J., and Eberhardt, E. 2004. Estimating landslide motion mechanism, travel distance and velocity.
66. Hürlimann, M., Rickenmann, D., Medina, V., and Bateman, A. 2008. Evaluation of approaches to calculate debris-flow parameters for hazard assessment. Engineering Geology, 102(3): 152-163.
67. Hynes-Griffin, M.E., and Franklin, A.G. 1984. Rationalizing the seismic coefficient method. Defense Technical Information Center.
68. Ishikawa, T., Sekine, E., and Ohnishi, Y. 2002. Shaking table tests of coarse granular materials with discontinuous analysis. Proc. of ICADD-5, BALKEMA: 181-187.
69. Jakob, M., and Hungr, O. 2005. Debris-flow hazards and related phenomena. Springer.
70. Jibson, R.W. 1993. Predicting earthquake-induced landslide displacements using Newmark's sliding block analysis. Transportation Research Record: 9-9.
71. Jibson, R. 2000. A method for producing digital probabilistic seismic landslide hazard maps. Engineering Geology.
72. Jibson, R.W. 2007. Regression models for estimating coseismic landslide displacement. Engineering Geology, 91(2-4): 209-218.
73. Jibson, R.W. 2011. Methods for assessing the stability of slopes during earthquakes—A retrospective. Engineering Geology, 122(1-2): 43-50.
74. Jibson, R.W., and Jibson, M.W. 2003. Java programs for using Newmark's method and simplified decoupled analysis to model slope performance during earthquakes. US Department of the Interior, US Geological Survey.
75. Jibson, R.W., Harp, E.L., and Michael, J.A. 1998. A Method for Producing Digital Probabilistic Seismic Landslide Hazard Maps: An Example from the Los Angeles, California, Area.
76. Kavazanjian, E., and Consultants, G. 1997. Design Guidance: Geotechnical Earthquake Engineering for Highways. Design Principles. Federal Highway Administration.
77. Kawai, T. 1977. A new discrete analysis of nonlinear solid mechanics problems involving stability, plasticity and crack. In the Symposium on Applications of Computer Methods in Engineering, Los Angeles, USA, pp. 1029-1038.
78. Kawai, T. 1978. New discrete models and their application to seismic response analysis of structures. Nuclear Engineering and Design, 48(1): 207-229.
79. Kawai, T., Takeuchi, N., and Kumeta, T. 1981. New discrete models and their application to rock mechanics. In ISRM International Symposium.
80. Kawai, T., Kawabata, Y., Kumagai, K., and Kondou, K. 1978. A new discrete model for analysis of solid mechanics problems. Numerical methods in fracture mechanics: 26-37.

81. Ke, T.C. 1996. The issues of rigid-body rotation in DDA. In First international forum on discontinuous deformation analysis (DDA) and simulations of discontinuous media, Berkeley, USA, pp. 318-325.
82. Keefer, D.K. 2000. Statistical analysis of an earthquake induced landslide distribution—the 1989 Loma Prieta, California event. Engineering Geology, 58(3): 231-249.
83. Kirby, M.J., and Statham, I. 1975. Surface stone movement and screen formation. Journal of Geology, 83(3): 349-362.
84. Kobayashi, Y., Harp, E., and Kagawa, T. 1990. Simulation of rockfalls triggered by earthquakes. Rock Mechanics and Rock Engineering, 23(1): 1-20.
85. Komodromos, P., Papaloizou, L., and Polycarpou, P. 2008. Simulation of the response of ancient columns under harmonic and earthquake excitations. Engineering Structures, 30(8): 2154-2164.
86. Kong, X., and Liu, J. 2002. Dynamic failure numeric simulations of model concrete-faced rock-fill dam. . Soil Dynamics and Earthquake Engineering 22(9–12): 1131–1134.
87. Koo, C.Y., and Chern, J.C. 1998. Modification of the DDA method for rigid block problems. International Journal of Rock Mechanics & Mining Sciences, 35: 683-693.
88. Kostaschuk, R. 1987. Identification of debris flow hazard on alluvial fans in the Canadian Rocky Mountains. Debris flows/avalanches: process, recognition, and mitigation, 7: 115.
89. Kramer, S.L. 1996. Geotechnical earthquake engineering. Prentice-Hall Civil Engineering and Engineering Mechanics Series, Upper Saddle River, NJ: Prentice Hall,| c1996, 1.
90. Kramer, S.L., and Smith, M.W. 1997. Modified Newmark model for seismic displacements of compliant slopes. Journal of Geotechnical and Geoenvironmental Engineering, 123(7): 635-644.
91. Krinitzsky, E.L. 1993. Fundamentals of earthquake-resistant construction. Wiley.com.
92. Lan, H., Derek Martin, C., and Lim, C. 2007. RockFall analyst: A GIS extension for three-dimensional and spatially distributed rockfall hazard modeling. Computers & Geosciences, 33(2): 262-279.
93. Latha, G.M., and Garaga, A. 2010. Seismic Stability Analysis of a Himalayan Rock Slope. Rock Mechanics and Rock Engineering, 43(6): 831-843.
94. Lee, K.L. 1974. Seismic permanent deformation in earth dams, University of California, Los Angeles, CA.
95. Leger, P., and Katsouli, M. 1989. Seismic stability of concrete gravity dams. Earthquake Engineering & Structural Dynamics, 18(6): 889-902.
96. Li, T. 1983. A mathematical model for predicting the extent of a major rockfall. Zeitschrift Fur Geomorphologie, 24: 473-482.
97. Lin, J.S., and Whitman, R.V. 1983. Decoupling approximation to the evaluation of earthquake-induced plastic slip in earth dams. Earthquake Engineering & Structural Dynamics, 11(5): 667-678.
98. Lin, C.T., Amadei, B., Jung, J., and Dwyer, J. 1996. Extensions of discontinuous deformation analysis for jointed rock masses. international Journal of Rock Mechanics and Mining Sciences & Geomechanics Abstracts, 33(7): 671-694.
99. Ling, H.I. 2001. Recent applications of sliding block theory to geotechnical design. Soil Dynamics and Earthquake Engineering, 21: 189-197.
100. Ling, H., and Leshchinsky, D. 1998. Effects of vertical acceleration on seismic design of geosynthetic-reinforced soil structures. Geotechnique, 48(3): 347-373.
101. Lomnitz, C. 1970. Casualties and behavior of populations during earthquakes. Bulletin of the Seismological Society of America, 60(4): 1309-1313.
102. Luan, M., Li, Y., and Yang, Q. 2000. Discontinuous deformation computational mechanics model and its application in stability analysis of rock slope. Chinese Journal of Rock Mechanics and Engineering, 3: 006.

103. Makdisi, F.I., and Seed, H.B. 1977. Simplified procedure for estimating dam and embankment earthquake-induced deformations. In ASAE Publication No. 4-77. Proceedings of the National Symposium on Soil Erosion and Sediment by Water, Chicago, Illinois, December 12-13, 1977.
104. Makdisi, F.I., and Seed, H.B. 1978. Simplified procedure for estimating dam and embankment earthquake-induced failures. Journal of the Geotechnical Division, ASCE, 104: 849–861.
105. Makris, N., and Roussos, Y. 2000. Rocking response of rigid blocks under near-source ground motions. Geotechnique, 50(3): 243-262.
106. Mankelow, J.M., and MURPHY, W. 1998. Using GIS in the probabilistic assessment of earthquake triggered landslide hazards. Journal of Earthquake Engineering, 2(4): 593-623.
107. Marcuson, W. 1981. Moderator's report for session on Earth Dams and Stability of Slopes under Dynamic Loads. In Proceedings, International Conference on Recent Advances in Geotechnical Earthquake Engineering and Soil Dynamics, Vol.3, p. 1175.
108. Marcuson III, W.F., and Franklin, A.G. 1983. Seismic Design, Analysis, and Remedial Measures to Improve Stability of Existing Earth Dams, DTIC Document.
109. Masuya, H., Amanuma, K., Nishikawa, Y., and Tsuji, T. 2009. Basic rockfall simulation with consideration of vegetation and application to protection measure. Natural Hazards and Earth System Science, 9(6): 1835-1843.
110. McDougall, S. 2006. A new continuum dynamic model for the analysis of extremely rapid landslide motion across complex 3D terrain, University of British Columbia.
111. McDougall, D. 2006. The distributed criterion design. Journal of Behavioral Education, 15(4): 236-246.
112. McDougall, S., and Hungr, O. 2004. A model for the analysis of rapid landslide motion across three-dimensional terrain. Canadian Geotechnical Journal, 41(6): 1084-1097.
113. Meunier, P., Hovius, N., and Haines, A.J. 2007. Regional patterns of earthquake-triggered landslides and their relation to ground motion. Geophysical Research Letters, 34(20).
114. Miles, S.B., and Ho, C.L. 1999. Applications and issues of GIS as tool for civil engineering modeling. Journal of Computing in Civil Engineering, 13(3): 144-152.
115. Miles, S.B., and Keefer, D.K. 2000. Evaluation of seismic slope-performance models using a regional case study. Environmental & Engineering Geoscience, 6(1): 25-39.
116. Mitchell, A.R., and Griffiths, D.F. 1980. The finite difference method in partial differential equations(Book). Chichester, Sussex, England and New York, Wiley-Interscience, 1980. 281 p.
117. Moretti, L., Mangeney, A., Capdeville, Y., Stutzmann, E., Christian Huggel, C., Schneider, D., and Francois Bouchut, F. 2012. Numerical modeling of the Mount Steller landslide flow history and of the generated long period seismic waves. Geophys. Res. Lett., 39(L16402).
118. Moriwaki, H., Yazaki, S., and Oyagi, N. 1985. A gigantic debris avalanche and its dynamics at Mount Ontake caused by the Nagano-ken-seibu earthquake, 1984. In Proc. 4th Int. Conf. Field Workshop on Landslides, pp. 359-364.
119. Newmark, N.M. 1965. Effects of earthquakes on dams and embankments. Géotechnique, 15: 139-159.
120. Ning, Y., and Zhao, Z. 2012. A detailed investigation of block dynamic sliding by the discontinuous deformation analysis. International Journal for Numerical and Analytical Methods in Geomechanics: 1-21.
121. Niwa, K., Kawai, T., Ikeda, M., and Takeda, T. 1984. Application of a new discrete method to fracture analysis of brittle materials. In the 3rd International Conference on Numerical Methods in Fracture Mechanics, Swansea, U.K., pp. 13-27.
122. Ochiai, H., Okada, Y., Furuya, G., Okura, Y., Matsui, T., Sammori, T., Terajima, T., and Sassa, K. 2004. A fluidized landslide on a natural slope by artificial rainfall. Landslides, 1(3): 211-219.
123. Ohnishi, Y., Chen, G., and Miki, S. 1995. Recent development of DDA in rock mechanics. Proc. ICADD, 1: 26-47.

124. Okura, Y., Kitahara, H., and Sammori, T. 2000. Fluidization in dry landslides. Engineering Geology, 56(3): 347-360.

125. Okura, Y., Kitahara, H., Sammori, T., and Kawanami, A. 2000. The effects of rockfall volume on runout distance. Engineering Geology, 58(2): 109-124.

126. Okura, Y., Kitahara, H., Ochiai, H., Sammori, T., and Kawanami, A. 2002. Landslide fluidization process by flume experiments. Engineering Geology, 66(1): 65-78.

127. Pal, S., Kaynia, A.M., Bhasin, R.K., and Paul, D.K. 2011. Earthquake Stability Analysis of Rock Slopes: a Case Study. Rock Mechanics and Rock Engineering.

128. Papaloizou, L., and Komodromos, P. 2009. Planar investigation of the seismic response of ancient columns and colonnades with epistyles using a custom-made software. Soil Dynamics and Earthquake Engineering, 29(11-12): 1437-1454.

129. Papantonopoulos, C., Psycharis, I.N., Papastamatiou, D.Y., Lemos, J.V., and Mouzakis, H.P. 2002. Numerical prediction of the earthquake response of classical columns using the distinct element method. Earthquake Engineering & Structural Dynamics, 31(9): 1699-1717.

130. Pekau, O.A, and Yuzhu, C. 2004. Failure analysis of fractured dams during earthquakes by DEM. Engineering Structures, 26(10): 1483-1502.

131. Pekau, O.A, and Yuzhu, C. 2004. Seismic collapse behaviour of Damaged dams. In 13 WCEE: 13 th World Conference on Earthquake Engineering Conference Proceedings.

132. Pfeiffer, T.J., and Bowen, T. 1989. Computer simulation of rockfalls. Bulletin of the Association of Engineering Geologists, 26(1): 135-146.

133. Pirulli, M. 2005. Numerical modelling of landslide runout. A continuum mechanics approach, Politecnico di Torino.

134. Poisel, R., Preh, A., and Hungr, O. 2008. Run Out of Landslides–Continuum Mechanics versus Discontinuum Mechanics Models. Geomechanics and Tunnelling, 1(5): 358-366.

135. Prevost, J.H. 1981. DYNA-FLOW: a nonlinear transient finite element analysis program. Princeton University, Department of Civil Engineering, School of Engineering and Applied Science.

136. Psycharis, I., Lemos, J., Papastamatiou, D., Zambas, C., and Papantonopoulos, C. 2003. Numerical study of the seismic behaviour of a part of the Parthenon Pronaos. Earthquake Engineering & Structural Dynamics, 32(13): 2063-2084.

137. Pyke, R. 1991. Selection of Seismic Coefficients for Use in Pseudo-Static Slope Stability Analyses. http://www.tagasoft.com/Discussion/article2_html.

138. Rathje, E.M., and Bray, J.D. 1999. An examination of simplified earthquake-induced displacement procedures for earth structures. Canadian Geotechnical Journal, 36(1): 72-87.

139. Rathje, E.M., and Bray, J.D. 2000. Nonlinear coupled seismic sliding analysis of earth structures. Journal of Geotechnical and Geoenvironmental Engineering, 126(11): 1002-1014.

140. Richards, R., and Elms, D.G. 1979. Seismic behavior of gravity retaining walls. Journal of the Geotechnical Engineering Division, 105(4): 449-464.

141. Richards, J., Elms, D., and Budhu, M. 1993. Seismic bearing capacity and settlements of foundations. Journal of Geotechnical Engineering, 119(4): 662-674.

142. Rickenmann, D. 1999. Empirical relationships for debris flows. Natural Hazards, 19(1): 47-77.

143. Rodriguez, C.E., Bommer, J., and Chandler, R.J. 1999. Earthquake-induced landslides 1980-1997. soil Dynamics and Earthquake Engineering, 18: 325-346.

144. Sarma, S.K. 1975. Seismic stability of earth dams and embankments. Geotechnique, 25(4): 743-761.

145. Sarma, S.K. 1981. Seismic displacement analysis of earth dams. Journal of the Geotechnical Engineering Division, 107(12): 1735-1739.

146. Sasaki, T., Hagiwara, I., Sasaki, K., Ohnishi, Y., and Ito, H. 2007. Fundamental studies for dynamic response of simple block structures by DDA. In In Proceedings of the eighth international conference on analysis of discontinuous deformation:

fundamentals and applications to mining & civil engineering, Beijing, China, pp. 141–146.
147. Sasaki, T., Hagiwara, I., Sasaki, K., Yoshinaka, R., Ohnishi, Y., and Nishiyama, S. 2004. Earthquake response analysis of rock-fall models by discontinuous deformation analysis. In In Proceedings of the ISRM international symposium 3rd ARMS, Kyoto, Japan, pp. 1267–1272.
148. Sassa, K. 1988. Motion of Landslides and Debris Flows: Prediction of Hazard Area: Report for Grant-in-aid for Scientific Research by Japanese Ministry on Education, Science and Culture (project No. 61480062). Disaster Prevention Research Institute.
149. Savage, S., and Hutter, K. 1989. The motion of a finite mass of granular material down a rough incline. Journal of Fluid Mechanics, 199(1): 177-215.
150. Sawada, T., Chen, W.F., and Nomachi, S.G. 1993. Assessment of seismic displacements of slopes. Soil Dynamics and Earthquake Engineering, 12: 357-362.
151. Saygili, G., and Rathje, E.M. 2009. Probabilistically based seismic landslide hazard maps: An application in Southern California. Engineering Geology, 109(3): 183-194.
152. Scheidegger, A.E. 1973. On the prediction of the reach and velocity of catastrophic landslides. Rock Mechanics, 5(4): 231-236.
153. Seed, H.B. 1973. Analysis of the Slides in the San Fernando Dams During the Earthquake of Feb. 9, 1971: Report to State of California Department of Water Resources, Los Angeles Department of Water and Power, National Science Foundation. College of Engineering, University of California.
154. Seed, H.B. 1979. Considerations in the earthquake-resistant design of earth and rockfill dams. Geotechnique, 29(3): 13-41.
155. Seed, H.B., and Martin, G.R. 1966. The seismic coefficient in earth dam design. Journal of Soil Mechanics & Foundations Div, 92(Proc. Paper 4824).
156. Serff, N. 1976. Earthquake induced deformations of earth dams. College of Engineering, University of California.
157. Shi, G.-H. 1988. Discontinuous Deformation Analysis A New Numerical Model for the Statics and Dynamics of Block Systems, University of California, Berkeley.
158. Shi, G. 2002. Single and multiple block limit equilibrium of key block method and discontinuous deformation analysis. In Proceedings of the 5th International Conference on Analysis of Discontinuous Deformation. Rotterdam: AA Balkema, pp. 3-43.
159. Shi, G.-H., and Goodman, R.E. 1985. Two dimensional discontinuous deformation analysis. International Journal for Numerical and Analytical Methods in Geomechanics, 9: 541-556.
160. Shi, G.-H., and Goodman, R.E. 1989. Generalization of two-dimensional discontinuous deformation analysis for forward modelling. International Journal for Numerical and Analyrucal Methods in Geomechanics, 13: 359-380.
161. Sousa, J., and Voight, B. 1991. Continuum simulation of flow failures. Geotechnique, 41(4): 515-538.
162. Stamatopoulos, C. 1996. Sliding system predicting large permanent co-seismic movements of slopes. Earthquake Engineering & Structural Dynamics, 25(10): 1075-1093.
163. Stewart, J.P., Blake, T.F., and Hollingsworth, R.A. 2003. A Screen Analysis Procedure for Seismic Slope Stability. Earthquake Spectra, 19(3): 697.
164. Taiebat, M., Kaynia, A.M., and Dafalias, Y.F. 2011. Application of an Anisotropic Constitutive Model for Structured Clay to Seismic Slope Stability. Journal of Geotechnical and Geoenvironmental Engineering, 137(5): 492.
165. Takahashi, T., Momiyama, A., Hirai, K., Hishinuma, F., and Akagi, H. 1992. Functional correlation of fetal and adult forms of glycine receptors with developmental changes in inhibitory synaptic receptor channels. Neuron, 9(6): 1155-1161.
166. Terzaghi, K. 1950. Theoretical Soil Mechanics.
167. Tsesarsky, M., Hatzor, Y., and Sitar, N. 2005. Dynamic displacement of a block on an inclined plane: analytical, experimental and DDA results. Rock Mechanics and Rock Engineering, 38(2): 153-167.

168. Varnes, D.J., Landslides, t.I.A.E.G.C.o., and Slopes, O.M.M.o. 1984. Landslide hazard zonation: a review of principles and practice.
169. Wartman, J., Asce, M., Bray, J.D., and Seed, R.B. 2003. Inclined Plane Studies of the Newmark Sliding Block Procedure. Journal of Geotechnical and Geoenvironmental Engineering, 129(8): 673-684.
170. Wasowski, J., Keefer, D.K., and Lee, C.-T. 2011. Toward the next generation of research on earthquake-induced landslides: Current issues and future challenges. Engineering Geology, 122(1-2): 1-8.
171. Wilson, R.C., and Keefer, D.K. 1983. Dynamic analysis of a slope failure from the 6 August 1979 Coyote Lake, California, earthquake. Bulletin of the Seismological Society of America, 73(3): 863-877.
172. Wu, J.-H. 2003. Numerical analysis of discontinuous rock masses using discontinuous deformation analysis, Kyoto University, Kyoto, Japan.
173. Wu, J.-H. 2010. Seismic landslide simulations in discontinuous deformation analysis. Computers and Geotechnics, 37(5): 594-601.
174. Wu, J.-H., and Chen, C.-H. 2011a. Application of DDA to simulate characteristics of the Tsaoling landslide. Computers and Geotechnics, 38(5): 741-750.
175. Wu, J.-H., and Tsai, P.-H. 2011b. New dynamic procedure for back-calculating the shear strength parameters of large landslides. Engineering Geology.
176. Wu, A., Ren, F., and Dong, X. 1997. A study on the numerical model of DDA and its preliminary application to rock engineering. Chinese Journal of Rock Mechanics and Engineering, 16(5): 411-417.
177. Wu, J., Lin, J., and Chen, C. 2009. Dynamic discrete analysis of an earthquake-induced large-scale landslide. International Journal of Rock Mechanics and Mining Sciences, 46(2): 397-407.
178. Yagoda-Biran, G., and Hatzor, Y.H. 2010. Constraining paleo PGA values by numerical analysis of overturned columns. Earthquake Engineering & Structural Dynamics, 39(4): 463-472.
179. Yegian, M.K. 1991a. Seismic risk analysis for earth dams. ASCE.
180. Yegian, M.K., Marciano, E.A., and Ghahraman, V.G. 1991b. Earthquake-induced permanent deformations: probabilistic approach. Journal of Geotechnical Engineering, 117(1): 35-50.
181. Yegian, M., Harb, J., and Kadakal, U. 1998. Dynamic response analysis procedure for landfills with geosynthetic liners. Journal of Geotechnical and Geoenvironmental Engineering, 124(10): 1027-1033.
182. Zhang, C., Pekau, O.A., Jin, F., and Wang, G. 1997. Application of distinct element method in dynamic analysis of high rock slopes and blocky structures. soil Dynamics and Earthquake Engineering, 16: 385-394.
183. Zhang, Y., Chen, G., Zheng, L., Wu, J., and Zhuang, X. 2012a. Effects of vertical seismic force on the initiation of the Daguangbao landslide induced by the Wenchuan earthquake. In The 8th Annual Conference of International Institute for Infrastructure, Renewal and Reconstruction, Kumamoto, Japan, pp. 530-539.
184. Zhang, Y., Chen, G., Zheng, L., and Li, Y. 2012b. Numerical analysis of the largest landslide induced by the Wenchuan earthquake, May 12, 2008 using DDA. In International Symposium on Earthquake-induced Landslides, Kiryu, Japan.
185. Zhang, Y., G. Chen, L. Zheng, Y. Li and J. Wu. 2013. Effects of near-Fault Seismic Loadings on Run-out of Large-Scale Landslide: A Case Study. Engineering Geology 166, 216-236.
186. Zhang, Y., G. Chen, L. Zheng, Y. Li, X. Zhuang. 2014. Effects of vertical seismic force on initiation of the Daguangbao landslide induced by the 2008 Wenchuan earthquake, Soil dynamics and earthquake engineering (In press).
187. Zhao, S.L., Salami, M.R., and Rahman, M.S. 1997. Discontinuous De-formation Analysis Simulation of Rock Slope Failure. In 9th International Conference on Computer Methods and Advances in Geomechanics, Wuhan, China
188. Zheng, L. 2010. Development of new models for landslide simulation based on discontinuous deformation analysis, Kyushu University, Fukuoka, Japan.

CHAPTER 8

Landslide Hazard Zonation in NH-1A in Kashmir Himalaya, India

R. K. Chingkhei, A. Shiroyleima, L. Robert Singh, Arun Kumar

Department of Earth Sciences, Manipur University, Imphal, India

ABSTRACT

The paper presents a methodology for landslide hazard zonation mapping using GIS and remote sensing data. The study has been carried out along NH-1A-Udhampur to Banihal in Kashmir Himalaya as this terrain is prone to the landslide hazards. The present study has been made to derive and identify the important terrain factors contributing to landslide occurrences in the region and corresponding thematic data layers are generated in GIS domain. These terrain data are collected from the topographic maps, satellite imageries, field visits and available published maps. A GIS database is prepared by digitizing these maps along with tabular data and their corresponding weightage is assigned using Analytic Hierarchy Process (AHP) based criteria ranking method for determining the standardized scores of criteria expressing their factor of importance for a given decision problem in terms of thematic parameters, categories and their normalized weights. Statistically integrating weightages from these thematic maps, a specific landslide hazard map was developed on a GIS platform. The resulting landslide hazard zonation map delineates the area into different zones of six classes of landslide hazard zones i.e., severe, very high, high, moderate, low and very low.

Keywords: Landslide Hazard Zonation; Landslide Hazard Management; GIS

1. INTRODUCTION

Landslides are among the most common natural hazards and are the most damaging, leading to a variety of human and environmental impacts. Though there are various steps involved in minimization of landslide hazard, there efforts are made to emphasize only landslide hazard zonation mapping. A Landslide Hazard Zonation (LHZ) map depicts division of land surface into zones of varying degree of stability based on the estimated significance of the causative factors in inducing instability. The LHZ maps have an important role in planning and implementation of development schemes in mountainous areas [1]. Landslide is the result of a wide variety of processes which include geological, geomorphologic and meteorological factors. The main factors which influence land sliding are discussed by Varnes [2] and Hutchinson [3]. Normally, the most important factors are bedrock geology (lithology, structure and degree of weathering), geomorphology (slope gradient, aspect, and relative relief), soil (depth, structure, permeability, and porosity), land use and land cover, and hydrologic conditions. Landslides are triggered by many causative factors. Most landslide-triggering factors can be divided into four main categories including geological, hydrological, topographical factors and loading conditions. Landslide hazard is a particular case of natural hazard defined as "the probability of occurrence within a specified period of time and within a given area of a potentially damaging phenomenon" [2]. From the experience of the USGS, the definition of a "landslide hazard map" includes "zonation showing annual probability (likelihood) of landslide occurring throughout an area". Over the past few years, Geographic Information System (GIS) has gained significant importance for spatial data analysis. It has been proved to a very powerful tool for landslide study. In recent years, GIS has been employed globally for spatial data analysis for landslide hazard zonation mapping [4-11]. Many parts of Jammu and Kashmir are vulnerable to landslide. The JammuSrinagar National Highway gets blocked at a number of places during the monsoons, due to landslides. It is worth mentioning that the Ramban-Banihal stretch of the highway has become one of the most affected portions.

The present paper deals with thematic data layer generation and their spatial analysis in GIS environment for landslide hazard zonation mapping along NH-1A-Udhampur to Banihal. This zone is known among the most landslides sensitive zones in the world. The major objectives of the study are 1) to prepare various thematic layers including lithology, geomorphology, lineament, fault, percent slope, slope morphology, slope aspect, drainage, landslide incidence, soil texture, soil depth, rock weathering, slope-dip relation, landuse/landcover from various sources in GIS domain; 2) to collect detailed field information about major landslides and soils of the area in order to generate a soil & terrain information system. GIS layer generation of all the thematic maps and their integration using ArcGIS software and 3) to prepare landslide hazard zonation maps through the integration of various thematic maps using Landslide Information System (LIS) software based GIS environment and to prepare landslide hazard management maps.

2. STUDY AREA

The study area is located along the National Highway NH-1A from Udhampur to Banihal. The extent of this study area lies between 32°53' to 33°28'N Latitude and 75°5' to 75°22'E Longitude (**Figure 1**). The study area falls under the Survey of India (SOI) toposheets No: 43P1, 4304, 4308, 4303and 4307. In the present study various zones have been marked as most hazardous that may cause severe disasters in future and may affect the local inhabitants in and around the area extending from Udhampur to Banihal on one hand and may cause serious disruption of traffic on strategic Jammu Srinagar Highway on the other. This zone is known among the most landslides sensitive zones in the world. The area shows a hilly topography with an average altitude of 2740 meters above MSL.

The area experiences hot summers and severe winters. Temperature is lowest between November & February when the minimum night temperature touches zero degree in the hill area and 3°C - 4°C in the outer plain area. Temperature rises from March onward. It becomes unbearable during May-June. Maximum day temperature in June touches sometime 47°C in the outer plain and about 30°C - 35°C in the hills. Most of the rainfall is received through the southwest monsoon which lasts from the last week of June to end of September. During the remaining period, rainfall is sporadic and scanty. Average annual rainfall is about 1052 mm.

3. METHODOLOGY

Different thematic layers corresponding to the causative factors that are responsible for the occurrence of landslides in a region were prepared from the Remote Sensing data and the Survey of India (SOI) topographic maps along with field data in GIS domain using ArcGIS and ERDAS. The integration of these thematic layers with weights assigned according to their relative importance in a GIS environment leads to the generation of a Zonal hazard map [11-14]. The detailed methodology adopted in the present study is represented in the form of a flow chart as below (**Figure 2**).

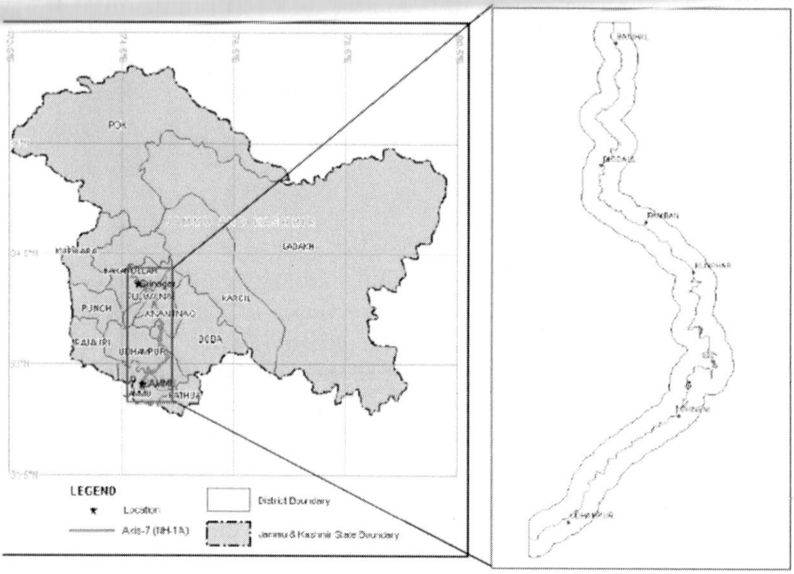

Figure 1. Location Map of the Study Area.

3.1. Preparation of Input Layers

Input Thematic layers were prepared in GIS domain using ARCGIS and ERDAS. These thematic layers are prepared from various sources like the SOI topographic maps (1:50,000 scale), Remote Sensing data (IRS P6L-3 satellite data, Google earth images etc.), Geological maps supplied by Geological Survey of India, soil maps supplied by National Bureau of Soil Survey and Landuse Planning, climate and earthquake data supplied by India Meteorological Department (IMD) and other related data, published literatures and field investigation data have been used during the course of this study. Altogether 20 thematic inputs have been prepared for the present study.

3.2. Digitization and GIS Layer Generation

The digitization/scanning and GIS layer generation work was carried out using ARCGIS. The codes or IDs of different thematic maps were assigned as per the standard code scheme worked out by Department of Space, Government of India in consultation with other leading Organizations, Institutions and Universities working in the field of terrain studies and landslide modeling in Himalayas. Trunk stream (main river) were depicted as single digitization. The base details pertaining to each individual map sheet were generated on 1:50,000 scale in three different layer viz. Base-poly, Baseline and Base-point. Roads were shown in Baseline layer, settlement was shown in Base-point layer and major rivers were shown in base poly layer in polygon feature type. The lineaments and faults were classified as per length. All the sheets were edge matched and in some of the themes the maps of entire area was generated and thereafter they have been clipped sheet wise in order to integrate maps of each individual map sheet.

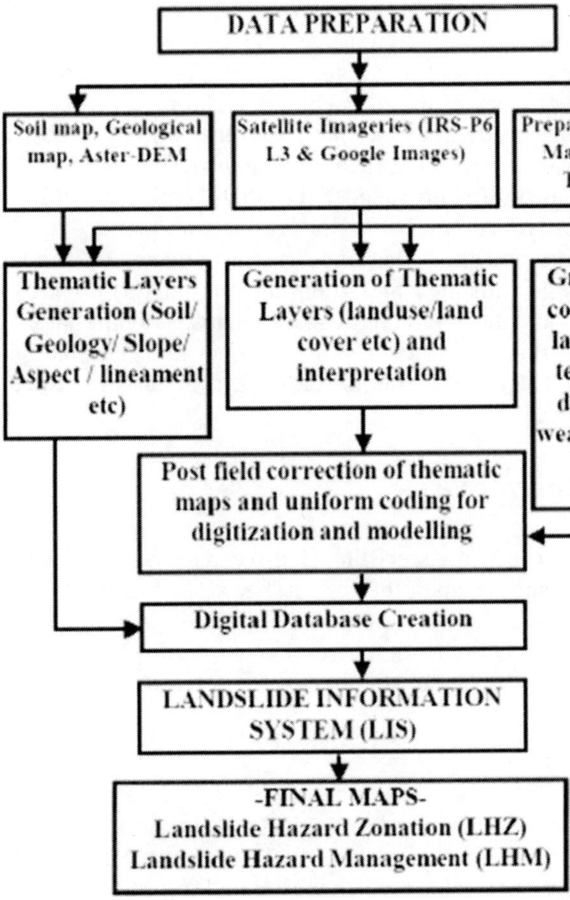

Figure 2. Flow chart of the steps involved in the present study.

4. RESULTS AND DISCUSSION

4.1. Thematic Data Layers

Selection of factors and preparation of corresponding thematic data layers are crucial components of any model for landslide hazard zonation and management mapping. The factors governing instability in a terrain are primarily geology, slope morphology, drainage, land use, anthropogenic activity, and climatic condition. These factors can be broadly grouped into two categories—the preparatory factors which make the slope susceptible to movement and the triggering factors which set off the movement [15]. The triggering factors such as rainfall and earthquakes, whose magnitude and temporal behavior are difficult to gauge, could not be considered here due to non-availability of past records regarding landslide occurrence. The factors considered in our study are essentially the preparatory factors for which pertinent data can be collected from available resources as well as from the field. These include slope, aspect, lithology, soil data, drainage density, lineament density, presence of faults, land use and landcover. The selection of these factors and their classes is primarily based on the field observations of existing landslides and their associated terrain factors. Based on the information collected from available maps, satellite data, and field investigations the thematic data layers are generated. The mapping was done at a scale of 1:125,000. The details of these layers are described in the following paragraphs.

4.2. DEM and Its Derivatives

In the present study the 30 m AsterDEM (**Figure 3**) has been used to generate the slope and aspect map of the region. The slope map is classified into 7 classes (0 - 15, 15 - 25, 25 - 30, 30 - 35, 35 - 40, 40 - 45, 45 - 60 and above 60 m) following the classification of earlier workers [9,11]. A slope aspect map, which shows direction of slope, was also derived from the DEM.

4.3. Lithology

The lithology map was derived from the geological map (scale 1:250,000) of the western Himalaya region compiled by V.C. Thakur and B.S. Rawat of Wadia institute of Himalayan Geology which was scaled to 1:25,000 for field investigation [16]. There are seven rock types present in the area. These are Boulder conglomerate, limestone, phyllite, granites, shale, volcanics and meta volcanics, and sandstone with shale and siltstone.

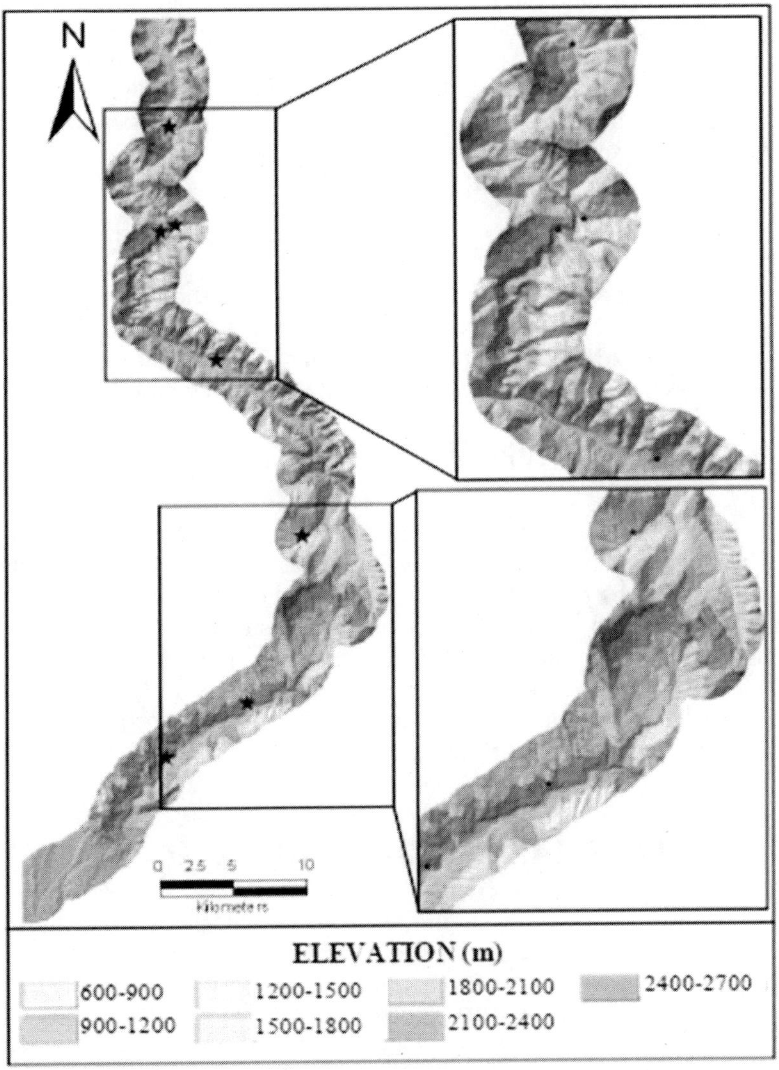

Figure 3. DEM showing the elevation classes in the study area. Black stars indicate the landslide sites in the study area.

4.4. Faults and Lineaments

Faults and lineaments present in the study area are extracted from the geological maps and the DEM. In order to identify linear topographic features from the DEM, eight shaded relief images were generated and the lineaments were generated from these shaded relief images using the line feature of ArcGIS. The lineaments are classified into minor lineament (length between 0 - 500 m), major lineament (length between 500 - 1500 m), and mega lineament (length greater than 1500 m). Further, faults, which are mega lineaments, showing offsets of ridges and drainage lines, were also identified.

4.5. Soil

The topsoil cover on a slope has an influence on landslide occurrence. For our study, the soil map was derived from a regional soil map at a 1:250,000 scale prepared by National Bureau of Soil Survey and Land Use Planning. The soils present in the area are loamy skeletal soil and loamy soil.

4.6. Landuse and Landcover

The landuse and landcover is one of the important parameters governing slope stability. The landuse map of the area was prepared from the remote sensing data. In the present study IRS P6 L3 and google images are used to extract various land use and land cover in terms of vegetation cover. The landuse map, thus prepared, classifies the area into dense vegetation, medium vegetation, degraded vegetation, agricultural land, scrub land/grass land, river sand and builtup area. The area is mostly covered by dense vegetation and agricultural land.

5. LANDSLIDE HAZARD ZONATION AND MANAGEMENT MAPPING

Various thematic maps have been prepared for generation of Landslide Hazard Zonation and the corresponding Management maps. Thematic maps/layers generated are lithology, geological structure/lineaments, geomorphology, drainage, slope angle, slope aspect, slope morphology, slope dip (bedding, joint) relation, landuse/landcover, rock weathering, soil texture, soil depth, landslide occurrence and anthropogenic factor. These thematic map layers are integrated using the LIS for generation of Landslide Hazard Zonation maps. The LIS software is capable of categorizing six hazard classes viz severe, very high, high, moderate, low and very low assigning higher wavelength colours to the high hazard classes and lower wavelength colours to the low hazard classes in the maps.

The criteria adopted for classification of six categories of landslide hazard zonation is based on ranking and weightage assigned to each categories in different thematic layers prepared for generation of landslide hazard zonation maps. Low and very low hazard zones indicating relative instability for each hazard zones. The tail ends of values obtained are classified into low and very low category of hazard zonation representing relative stability of these regions. The higher end of values obtained are classified into three different categories viz. severe, very high and high hazard zones indicating relative instability of these regions, whereas middle values which are distributed uniformly over area is classified as moderate hazard zone indicating moderate instability of the area. In the present study it is found that the area is dominantly occupied by very high, high, moderate and low hazard zones (**Figure 4**).

Very high hazard zone areas are prone to landslide activity which depicts steep slopes, scarp slopes higher weathering, higher dissection and denudation processes, torrential nature of streams and high velocity of runoff carrying huge debris materials, accelerated anthropogenic activities destabilizing the slopes. The rock formations in the area is folded, faulted, fractured & displayed at many places. The area need urgent attention in the form of mitigation measures like slope stabilization, bio engineering measures including regeneration of natural vegetation, reforestation, drainage correction etc. unless immediate action plans are implemented this zone will soon deteriorate further to the severe category. The settlement is to be avoided and permitted only in localized safe area.

High hazard zone area is also prone to landslide activity. The drainage follows structural features, the rocks are dissected, fractured and displaced, and denudation and weathering processes are active. The springs found along linear features are also responsible for natural hazards in the area. Hydrological and slope stabilization measures are required to be taken for mitigation.

Moderate hazard zone in which landsliding activity destabilize the area, are to be planned systematically so as to maintain its present status however, as natural drainage is disrupted and slope modified, landslide could be triggered. Mainly sheared zones of schist, genesis reveals the moderate slopes zone. The overall analysis reveals that moderate slopes have contributed more landslides in the past rather than very steep slopes.

Low hazard zone is occupied by an area of lower valley bottoms. The gentle middle spurs also come in the low hazard zone. However, many slopes falling within this zone could be destabilized by uncontrolled erosion, improper landuse practices and other developmental activities. Low hazard zone areas are relatively stable and safe from mass movements by virtue of its present environmental set up.

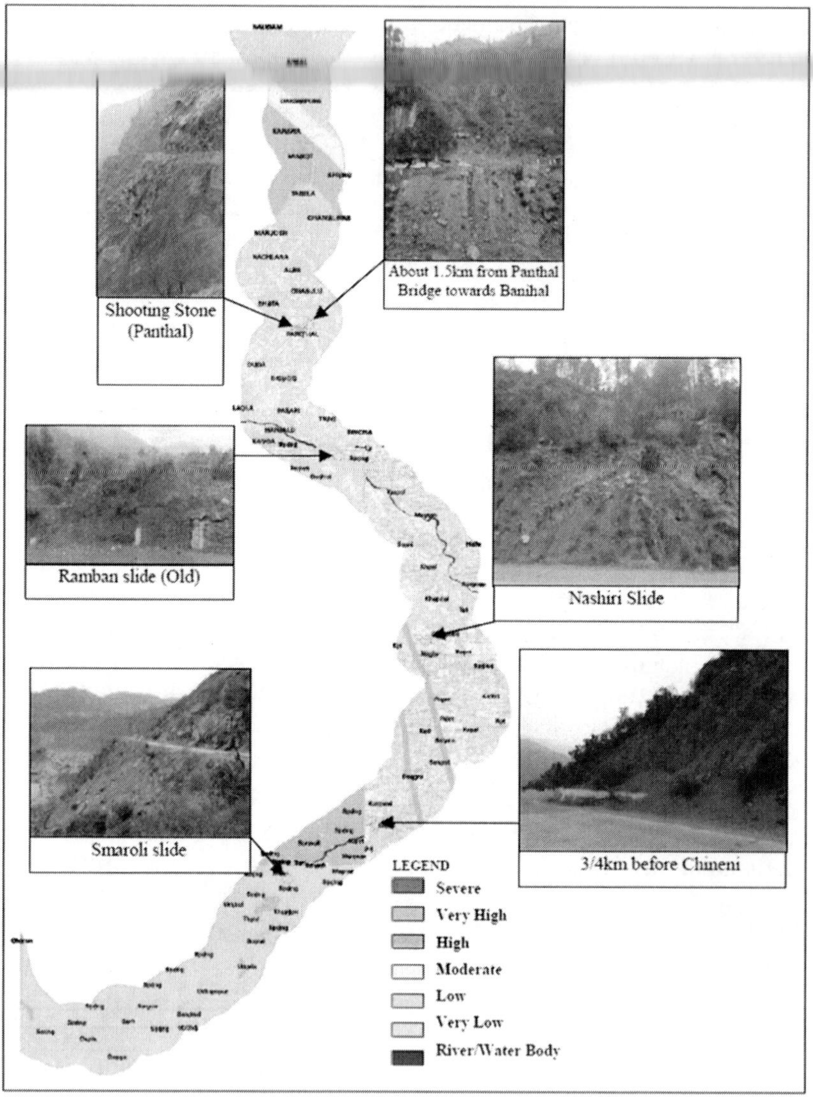

Figure 4. Landslide Hazard Zonation Map of the study area along with the field verification photographs.

Figure 5. Landslide Management Map of the Study Area Showing the Suggested Management Practices.

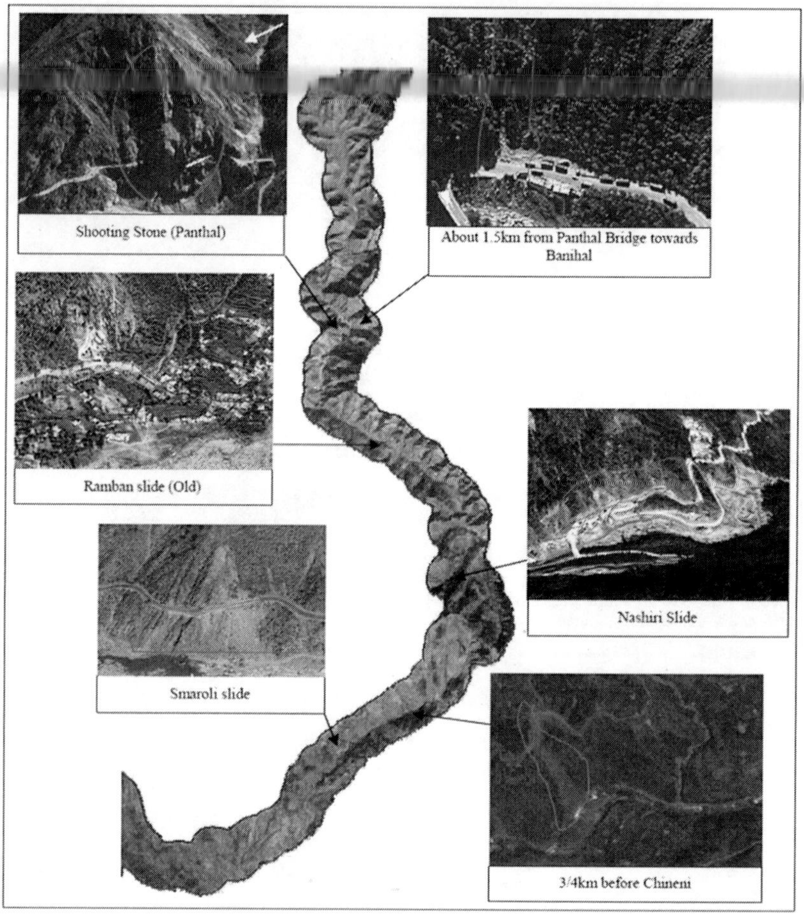

Figure 6.Google Earth image of the study area showing the landslide sites.

Map sheet wise area extent of each hazard zone, given in the map, is also generated taking into consideration the different hazard classes. As the landslides affect mainly the land use/land cover, making a great loss to life and property stress has been given to formulate the management rules taking into consideration mainly the landuse the management maps are generated. The management maps show 15 suggested management practice, having various combinations of management (**Figure 5**). After generating the landslide zonation map, it is validated by comparing with the field data and show a good correlation (Figures 4 and 6).

6. CONCLUSION

A methodology for landslide hazard zonation using remote sensing data and GIS has been successfully implemented in the present study area which can be further applied to other regions as well. The use of 30 m AsterDEM in generating various terrain units has also reduced the time quite efficiently. Landslide hazard zonation maps along with the corresponding management maps are of immense importance in hilly areas, like the present study area, to assess the potentiality of landslide occurrence and their management practices. In the present study, with the help of available information about the geology, topography of the area and using remote sensing data, considerable estimation of landslide hazards could be obtained to identify the broad trends of landslide hazard zonation. Last but not the least, since the entire database is created in digital format, the monitoring and assessment of landslides in the region can also be done at ease with necessary updates from time to time.

7. ACKNOWLEDGEMENTS

R. K. Chingkhei, A. Shiroyleima, L. Robert Singh, Arun Kumar The authors thankfully acknowledge the financial assistance given by the DTRL, New Delhi, Government of India.

REFERENCES

1. R. Anabalagan, "Landslide Hazard Evaluation and Zonation Mapping in Mountainous Terrain," Engineering Geology, Vol. 32, No. 4, 1992, pp. 269-277.
2. D. J. Varnes, "Landslide Hazard Zonation: A Review of Principles and Practice," Natural Hazards 3, Commission on Landslides of the IAEG, UNESCO, Paris, 1984, 63p.
3. J. N. Hutchinson, "Keynote Paper: Landslide Hazard Assessment," In: Bell, Ed., Landslides, Balkema, Rotterdam, 1995, pp. 1805-1841.
4. R. P. Gupta and B. C. Joshi, "Landslide Hazard Zoning Using the GIS Approach—A Case Study from Ramganga Catchment Himalayan," Engineering Geology, Vol. 28, No. 1-2, 1990, pp. 119-131.
5. A. Carrara, M. Cardinali, F. Detti, F. Guzzetti, V. Pasqui and P. Reichenbach, "GIS Techniques and Statistical Models in Evaluating Landslide Hazard," Earth Surface Processes and Landforms, Vol. 16, No. 5, 1991, pp. 427-445.
6. C. J. van Westen, "GIS in Landslide Hazard Zonation: A Review, with Examples from the Andes of Colombia," In: M. F. Price and D. I. Heywood, Eds., Mountain Environments and Geographical Information Systems, Taylor & Francis Publishers, London, 1994. pp. 135-165.
7. C. F. Chung, A. G. Fabbri and C. J. van Westen, "Multivariate Regression Analysis for Landslide Hazard Zonation," In: A. Carrara, and F. Guzzetti, Eds., Geographical

Information Systems in Assessing Natural Hazards, Kluwer Academic Publishers, Dordrecht, 1995, pp. 107-134.

8. R. Nagarajan, A. Mukherjee, A. Roy and M. V. Khire, "Temporal Remote Sensing Data and GIS Application in Landslide Hazard Zonation of Part of Western Ghat, India," International Journal of Remote Sensing, Vol. 19, No. 4, 1998, pp. 573-585.
9. A. S. Dhakal, T. Amada and M. Aniya, "Landslide Hazard Mapping and Its Evaluation Using GIS: An Investigation of Sampling Schemes for a Grid-Cell Based Quantitative Method," Photogrammetric Engineering & Remote Sensing, Vol. 66, No. 8, 2000, pp. 981-989.
10. A. K. Saha, R. P. Gupta and M. K. Arora, "GIS-Based Landslide Hazard Zonation in the Bhagirathi (Ganga) Valley, Himalayas," International Journal of Remote Sensing, Vol. 23, No. 2, 2002, pp. 357-369.
11. S. Sarkar and D.P. Kanungo, "An Integrated Approach for Landslide Susceptibility Mapping Using Remote Sensing and GIS," Photogrammetric Engineering & Remote Sensing, Vol. 70, No. 5, 2004, pp. 617-625.
12. D. P. Kanungo, M. K. Arora, S. Sarkar and R. P. Gupta, "Comparative Study of Conventional, ANN Black Box, Fuzzy and Combined Neural and Fuzzy Weighting Procedures for Landslide Susceptibility Zonation in Darjeeling Himalayas," Engineering Geology, Vol. 85, 2006, pp. 347-366.
13. J. Mathew, V. K. Jha and G. S. Rawat, "Weights of Evidence Modelling for Landslide Hazard Zonation Mapping in Part of Bhagirathi Valley," Uttarakhand Current Science, Vol. 92, No. 5, 2007, pp. 628-638.
14. R. P. Gupta, A. K. Saha, M. K. Arora and A. Kumar, "Landslide Hazard Zonation in a Part of Bhagirathi Valley, Garhwal Himalayas, Using Integrated Remote Sensing—GIS," Journal of Himalayan Geology, Vol. 20, No. 2, 1999, pp. 71-85.
15. M. J. Crozier, "Landslides: Causes, Consequences and Environment," Croom Helm Australia Pty. Ltd., London, 1986, 252 p.
16. V. C. Thakur and B. S. Rawat, "Geological Map of the Western Himalaya," Wadia Institute of Himalayan Geology, Dehradun, 1992.

CHAPTER 9

Landslide Cartography at the Region of Nabeul-Hammamet Based on Geographic Information System and Geomatic

Romdhan Haddad[1,2], Mariam Labiad[1,2], Mohamed Khaled Bouzid[1,2], Rym Rabia[1,2], Dallel Makloufi[1], Mohamed Chedly Rabia[1]

[1]*UR11ES41 Geomatics and Geosystems, University of Manouba, Manouba, Tunisia*

[2]*Faculty of Sciences Bizerte, Zarzouna, University of Carthage, Carthage, Tunisia*

ABSTRACT

Landslides are one of the most significant natural damaging disasters in hilly environ- ment [1] . The location of our study area is to the north of Tunisia, home to several manifestations of land instabilities, we bring to study this area of interest by GIS and geomatic approach to reduce social economic losses due to landslides. The performance of a cartographic data base for the landslide study in the Cap-Bon region was realized through studying geologic 1/50,000 and topographic 1/25,000 maps, aster optic Remote Sensing, land observation, and climatologic seismic data. These data will be digitalized, georeferenced, vectorized, spatially analyzed, classified and geotreated in order to produce a landslides card. The findings have shown that fields with smooth and friable lithology which are located at rather important seismic zones (more than 4 at Richter's scale) have some stability. However, the most endangered zones are in the North-West around Oued El Kbir and El Ain. Realizing this work helps to determine the most hazardous zones so that policy makers have an effective field intervention.

Keywords: Cartography, GIS, Geomatic, Landslide, Nabeul-Hammamet

1. INTRODUCTION

The Cap Bon region has an interesting economical importance for Tunisia. It contains about 30% of the touristic resorts of Tunisia and is having an important urban development. For these reasons, this work is focusing on cartography of instability hazards of land for the region lying between the town Beni Khiar and Hammamet. Landslides are a common hazard, especially during the rainy season [2] [3] . For these reasons, a data collection study about land instability has been conducted. By analyzing and integrating it, a "SIG" allowed us to realize a land instability map which could be used as a helping tool for decision makers for better urban and environmental management.

2. THE STUDY METHOD

The cartography of land instability areas in the regions of Nabeul and Hammamet compels us to analyze the process of realizing this map step by step (Figure 1).

2.1. Geographic Level

Nabeul-Hammamet region is located in the North-East of Tunisia between the meridiens 641.963 and 667.000 and the parallels 4.028.304 and 4.054.072. It lies on the south west of the Cap Bon peninsula (Figure 2).

2.2. Definition of Land Instability

Land instability means wrench, displacement on a slope and along an identifiable rupture of a surface of ground and rockets under the effect of gravity [4] . It depends mainly or the type of land morphology, the geologic structure, climate conditions, the seismicity, the hydrographic system, underground hydraulic pressure and anthropic activities [5] . These phenomena differ according to their extent, to their movement speed, to the land lithology and to the geometry of their braking surfaces.

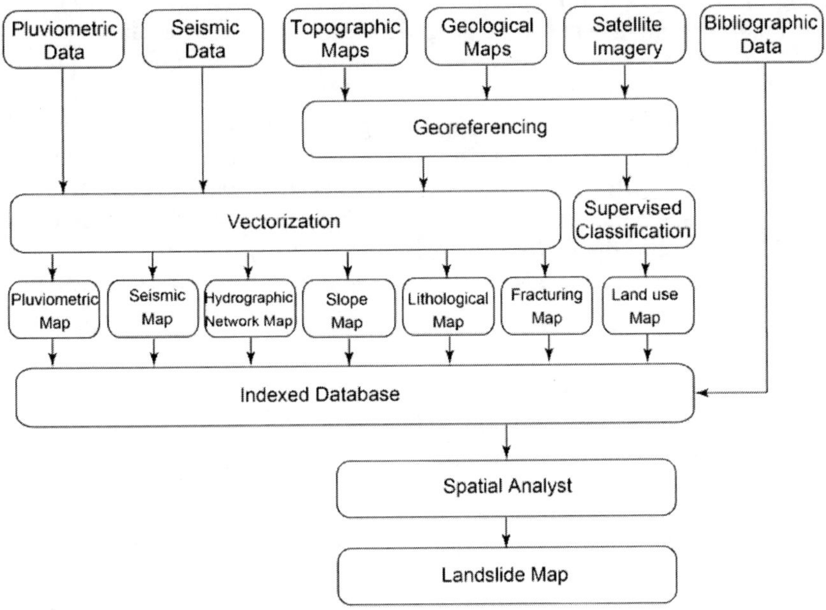

Figure 1. Methodology adopted for the implementation of the hazard map.

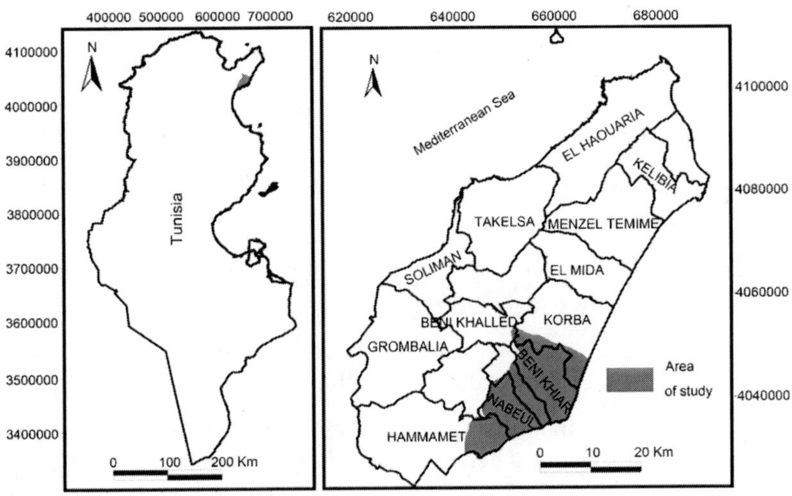

Figure 2. Geographical setting.

2.3. Identifying the Cartographic Data Base for Land Instability

Geographic Information System (GIS) modelling of landslide phenomena has taken precedence in recent time. Geospatial technologies like the use of GIS, and Remote Sensing are useful in the hazard assessment, risk identification, and disaster management for landslides [6] . For it while performing this study we have used image processing software (Envi 4.7) and SIG (Arcviews 3.1) so that to set a geographic data base (GDB) which allows us to a better understanding to the risks of land instability in the region of Nabeul-Hammamet. Realizing the GDB needed a collection of multidisciplinary data, its numerisation, and georeferencement and vectorisation so that produce the necessary themes for the targeted maps.

2.3.1. Lithologic Maps

Our lithologic map was realized according to a vectorization of two geologic cards at 1/50,000 of Nabeul and Hammamet. With regard to the works of [7] - [10] , the studied land is composed mainly of sand, clay and sandstone (Figure 3).

- -The Somaa formation, it is composed yellow or red sands intercalated with cong- lomeratic levels.
- -The Beni khiar formation, presents a progressive passage of continental sandstones to oolitic limestone.
- -The Oued El Bir formation is composed of sandstones, sand and azoic clays.
- -The pottery clay is plastic with a grey color and bleu patina.
- -The yellow sand of Nabeul is characterized by a varying granulometry with some conglomeratic levels.
- -The sand and sandstones of Hammamet are yellow fin sand, sometimes containing clay.
- -The Sicilian, its levels are composed generally by conglomerates poorly consolidated and dark grey sand.
- -The Tyrrhenian outcrops a cord formed by sandstone and limestone and sandstone at the base dune large oblique stratifications at the top.

The lithologic nature and the consolidation state of each formation allowed us to predict the mechanical behavior of each formation.

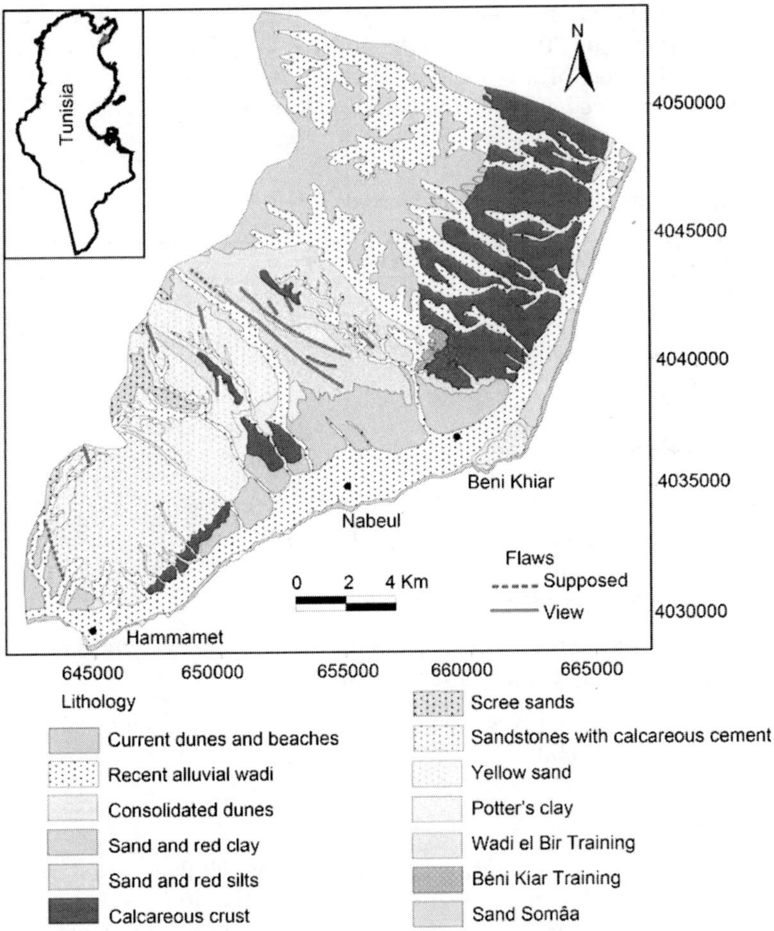

Figure 3. Map of lithology and fracturing.

This is checked out by the work of [11] that had proved the direct relation with the friction angle φ and the horizontal displacement of materials:

$$\sigma h = Ka \cdot \sigma v$$

σh: horizontal stress.

σv: vertical stress.

Ka: thrust coefficient of land.

$$Ka = \tan^2\left(\frac{\pi}{4} - \frac{\varphi}{2}\right)$$

Applying this equation allowed us to classify the lithologic unites into 3 classes depending on their aptitude to the instability. To facilitate implementing this classification in the data base, some proportional indexes at each instability have been affected to the surface lithology (Table 1).

Table 1. Classification of evidence of hazards lithological.

Land state	Stable	moderately stable	unstable
Hazard index	1	2	3
Lithology	Sandstone and limestone	Consolidated dune	Dune, Alluvium, Sand and Clay

2.3.2. Fracture Map

It includes the estimated and identified breaks at the level of our section. These breaks are generally oriented N120 to N170 (Figure 3). According to [12] , 1981 who affirm that break disposition is the most important criteria in our thematic, we have chosen to attribute:

-Index 1: to the estimated faults that affected sand and clayful lands or those which are not reactivated for a long time.

-Index 3: for the faults that represent a dip less to that designed breaks and also for the conjugated faults.

2.3.3. Slopes Map

The obtained slopes map is a derived map through a digital elevation model (DEM) product following the extraction and processing of contour to 25 m not included on topographic maps at the scale 1/25,000 in the Fevilles of Hammamet and Nabeul NE, NW, SE and SW.

It shows that the study area is a peneplain except the Nabeul-Hammamet monoclinal where the slopes can exceed the value of 30% (Figure 4). The works of [13] [14] have showed that stability of slopes is maintained thanks to a balance between the propulsive forces and the resistance forces.

$$F = \frac{\text{maximum resistant forces}}{\text{Driving forces}} = \frac{\tan \varphi}{\tan \alpha}$$

φ: friction angle, α: slope.

Applying this equation, which is in relation with the slope, allowed us to classify the topography into three classes depending on their contribution to instability (Table 2).

2.3.4. Hydrographic Map

The same previous topographic maps were used as a data base during the vectorization of the hydrographic network illustrated in the study area (Figure 5). The result has shown that the majority of the networks are in the form of trills which prove the preserve of permeable and friable land where water could dig and mobilize the particules easily [15] . In consideration with Strahler Classification to the whole network, land observation has proved that land instability is proportional is proportional to the order of the drain.

Table 2. Classification of topographic indices hazards.

Slope %	Land state	Instability index
≤15	stable	1
between 15 and 30	moderately stable	2
>30	unstable	3

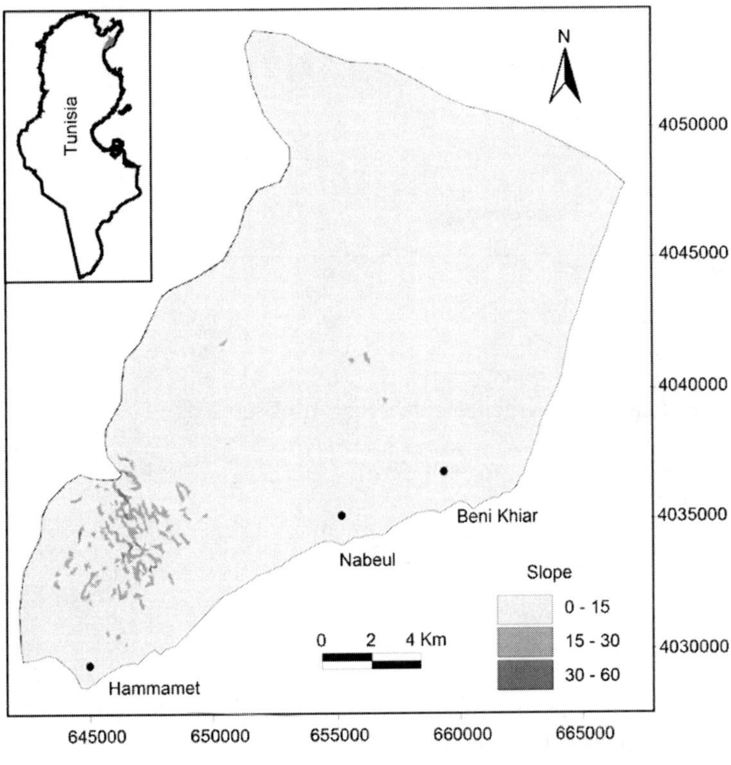

Figure 4. Slope of map.

This is confirmed by the progressive increase of the solid charge in the drains from upstream to downstream. Hence the choice of indices is for different drains classes (Table 3).

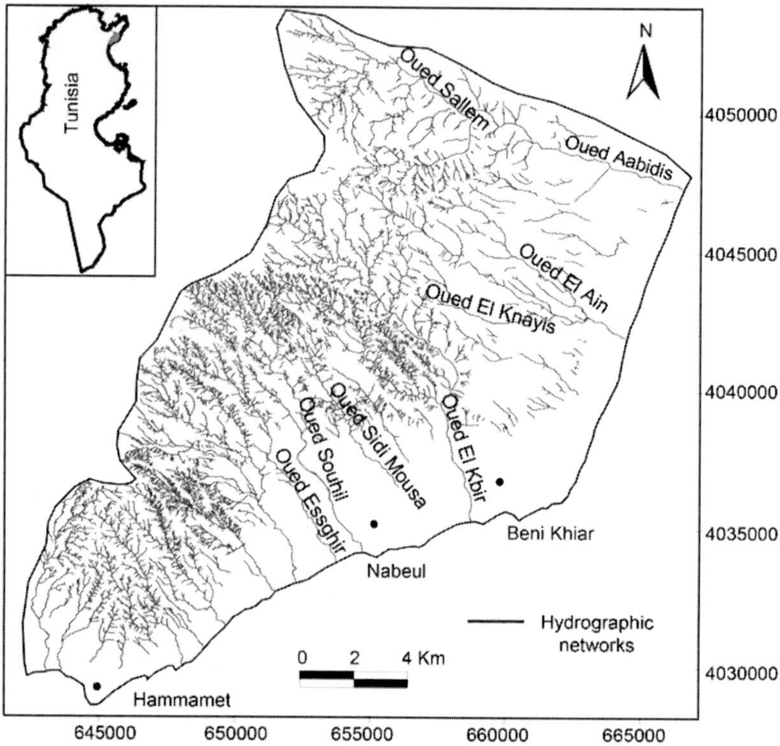

Figure 5. Hydrographic map.

Table 3. Classification of evidence of hydrographic hazards.

Order of drain	Land state	Instability index
1, 2 and 3	Stable	1
4 and 5	moderately stable	2
>5	unstable	3

2.3.5. Seismic Map

The seismic map (Figure 6) of the average change in the intensity of earthquakes that affected the southern part of Cap Bon for the last thirty years which were registered by many geophones situated in Nabeul-Hammamet, Dar Chabaan, Bou Argoub and Grombalia. These data issued by

National Meteorology Institute represented in the raster mode prove that the Nabeul-Hammamet region is seismically stable. In fact the maximum magnitude recorded was at the scale of 4.1.

The following Table 4 illustrates the adopted classification for the seismic factors.

2.3.6. Rainfall Map

The rainfall map (Figure 7) shows the special distribution of the interannual average rainfall for the last decade. The data base used in realizing this map is the monthly precipitations measured in the meteorological stations: Nabeul, Oued Souhil, Hammamet, Khangt El Hagag and Bou argoub.

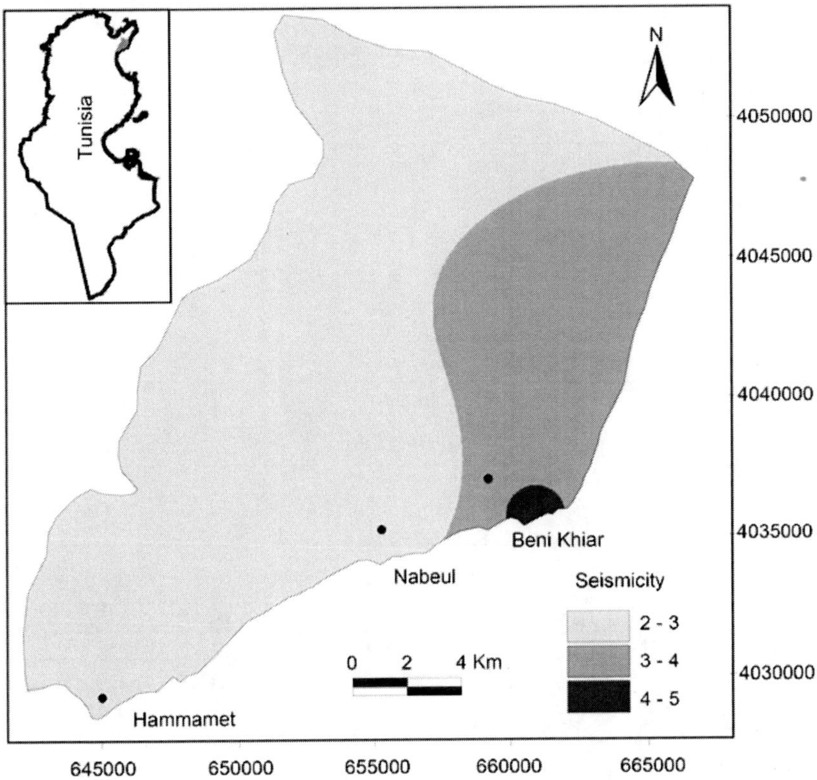

Figure 6. Seismic map.

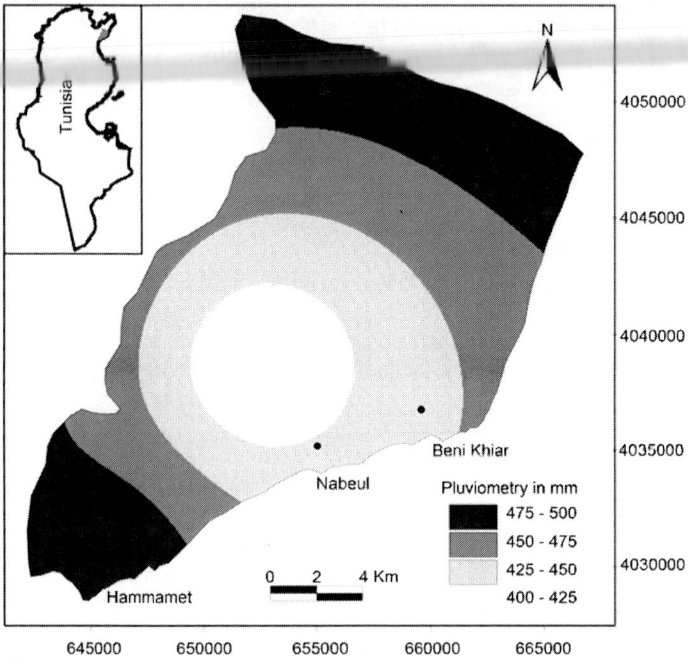

Figure 7. Rainfall map.

Table 4. Classification clues seismic hazards.

Seismic magnitude	Land state	Instability index
<3	stable	1
>3 and <5	moderately stable	2

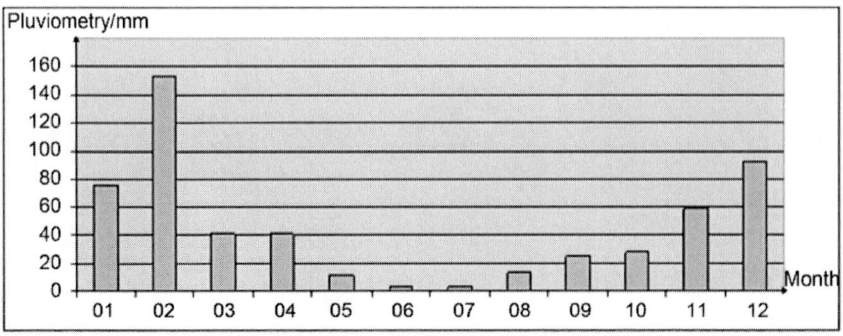

Figure 8. Average monthly precipitation the past decade.

Belonging to the Mediterranean climate which is characterized by its precipitations of low frequency and high intensity (Figure 8), it allowed us to attribute the instability index 2 to the whole area.

2.3.7. Soil Exploitation Map

The land use map was created following the supervised classification of satellite Aster high resolution image acquired April 16, 2002, which essentially shows five classes of land cover (Figure 9). It is essentially exploited by agricultural land that covers more than 65%, the urban sites (Nabeul, Hammamet, Dar Chaabane and Beni Khiar), the forest of Jebel Nabeul-Hammamet, Jebel Ksous and waste Lands.

Every type of exploitation contributes to the land instability in a special way. This is illustrated through the following Table 5.

2.4. Spatial Analyst

2.4.1. Applying Tampon Areas to the Break Theme

[16] has proved that the impact of fracturation sector on the land instability is generally localized at the level of a nearly 100 m thick corridor. This offered us a basis for creating "tampon" areas which are at the level of 50 m and an instability index to the present breaks in our data base.

2.4.2. Applying Tampon Areas to the Hydrographic Factor

In the case of hydrographic network this operation is a bit more complicated because we have attributed a specific distance for each chosen index:

- 25 meters for the index 1 drains
- 50 meters for the index 2 drains
- 75 meters for the index 3 drains

Table 5. Soil types classification according to its contribution to land instability.

Soil exploitation	Land state	Instability index
Urban sites and water plan	stable	1
Agricultural lands and forests	moderately stable	2
Bare areas	unstable	3

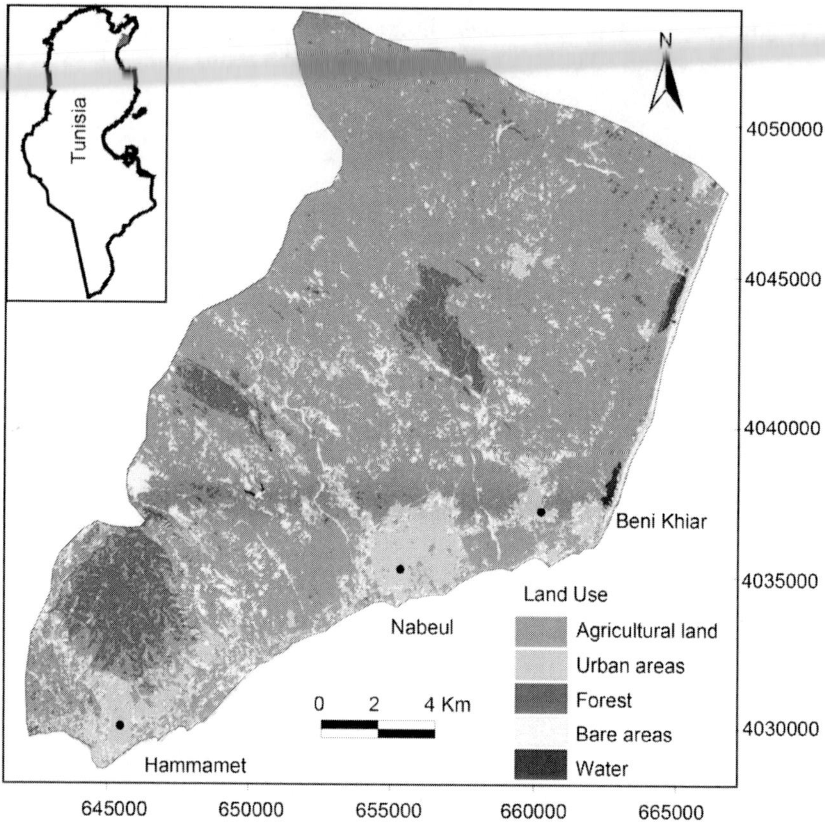

Figure 9. Land use map.

This choice was taken through land observation of the most sensitive areas to land instability around oued Essghir, El Kbir and Abid.

2.4.3. Themes Union

This operation associates the entities of an introductory these with the polygons of a recovering theme to produce a closing theme containing the attributes and the general view of the two themes.

As a result of our choice, we elaborated the synthesis map by working in vector mode by converting maps of the slopes, the soil exploitation, rainfall, seismicity and the tampon areas in vector files.

The first union operation was realized through lithologic and break themes, the result was afterwards combined with the slope theme.

Analogue operations were performed by integrating theme after theme the instability indexes. These were covered through:

- Rainfall,
- seismicity,
- drain classes,
- Soil exploitation.

After each association operation we have paid tribute to the data base to multiply the indexes of instability for the introductory theme by the recovering theme (Table 6).

Table 6. Multiplication of the proportional indexes to the lithologic theme and to the break theme.

Faults index X Lithology index	1	2	3
1	1	2	3
3	3	6	9

The above box shows the use of new indexes produced through multiplying the above indexes in box 8 by the instability indexes linked to the slope factor (Table 7).

Table 7. Second multiplication of instability indexes.

Slope index X Index Litho + Faults	1	2	3	6	9
1	1	2	3	6	9
2	2	4	6	12	18
3	3	6	9	18	27

The final result of these special analysis operations and data combination is illustrated through the instability map of Nabeul-Hammamet (Figure 10).

2.5. Interpreting the Cartographic Maps

The realized land instability map contains 11 indexes representing the state of the land instability. The lowest indexes 4, 6, 12 and 16 cover 87.08% of the total surface of the study area. This indicates that the studied area is generally stable despite the abundance of crumbly lands. The impact of topographic, tectonic, seismic and hydrographic factors have generated very important indexes 24, 36 and 48 which proved the increase of surfaces instability. These indexes are mainly in the North-East of the studied area at the level of the seismic aureole of the index 2 or on the tampon areas where rivers and breaks are present.

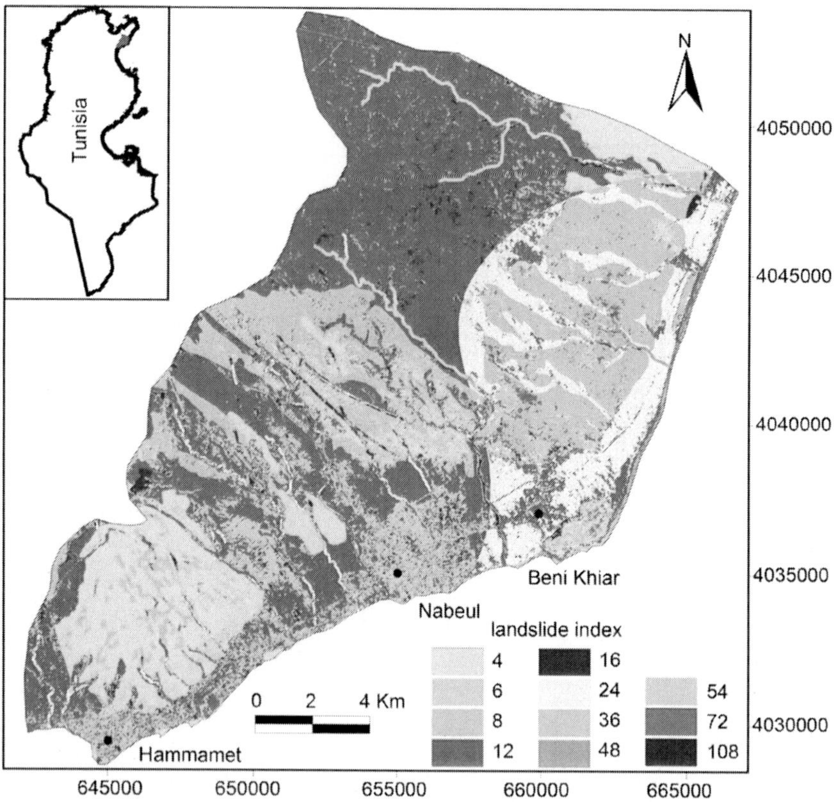

Figure 10. Map of field instability hazard.

The most important indexes which represent the most concerned areas by land instability are only 0.3% of the total surface, but they cover 103 hectares. These indexes are localized on the bank of the wadi El Kbir and El Ain. They cross loose lands which are influenced by seismic factors aureole of the index 2.

The highest values on the map (Figure 10) are manifested on land in different forms:

- Rational landslides that caused the fracture and detachment (Photo 1) and ever the fall (Photo 2) of some defensive walls.
- Some fractures and collapses due to the karst of limestone crusts on the bank of Wadi El Kbir (Photo 3).

Photo 1. Fracturing and detachment of the wall located on the edge El Wadi Ain.

Photo 2. Fall of a portion of the wall at the bridge El Kebir wadi.

Photo 3. Fracturing and collapse of limestone crusts.

2.6. Conclusions

Through this study, a hazard instability land map has been realized. This geo-carto- graphic multisource document is designed to attract the attention of planners to the potential or real risks of some lands in relation with nature and the specificity of each area.

Producing up-to-date and accurate landslide susceptibility maps can ensure safety to people and property at risk and avoid extensive economic loss (Kavzoglu et al., 2013).

The SIG techniques have facilitated the use and management of the data base for this work. The realized map has shown that in addition to the lithologic and topographic factors, land instability might be conditioned by hydrographic and seismic factors.

CITE THIS PAPER

Haddad, R., Labiad, M., Bouzid, M.K., Rabia, R., Makloufi, D. and Rabia, M.C. (2016) Landslide Cartography at the Region of Nabeul-Hammamet Based on Geographic Information System and Geomatic. International Journal of Geo- sciences, 7, 1088-1101.

REFERENCES

1. Ayala, I.A., Chavez, O.E. and Parrot, J.F. (2006) Landsliding Related to Land-Cover Change: A Diachronic Analysis of Hillslope Instability Distribution in the Sierra Norte, Puebla, Mexico. CATENA, 65, 152-165.
2. Khan, Y.A., Lateh, H., Baten, M.A. and Kamil, A.A. (2012) Critical Antecedent Rainfall Conditions for Shallow Landslides in Chittagong City of Bangladesh. Environmental Earth Sciences, 67, 97-106.
3. Ahmed, B. and Rubel, Y.A. (2013) Understanding the Issues Involved in Urban Landslide Vulnerability in Chittagong Metropolitan Area, Bangladesh. The Association of American Geographers (AAG), Washington DC. http://discovery.ucl.ac.uk/1418959/
4. Rouiller, J.-D., Jaboyedoff, M., Marro, C., Phlippossian, F. and Mamin, M. (1998) Pentes instables dans le Pennique valaisan. Rapport final PNR31. VDF, Zürich, 239 p.
5. Borgatti, L. and Soldati, M. (2010) Landslides as a Geomorphological Proxy for Climate Change: A Record from the Dolomites (northern Italy). Geomorphology, 120, 56-64.
6. Ahmed, B. (2014) Landslide Susceptibility Mapping Using Multi-Criteria Evaluation Techniques in Chittagong Metropolitan Area, Bangladesh. Landslides, 12, 1077-1095.
7. Bensalem, H. (1992) Contribution à la connaissance de la géologie du Cap Bon: Stratigraphie, tectonique et sédimentologie. Thèse de 3ème cycle, Fac. des Sciences de Tunis.
8. Colleuil, B. (1976) Etude stratigraphique et néotectonique des formations néogènes et quaternaire de région de Nabeul-Hammamet (Cap Bon Tunisie). Mémoire de D.E.S. Univ., Nice.
9. El Euch, N. (2002) Signature tectonique et eustatique des séries du Miocène moyen du Cap Bon. D.E.A. Univ. De Tunis II. Fac. Sc. De Tunis.
10. Paskoff, R. and Sanlaville, P. (1983) Les c?tes de la Tunisie. Variation du niveau marin depuis le Tyrrhénien. Coll. Maison de l'Orient Médit. de Lyon, n 14. série géogr.et préhist.
11. Philipponnat, G. and Hubert, B. (1997) Fondations et ouvrages en terre. http://www.eyrolles.com/BTP/Livre/fondations-et-ouvrages-en-terre-9782212072181
12. Hoek, E. and Bray, J.W. (1981) Rock Slope Engineering. Revised 3rd Edition, The Institution of Mining and Metallurgy, London, 341-351.
13. Jeannin, M. (2006) Etude des processus d'instabilités des versants rocheux par prospection géophysique. Apport du radar géologique. Thèse de 3ème cycle de l'Université Joseph Fourier, Grenoble.
14. Robitaille, V. and Tremblay, D. (1997) Mécanique des sols théorie et pratique. Modulo ed., Québec.
15. Sgahaier, A. (2003) Analyse morphostructurale de la région de Mehdia-El Jem-Jebeniana (Tunisie orientale). Mémoire de DEA. Fac. Sci. Tunis.
16. Guillande, R., Gelugne, P., Bardintzeff, J., Brousse, R., Chorowicz, J., Deffontaine, B. and Parrot, J. (1993) Cartographie automatique de zones à aléas de mouvements de terrain sur l'?le de Tahiti à partir de données digitales. Bull. Soc. géol. Fr., n 4.

CHAPTER 10

Converting Tsunami Wave Heights to Earthquake Magnitudes

Nils-Axel Mörner

Paleogeophysics & Geodynamics, Stockholm, Sweden

ABSTRACT

There is a fairly strict relation between maximum tsunami wave heights and causation earthquake magnitudes. This provides a new tool for estimating the magnitude of past earthquakes from the observed wave heights of related paleo-tsunami events. The method is subjected to a test versus two paleoseismic events with multiple independent estimates of corresponding earthquake magnitude. The agreement to the tsunami wave height conversion is good, confirming very high magnitudes of M 8.5 - 9.0 and M 8.4 - 8.5. Applying the same method to two Late Holocene events of methane venting tectonics indicates a ground shaking of forces equivalent to a M 8.0 earthquake, seriously changing previous long-term crustal hazard assessments.

Keywords: Tsunamis, Wave Height, Earthquakes, Magnitudes, Paleo-Tsunamis, Methane Venting Tectonics, Hazard Assessment

1. INTRODUCTION

Major tsunami events are primarily generated by submarine earthquakes [1] [2] . In principle, there is a relation between earthquake magnitude and tsunami wave height. During the last 13 years, there has been eight high-amplitude tsunami wave events |3]; viz. 1) the 2004 Indian Ocean event with a tsunami wave height of 20 m (a statement of a 30 m height is considered to represent a run-up height, not an actual wave height) and a Mw 9.1 earthquake magnitude, 2) the Java 2006 event with a wave height of 8.6 m and a magnitude of Mw 7.7, 3) the Benkula 2007 event with a wave height of 1.65 m and a magnitude of Mw 8.5, 4) the Peru 2007 event with a wave height of 3 - 4 m and a magnitude of Mw 8.0, 5) the Samoa 2009 event with a wave height of 11.9 m and a magnitude of Mw 8.1, 6) the Mentawai 2010 event with a wave height of 10 m and a magnitude of Mw 7.7, 7) the Chile 2010 event with a wave height of 17.2 m and a magnitude of Mw 8.8 and 8) the Tihoku-oki 2011 event with a wave height of 19.5 m and a magnitude of Mw 9.0. Because these 8 events are known both to tsunami wave height and causation earthquake magnitude, they can be used to establish the relationship between tsunami wave height and seismic magnitude [3] [4] . This is illustrated in Figure 1 (modified from [4]). Events 3 and 4 obviously never reached a maximum wave height, and are excluded in establishing the red line relation. Value 1a is superseded by the better 1b value.

2. PALEO-TSUNAMIS

Paleo-tsunamis can rarely be evaluated with respect to earthquake magnitudes [5] [6] [7] [8] . Having established the tsunami height/seismic magnitude relation in Figure 1, we now have a new tool for estimating the corresponding earthquake magnitudes [4] .

Even the reverse may apply; i.e. knowing the seismic magnitude of a paleoseismic event, one may estimate the maximum tsunami height. This is the case with the Crete AD 365 paleoseismic event; its magnitude has been estimated to "at least 8.5" [6] or "8.3 - 8.5" [7] . It set up a major tsunami, which hit and destroyed the Library at Alexandria, destroyed 50,000 homes, and killed about 5000 persons [7] . It also left a "mega-turbidite" [7] . The tsunami height is less well known. From Figure 1, it is easy to read that an Mw 8.5 earthquake may set up a tsunami with a maximum wave height of about 15 m. This seems to fit well with observed records [7] .

Today Sweden is an area of low to moderately low seismic activity. Due to the very high rate of glacial isostatic uplift at the time of deglaciation it was,

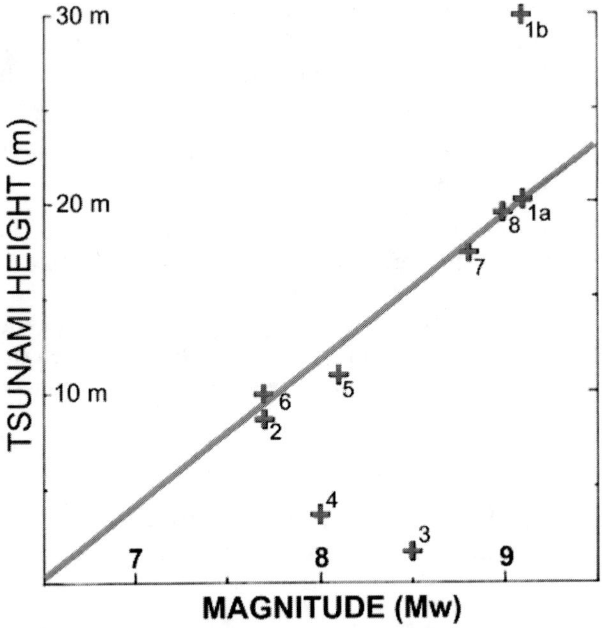

Figure 1. Relation between observed maximum tsunami heights and magnitudes of causational earthquakes (events 1 - 8). Having established this relation, observed tsunami heights of paleoseismic events can be converted to corresponding earthquake magnitudes [4] . The red line gives a ratio of 0.133 Mw per 1.0 m tsunami wave height.

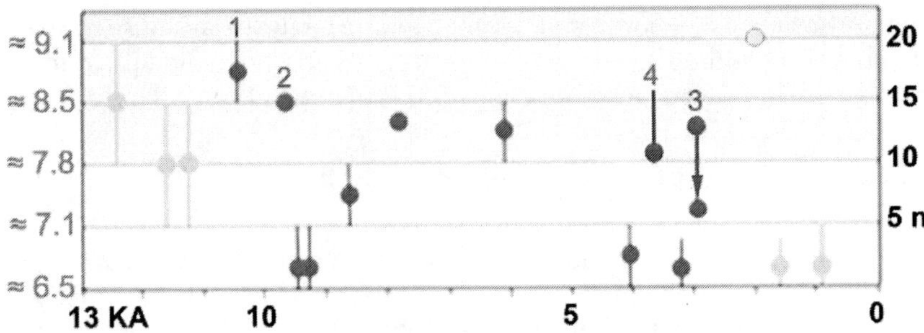

Figure 2. The Swedish paleoseismic database [10] [11] includes 17 tsunami events [4] [14] ; 5 in the Kattegatt Sea (green) and 12 in the Baltic (blue) plotted chronologically with respect to observed wave heights. Purple figures to the left give corresponding earthquake magnitudes as read from Figure 1 relation (modified from [4]). The four events further discussed in section 3 are marked in red (1 - 4).

at that time, an area of very high paleoseismic activity in frequency as well as in magnitude [8] - [13] . In total, 62 paleoseismic events have been documented [13] , out of which 17 events generated tsunamis [5] [11] [12] [13] [14] .

In Figure 2, the Swedish database of tsunami events and wave heights is converted to magnitudes using Figure 1 relations.

Figure 2 demonstrates that the corresponding paleoseismic event must have been of considerable magnitudes (viz. 7 between M 6.5 - 7.5, 6 between M 7.6 - 8.4, and 3 of M > 8.4). In some of the events (1 and 2 in Figure 2), there are independent magnitude estimates from liquefaction, fault movements and bedrock deformation [8] [10] [11] , which can be used for testing the relations among wave heights and magnitudes as given in Figure 1.

3. APPLICATION ON SOME SWEDISH PALEO-TSUNAMI EVENTS

The first Paleoseismic Catalogue of Sweden [10] included 52 paleoseismic events, and the second Paleoseismic Catalogue 62 events [11] . All events entering the catalogues were documented by multiple criteria [8] [10] [11] and well dated; often by varve chronology with an annual resolution [10] [13] (varve ages are assigned vBP, for distinction to conventional C14-dates in BP). All the events in the catalogues were assigned a magnitude estimate [10] [11] , based on a number of different criteria [5] .

Two of the Swedish paleoseismic events will be analyzed as a test of Figure 1 tsunami/magnitude relations in comparison with independent estimates based on other criteria presented in [5] [8] [10] and [11] .

3.1. The 10,430 vBP Paleoseismic Event

Stockholm is traversed by an old Permian fault that extends in west-east direction for about 400 km, and which may continue into the Bay of Finland and Lake Ladoga for another 300 - 400 km. The fault was reactivated in deglacial time some 10,500 - 10,400 vBP. The frequency of paleoseismic events was very high with 7 independent events recorded and dated with in 102 varve years from 10,490 to 10,388 vBP [8] .

In varve year 10,430 vBP a giant earthquake occurred [10] . A lateral to sympathetic fault located 1 km north of the main fault was displaced by 6 - 8 m, indicating a very high magnitude of the causation earthquake. Heavy bedrock fracturing is documented over an area of 50 × 100 km. Liquefaction has been recorded over an area of 320 km, which is indicative of an earthquake magnitude of about M 9.1 [5] [10] [11] . Liquefaction of gravel is another indicator of a very high magnitude event. An intra-varve turbidite is recorded over an area of 200 × 320 km. Magnetic grain rotation is recorded over an area of 500 × 600 km [10] [15] . At present we have no means of converting this value into a magnitude estimate. One thing is clear, however, the magnitude must have been very high, i.e. ? 8 [5] [8] .

This event also set up a gigantic tsunami event, which invaded several lake basins and washed the strait across southern Sweden (the so-called Närke Straight) free of pack-ice and ice-bergs so that entire Baltic became marine (the Yoldia Stage) within one year [10] [12] [16] . The tsunami wave must have had a wave height of 15 - 20 m.

In Figure 3, we compare the magnitude estimated from the observed tsunami wave height, with the estimates obtained previously on the bases of other criteria [5] [8] [10] [11] .

In principle, there is a very good agreement between all the different and independent means of estimating seismic magnitude. No doubt, the 10,430 vBP paleoseismic event was a very strong event of a magnitude of about M 8.5 - 9.0, which implies that it was a "mega-earthquake". It also means that the test of the tsunami vs. magnitude relation of Figure 1 seems to work very well.

3.2. The 9663 vBP Paleoseismic Event

This paleoseismic event is known as the Hudiksvall or Iggesund paleoseismic event [10] . It occurred in varve 9663 vBP (or 9150 C14-years BP). It is one of the ever best investigated paleoseismic events [10] , documented by primary fault, bedrock deformations over an area of 50 × 50 km (with 49 sites investigated in details), recorded and dated in 31 varve sequences with a distinct turbidite extending for 320 km along the coast, recorded by liquefaction in 15 sited over an area of 40 × 80 km, and documented as a tsunami event in 14 lakes (and 43 sediment cores) covering an area of 30 × 125 km. At two sites 35 km apart, the liquefaction event is composed of 5 different phases, interpreted as shock and after-shock signals, which calls for a very strong event [5] [8] [10] [11] .

The tsunami wave height can be very closely fixed at a minimum of 15 m [10] [12] [14] , providing a magnitude of about M 8.5 according to Figure 1 relation [5] .

The various independent means of estimating seismic magnitude are compared in Figure 4 to test the new tsunami vs. magnitude relation, and to provide a combined estimate of the corresponding magnitude.

Even for this event, there is a reasonably good agreement between all the different and independent means of estimating seismic magnitude. No doubt, the 9663 vBP paleoseismic event was a very strong event of a magnitude of about M 8.4 - 8.5. The test of the tsunami vs. magnitude relation of Figure 1 seems to work very well.

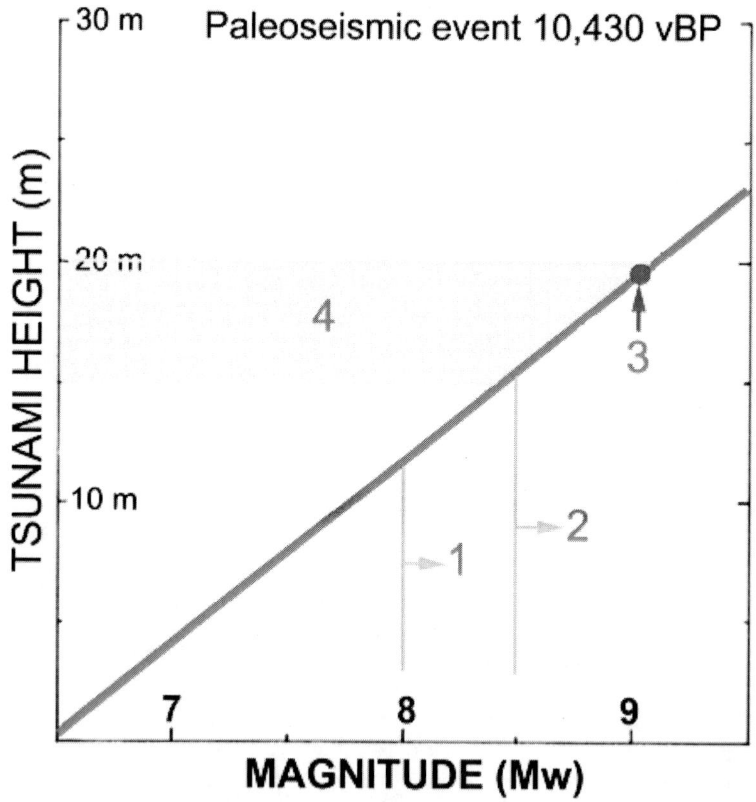

Figure 3. Comparision between previous magnitude estimates (1 - 3) and the present one (4) obtained via the tsunami wave height and earthquake magnitude relation in Figure 1. Group 1 refers to M > 8 suggested by liquefaction of gravel, fracturing opening of 10 - 20 cm, and turbidite spread. Group 2 refers to M ? 8 suggested by 6 - 8 m lateral fault displacement 1 km away from the main fault, spatial distribution of bedrock fracturing, seismic recurrence frequence, and rotation of magnetic grains over an immense area. Point 3 refers to the magnitude (M ~ 9.0) obtained from the spatial distribution of liquefaction. Point 4 represents the magnitude (M 8.5 - 9.0) converted from the tsunami height.

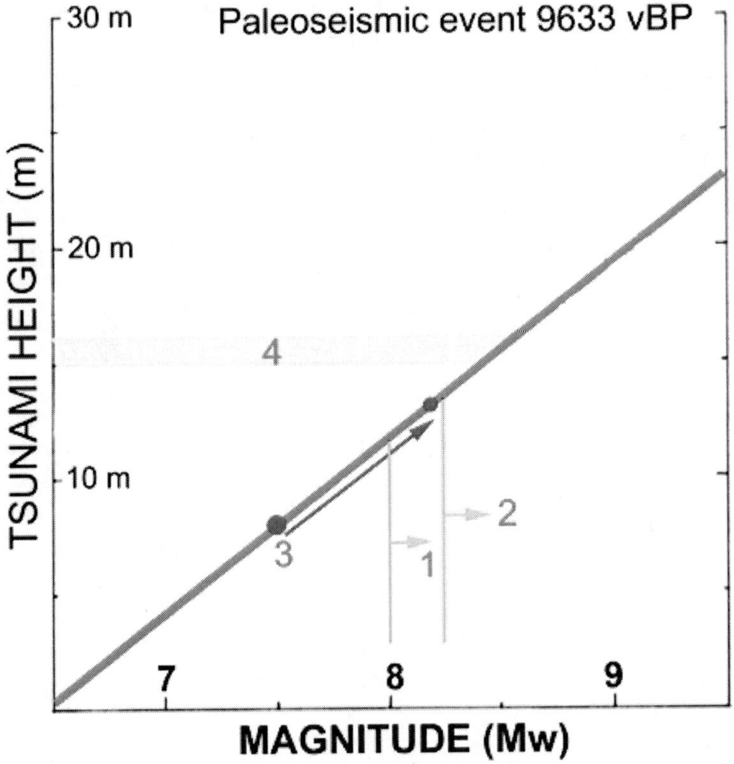

Figure 4. Comparison between previous magnitude estimates (1 - 3) and the present one (4) obtained via the tsunami wave height and earthquake magnitude relation in Figure 1. Group 1 refers to M > 8 suggested primary fault displacement, mode of bedrock deformations, slide volumes, spatial distribution of turbidities, liquefaction of gravel, and methane venting. Group 2 refers to M ? 8 suggested by spatial distribution of bedrock deformation, and mode of liquefaction in 5 phases. Point 3 refers to spatial distribution of liquefaction. Point 4 represents the magnitude (M 8.5 - 8.6) converted from the tsunami height.

3.3. The 2900 BP Methane Venting Tectonics Event

Methane venting tectonics is a novel factor, which it has taken time to evolve from the first idea [17] [18] [19] to observational confirmation [10] and finally to full presentation [20] [21] , summarized in [22] . It implies the sudden phase transition from methane hydrate stored in voids and fractures in the bedrock to methane gas venting explosively to the surface, by that causing severe bedrock deformation.

This event set up a major tsunami, recorded in nearby bogs and lakes [10] [12] . It was later found to be coincidental with a tsunami event recorded 100 km to the south [4] [14] with an age of 2900 C14-years BP. At the nearby sites the tsunami wave must have had a height of at least 12 m, and at southern sites a height of at least 6 m. At both sites, the wave had a submarine trimming depth of at least 18 m.

The bedrock is severely deformed in a huge cone of angular block torn out of the bedrock beneath. The cone is 20 - 25 m high and 100 × 150 m wide (i.e. elliptic). It is surrounded by a depression (compensating the rock masses of the cone). At the top of the cone, there are gigantic blocks of 10 m to >10 m diameters.

Methane venting tectonics is, of course, something quite different from earthquake deformational magnitudes. The deformation was so violent, however, that it was compared to a magnitude M ~7 earthquake event [10] [11] .

In Figure 5 the original magnitude estimate is compared to the magnitude es timate according to Figure 1 relation between tsunami height and earthquake magnitude. It reveals that the 2900 BP methane venting event set up a ground shaking comparable to a magnitude M 8.0 earthquake. This is, of course, quite remarkable because it seems to verify that methane venting tectonics imply ground deformations comparable to very high magnitude earthquakes.

3.4. The 3000 - 4000 BP Methane Venting Tectonic Event

A major methane venting tectonics event is recorded south of Stockholm [20] . It occurred sometime between 4000 and 3000 BP. The deformational structures indicate that the event must have been quite violent [20] [21] ; a 25 m high cone of 150 × 230 m width, and with gigantic blocks at the top. It seems that a tsunami event with a run-up of about 11 - 21 m can be associated with this event [10] [20] . The field data referring to the tsunami need to be revisited and checked before a serious magnitude estimate can be done. If an 11 m height is taken as wave height, we would be dealing with a ground-shaking magnitude of M 7.9 by applying Figure 1relations. The situation is quite similar to that of the 2900 BP event (Figure 5), in basic structure as well as in tsunami height.

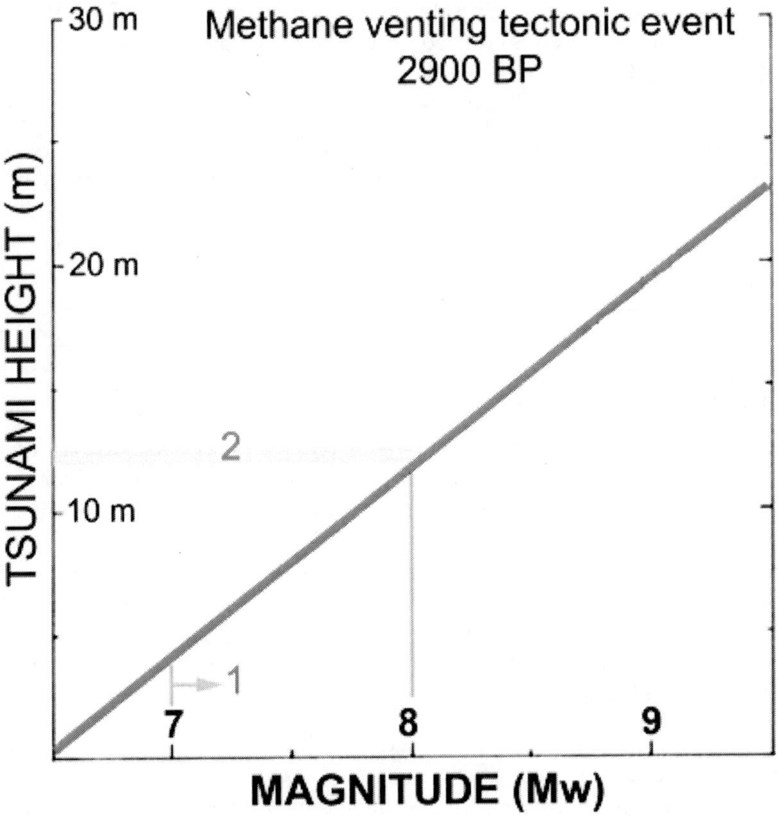

Figure 5. Comparing estimated earthquake magnitude equivalent (1) and observed tsunami wave height (2) of the methane venting tectonic event at 2900 BP [10] [14] [20] , indicating that the ground shaking associated with methane venting tectonics may reach very high magnitudes comparable to M 8.0 earthquake magnitudes. A quite similar situation applies to the 3000 - 4000 BP methane venting tectonic event documented south of Stockholm [20] [21] .

4. DISCUSSION

Figure 1 relations between tsunami wave height and earthquake magnitude [3] [4] implies an improved and simplified tool of converting observed tsunami wave height into corresponding earthquake magnitude as compared to preceding graphic relations [23] .

Sweden has a database of 17 postglacial tsunami events [10] [12] [14] . The wave heights documented is consistent with a very high seismicity, not only in Late Glacial time, but also in Mid-Holocene and Late Holocene time (Figure 1).

Two paleoseismic events were selected for a test of the significance of Figure 1 relations established; viz. the 10,430 vBP and the 9663 vBP events, because both of these event were established by means of a multiple criteria [10] [11] , besides they both represent very strong events with well-established tsunami heights. This implies the comparison between multiple independent parameters. The data referring to the 10,430 vBP event are consistent with a mega-event of a magnitude of about M 8.5 - 9.0 (Figure 3) and the data referring to the 9663 vBP event to an event of magnitude of M 8.4 - 8.5 (Figure 4).

In the Late Holocene, there were two major tsunami events (Figure 1). Both of those events were generated by methane venting tectonics, however [10] [20] . Hence, there is no straightforward application of Figure 1 relations. In order to obtain some sort of quantification of the forces involved and magnitude of the ground shaking, Figure 5 was drawn. The tsunami height of the 2900 BP event is consistent to a magnitude (rather magnitude equivalent) ground shaking of M 8.0, which seems to harmonize with the violence of the structures observed (Figure 5). The 3000 - 4000 BP event from the Stockholm area give a similar picture.

The process of methane venting tectonics [20] has by this (Figure 5) obtained a first serious quantification as to corresponding ground shaking, and both events were found to be consistent with a magnitude M 8.0 equivalent. This makes it a very dangerous factor for long-term stability and must be considered seriously in hazard assessment [5] [11] .

5. CONCLUSIONS

A new tool for the conversion of observed tsunami heights to corresponding causation earthquake magnitude is presented (Figure 1).

It is quite successfully tested on multiple independent magnitude estimates from two paleoseismic events in Sweden (Figure 3, Figure 4). The general agreement is very good.

Methane venting tectonics is a novel process of severe bedrock deformation [10] [20] . Big tsunami events were generated at an event occurring 2900 BP (the Skålboberget event in central Sweden) and at 3000 - 4000 BP (the Kvarnberget event in the Stockholm area). The tsunami heights documented correspond to an earthquake equivalent magnitude of about M

8.0 (Figure 5), indicating that methane venting tectonics implies violent deformation and ground shaking equivalent to high-magnitude seismic events.

CITE THIS PAPER

Mörner, N.-A. (2017) Converting Tsunami Wave Heights to Earthquake Magnitudes. Open Journal of Earthquake Research, 6, 89-97. https://doi.org/10.4236/ojer.2017.62005

REFERENCES

1. Mörner, N.-A. (2013) Tsunamis and Tsunamites: Origin and Characteristics. 4th International INQUA Meeting on Paleoseismology, Active Tectonics and Archeoseismology, 9-14 October 2013, Aachen, 165-168.
2. Mörner, N.-A. (2018) Tsunami Deposits. In: Finkle, C., Ed., Encyclopedia of Coastal Science, Springer, Dordrecht. (In Printing)
3. Lario, J., Bardaj, T., Silva, P.G., Zazo, C. and Goy, J.L. (2016) Improving the Coastal Record of Tsunamis in the ESI-07 Scale: Tsunami Environmental Effects Scale (TEE-16 Scale). Geologica Acta, 14, 179-193.
4. Mörner, N.-A. (2016) Seismic Hazard Assessment: A Challenge for Science and Geoethics. International Journal of Earthquake Engineering and Hazard Mitigation, 4, 64-70.
5. Fonseca, J.D. (2005) 1755 o terramoto de Lisboa—The Lisbon Earthquake. Argumentum. [Citation Time(s):10]
6. Stiros, S.C. (2010) The 8.5+ Magnitude, AD 365 Earthquake in Crete: Coastal Uplift, Topographic Changes, Archaeological and Historical Signature. Quaternary International, 216, 54-63.
7. Polonia, A., Bonatti, E., Camerlenghi, A., Lucchi, R.G., Panieri, G. and Gasperini, L. (2013) Mediterranean Megaturbidite Triggered by the AD 365 Crete Earthquake and Tsunami. Scientific Report, 3, Article No. 1285.
8. Mörner, N.-A. (2011) Paleoseismology: The Application of Multiple Parameters in Four Case Studies in Sweden. Quaternary Research, 242, 65-72.
9. Mörner, N.-A. (1991) Intense Earthquakes as a Function of Glacial Isostasy. Tectonophysics, 188, 407-410.
10. Mörner, N.-A. (2003) Paleoseismicity of Sweden—A Novel Paradigm. 16th International INWQUA Congress, Reno, 1-320. [Citation Time(s):25]
11. Mörner, N.-A. (2013) Patterns in Seismology and Palaeoseismology and Their Application in Long-Term Hazard Assessment—The Swedish Case in View of Nuclear Waste Management. Pattern Recognition in Physics, 1, 75-89.
12. Mörner, N.-A. and Dawson, S. (2011) Traces of Tsunami Events in Off- and on-Shore Environments. Case Studies in the Maldives, Scotland and Sweden. In: Mörner, N.-A., Ed., The Tsunami Threat: Research and Technology, InTech, Rijeka, 371-388.
13. Mörner, N.-A. (2013) Drainage Varves, Seismites and Tsunamites in the Swedish Varve Chronology. GFF, 135, 308-315.
14. Mörner, N.-A. (2016) Tsunamis in Sweden: Occurrence and Characteristics. In: Mokhtari, M., Ed., Tsunami, Chapter 6, InTech, Rijeka, 115-133.

15. Mörner, N.-A. and Sun, G. (2008) Paleoearthquake Deformations Recorded by Magnetic Variables. Earth Planetary Science Letters, 267, 495-503.

16. Mörner, N.-A. (1995) The Baltic Ice Lake—Yoldia Sea Transition. Quaternary International, 27, 95-98.

17. Björklund, A. (1990) Methane Venting as a Possible Mechanism for Glacial Plucking and Fragmentation. GFF, 112, 329-333.

18. Mörner, N.-A. (1993) Methane Dehydration Tectonics. Bulletin of the INQUA Neotectonic Commission, 16, 71-72.

19. Sjöberg, R. (1994) Bedrock Caves and Fractured Rock Surfaces in Sweden: Occurrence and Origin. PhD Thesis, Stockholm University, Stockholm, Vol. 7, 1-110.

20. Mörner, N.-A. (2017) Methane Hydrate in Crystalline Bedrock and Explosive Methane Venting Tectonics. Earth Science Review. (In Press) [Citation Time(s):9]

21. Mörner, N.-A. (2017) Supplementary Material to: Methane Hydrate in Crystalline Bedrock and Explosive Methane Venting Tectonics. Earth Science Review. (In Press) [Citation Time(s):3]

22. Mörner, N.-A. (2016) Methane Venting and Methane Venting Tectonics. 13rd International Conference on Gas in Marine Sediments, Tromso, 19-22 September 2016. https://www.researchgate.net/publication/308306076

23. Silva, P.G., Rodriguez-Pascual, M.A., Perez-Lopez, R., Bardaj, T., Lario, J., Alfaro, P., Martinez-Diaz, J.J., Reicherter, K., Gimenez, J., Giner, J., Azanon, J.M., Goy, J.L. and Zazo, C. (2007) Catalogacion de los efectosgeologicos y ambientales de los terremotos en Espana en la Escala ESI. 2007 y suaplicacion a los studio paleoseismlogicos. Geo-Temas, 10, 1063-1066.

CHAPTER11

The Calatabiano Landslide (Southern Italy): Preliminary GB-Insar Monitoring Data and Remote 3D Mapping

Luca Lombardi[1], Massimiliano Nocentini[1], William Frodella[1], Teresa Nolesini[1], Federica Bardi[1], Emanuele Intrieri[1], Tommaso Carlà[12], Lorenzo Solari[12], Giulia Dotta[12], Federica Ferrigno[1], Nicola Casagli[1]

[1]*Department of Earth SciencesUniversity of FirenzeFlorenceItaly*

[2]*Regional Doctoral School of Earth SciencesUniversity of FirenzeFlorenceItaly*

ABSTRACT

On 24 October 2015, following a period of heavy rainfall, a landslide occurred in the Calatabiano Municipality (Sicily Island, Southern Italy), causing the rupture of a water pipeline supplying water to the city of Messina. Following this event, approximately 250,000 inhabitants of the city suffered critical water shortages for several days. Consequently, on 6 November 2015, a state of emergency was declared (O.C.D.P. 295/2015) by the National Italian Department of Civil Protection (DPC). During the emergency management phase, a provisional by-pass, consisting of three 350-m long pipes passing through the landslide area, was constructed to restore water to the city. Furthermore, on 11 November 2015, a landslide remote-sensing monitoring system was installed with the following purposes: (i) analyse the landslide geomorphological and kinematic features in order to assess the residual landslide risk and (ii) support the early warning procedures needed to ensure the safety of the personnel involved in the by-pass construction and the landslide stabilization works.

The monitoring system was based on the combined use of Ground-Based Interferometric Synthetic Aperture Radar (GB-InSAR) and terrestrial laser scanning (TLS). In this work, the preliminary results of the monitoring activities and a remote 3D map of the landslide area are presented.

Keywords: GB-InSAR Laser scanning Landslides Emergency management 3D mapping

INTRODUCTION

Landslides represent one of the most frequent geo-hazards. They represent serious threats to human life and can cause serious socioeconomic losses on the order of billions of Euros, in terms of damage to property, infrastructure and environmental degradation (Kjekstad and Highland 2009; Petley 2012). Recent events show a significant increase in the number of disasters with natural and/or technological causes, and these disasters can have potentially serious consequences for critical infrastructure (CI) (Murray and Grubesic 2007). Where this infrastructure tends to fail or to be destroyed, the resulting cascade effects (chain of events) can lead to catastrophic damage and negative effects on people, the environment and the economy (Geertsema et al. 2009; Kadri et al. 2014). In countries characterized by a geologically young and tectonically active territory, the susceptibility to instability phenomena is further increased. In Italy in particular, approximately 70 % of the country can be considered exposed to landslide risk (http://www.protezionecivile.gov.it/) and the related estimated socioeconomic losses range from 1 to 2 billion dollars annually (Canuti et al. 2004). Currently, these values are certainly underestimated.

In the field of landslide detection, mapping, monitoring and management, the availability of advanced remote sensing technologies, which allow the systematic and easily updatable acquisition of data, may enhance the implementation of near-real-time monitoring activity and the production of landslide maps, optimizing field work (Guzzetti et al. 2012; Frodella et al. 2014). In recent years, the *Ground-Based Interferometric Synthetic Aperture Radar* (GB-InSAR) technique has been widely used to monitor ground displacements in research on landslides (Tarchi et al., 2003; Herrera et al. 2009; Barla et al. 2010; Schulz et al. 2012; Corsini et al. 2013; Bardi et al. 2014), volcanoes (Bozzano et al. 2011; Nolesini et al., 2013; Di Traglia et al. 2014) and sinkholes (Intrieri et al. 2015). *Terrestrial laser scanning* (TLS) has become a widely applied technique for the detection and characterization of several types of mass movements as it allows the rapid collection of detailed and highly accurate 3D ground representations (Abellan et al. 2006; Oppikofer et al. 2008; Jaboyedoff et al. 2009; Fanti et al. 2012; Gigli et al. 2014). The intrinsic characteristics of the abovementioned techniques, such as (i) producing near-real-time displacement maps without physical access to the analysed area; (ii) observing the investigated scenario 24 h per day and in all weather conditions; (iii) generating high-resolution images, especially for analysis of local-scale phenomena; and (iv) providing high versatility and transportability, represent consistent advantages with respect to traditional methods.

This work presents an example of the advantages provided by the integrated use of GB-InSAR and TLS to monitor and manage the post-emergency phase associated with a landslide located in the Catania Province (Sicily Island, Southern Italy). The landslide occurred on 24 October 2015 following intense rainfall that occurred during the previous days (a total of approximately 195 mm measured between 21 and 24 October 2015; data provided by Osservatorio delle Acque—Regione Siciliana) and ruptured a water pipeline that was part of the Messina city aqueduct. Consequently, a considerable lack of water occurred for a large number of the city inhabitants. A provisional by-pass, consisting of three 350-m long pipes passing through the landslide area, was implemented to restore water to the city during the emergency management phase. An integrated monitoring network was also implemented to assess the residual risk by analysing the geomorphological and kinematic features of the landslide and to support the early warning procedures needed to ensure the safety of the personnel involved in the by-pass construction and the long-term landslide stabilization works. The workflow in Fig. 1 explains the organization of the performed activities.

GEOLOGICAL AND GEOMORPHOLOGICAL SETTING

The study area is located in the Calatabiano municipality (Catania Province, Sicily Island), approximately 50 km southwest of the city of Messina and approximately 3 km inland from the Ionian coastline (Fig. 2). From a geomorphologic perspective, the study area is located in the Alcantara River valley, which is delimited to the north by the Peloritani Mountains, to the west by the Nebrodi Mountains and to the south by the north-eastern slopes of the Etna Volcano (Fig. 2).

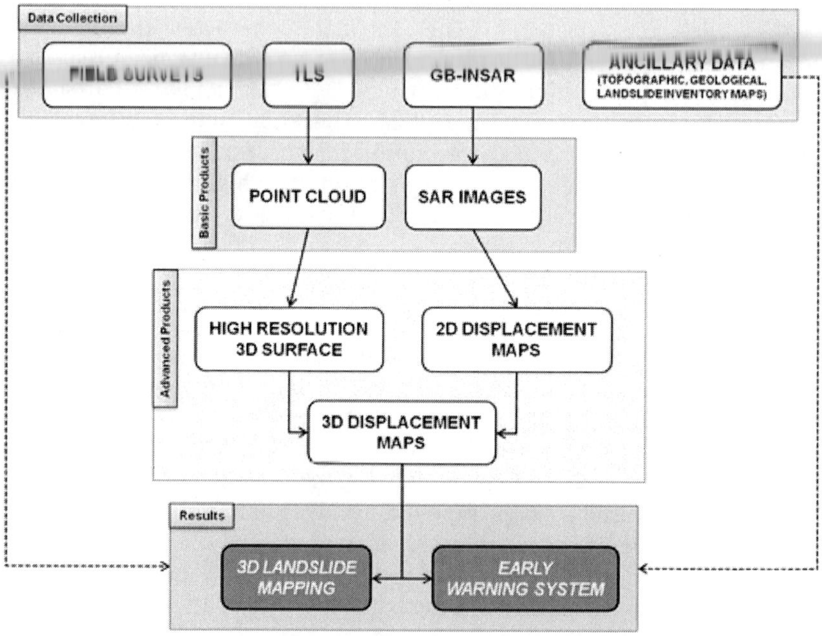

Figure 1. Logical scheme of the applied operative procedure

The study area is located on the right side of the lowermost Alcantara river alluvial plain and is characterized by hilly terrain ranging from approximately 60 to 250 m a.s.l. The landslide (lat. 37° 49′ 39″; long. 15° 13′ 26″), which featured a 100-m elevation range and was approximately 110 m in maximum length and 65 m in width, affected a large portion of the east-facing slope overlooking the medieval Calatabiano Castle (Figs. 3 and 4). Its topmost sector is characterized by a wide crown area, where a rotational movement (Cruden and Varnes 1996) occurred, displaying a retrogressive behaviour and almost reached the slope ridge at approximately 200 m a.s.l. The landslide body and toe are affected by minor scarps, while two erosional channels, formed by mud flows that almost reached the creek valley at approximately 100 m a.s.l., border the right and left flanks of the landslide (Figs. 3 and 4). The creek valley is located just a few hundred metres from the northern sector of the inhabited area of Calatabiano and represents a high risk for the local population in case of possible landslide reactivation. The landslide-affected slope is characterized by olive trees, shrubs and rocky scattered outcrops, and field surveys have revealed that the mass movement involved detrital slope deposits a few metres thick and consisting of heterogeneous clasts in a coarse sandy matrix with a small percentage of loam (Fig. 4).

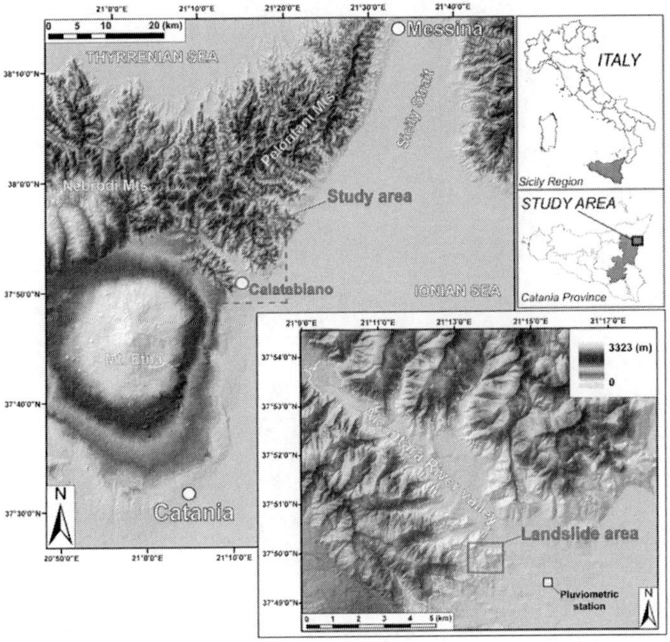

Figure 2. Study area setting and landslide area location, including the rain gauge station

Figure 3. a Image of the Calatabiano landslide acquired on 10 November 2015, from the monitoring system installation point. The *dashed white line* represents the main landslide crown, the *dashed red line* represents the interrupted pipeline (**b**), and the *black line* represents the new pipeline by-pass (**c**). P1–P8 represent the GB-InSAR monitoring points

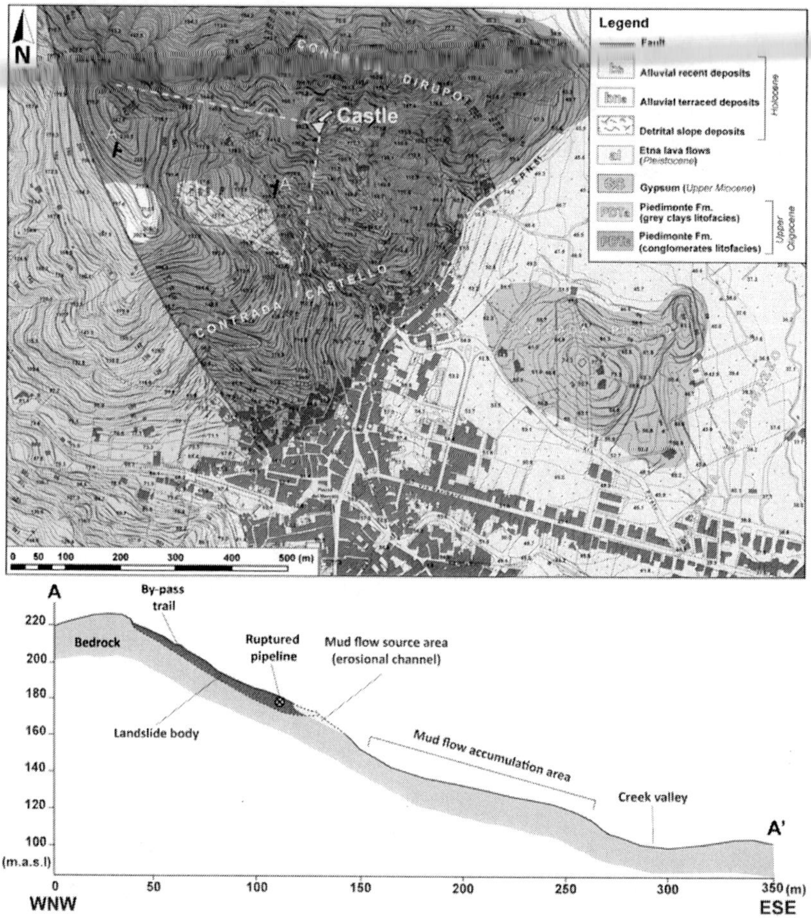

Figure 4. Geological framework of the landslide area and schematic cross section (*red oval*highlights the landslide area)

These deposits were formed by the weathering and erosion of the bedrock, represented by the arenaceous-conglomeratic facies (PDT_c) of the Piedimonte Formation (Upper Eocene-Lower Oligocene) (Catalano et al. 2010). This lithofacies is composed of irregular alternations of conglomerates and coarse sandstones. The conglomerate layers present thicknesses of up to 10 m, while the sandstone strata rarely exhibit thicknesses greater than 1 m (Catalano et al. 2010). In the upper part of the slope (Fig. 4), a tectonic thrust contact that erased the original heteropic succession of the two lithofacies represents the transition to the grey clays lithofacies (PDT_a) of the Piedimonte Formation (Cassola et al. 1991).

EMPLOYED REMOTE SENSING MONITORING TECHNIQUES

GB-InSAR monitoring

The employed radar system is composed of a coherent microwave transceiver unit operating on the Ku band with a bandwidth of 200 MHz and a central frequency of 17.2 GHz. Synthetic aperture is achieved by moving a motorized sled hosting the radar head along a 3-m long straight rail (parallel to the azimuth direction; Fig. 5a). The working principle of the GB-InSAR technique involves radiating microwaves toward the investigated area and measuring the backscattered signal, obtaining a SAR image. The SAR image is created by combining the spatial resolution along the direction perpendicular to the rail (range resolution; ΔRr) and the one parallel to the synthetic aperture (azimuth, or cross-range resolution; ΔRaz) (Luzi 2010), containing amplitude and phase information of the backscattered echo from the investigated scenario objects.

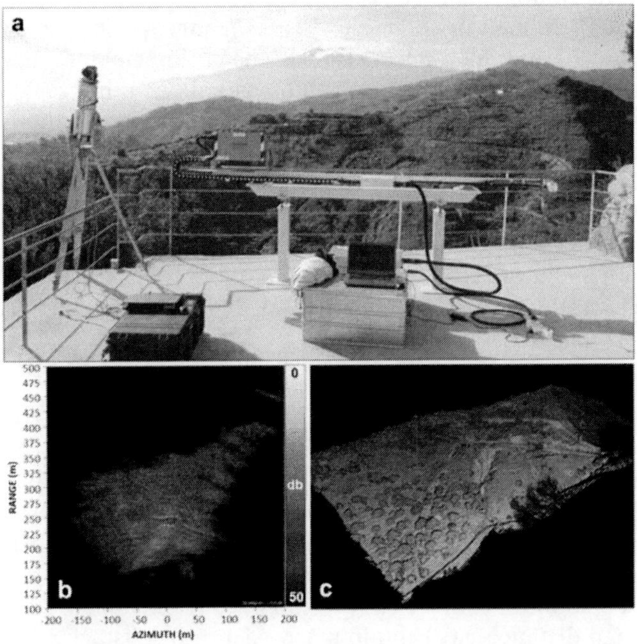

Figure 5. a Ground-based radar and TLS positioned on the roof of Calatabiano Castle; **b** power image of the GB-InSAR surveyed slope; **c** point cloud coloured with intensity values of the reflected laser beam.

By evaluating the phase difference, pixel by pixel, between two pairs of averaged sequential SAR images of the same scenario, it is possible to obtain a 2D displacement map of the investigated area, which constitutes an interferogram (Luzi et al. 2004; Monserrat et al. 2014). It is important to note that the system is able to measure only the movement component parallel to the instrument line of sight (L.O.S.—direction connecting the sensor and the investigated object), thus displacements that occur in the direction perpendicular to the sensor are missed. The cross-range resolution depends on the target distance, and good acquisition geometries thus depend on the observed scene. The principal limitations of the GB-InSAR technique are due to temporal decorrelation and atmospheric noise (Luzi et al. 2010).

TLS survey

A TLS device consists of a directional, coherent and in-phase optical beam (or discrete pulses) transmitter, a back-scattered signal receiver and a scanning mechanism (e.g., a rotating mirror). The main product of a long-range laser scanning technique is a high-resolution matrix of points (called a point cloud), defined in polar coordinates (range, horizontal and vertical scanning angle), obtained by measuring the scanner-object distance with great accuracy (on the order of millimetres or centimetres). Given the range distance and the scanning angle, the Cartesian coordinates of each point can be obtained. The high acquisition rate (up to hundreds of thousands of points per second) makes the detailed 3D shape of the object immediately available. Furthermore, for each point, the intensity of the reflected signal is acquired. The intensity data can provide some information about the type of material and the soil moisture content of the targets (Pesci and Teza 2008; Voegtle et al. 2008; Franceschi et al. 2009).

The monitoring system

In the post-landslide event phase, on 11 November 2015, a GB-InSAR system was installed on the terrace roof of the Calatabiano Castle (Fig. 5a, c). With the aim of optimizing the monitoring system efficiency, the following installation criteria were applied: (i) an adequate sensor-target distance (approximately 500 m), aimed at guaranteeing an azimuth resolution spanning from 40 cm (at 150 m range distance) to 106 cm (at 400 m range distance); (ii) minimum presence of obstacles between the radar sensor and the investigated objects; and (iii) a stable radar location (Fig. 5a). The radar system produces interferograms every 2 min, but cumulative displacement maps can be generated using longer (monthly) temporal baselines to detect both rapid and long-term movements. To obtain a high-resolution 3D surface of the post-event landslide area and to geo-rectify the acquired radar images for a better interpretation of the radar displacement data, TLS surveys were performed on 11 November from the radar system position (Fig. 5a). The employed terrestrial laser scanner is a long-range and high-accuracy 3D time-of-flight instrument (model Riegl LMS-Z420i; RIEGL 2010). This device is able to acquire the position of up to 12,000 points/s by calculating the round-trip travel time of a pulsed laser beam (near-infrared wavelength) from the instrument to the scanned object, with a maximum angular resolution of 0.008°, an accuracy of ±10 mm (one σ at 50 m range under RIEGL test conditions) and from a maximum distance of 800 m (Fig. 5d). The acquired TLS point cloud (approximately 14.5 million points) was linked to a global reference system by performing a GPS survey of 12 laser reflectors placed on the investigated slope. The raw data were subsequently cleared of vegetation, the data points were triangulated, and a 3D mesh was created to produce a continuous surface for the slope.

RESULTS

Monitoring data

The analysed preliminary radar data spans from 11 November to 31 December 2015, when the crucial landslide restoration works and by-pass construction was performed. The logistics of the GB-InSAR system installation provided a good spatial coverage of the landslide-affected area, including its main features (crown, minor scarps, erosional channels) and the pipeline by-pass and bulldozer trails related to the earthworks (Fig. 5). For the monitored site, the cumulative displacement maps were less affected by atmospheric noise than the single interferograms. Therefore, the cumulative displacement maps were considered more suitable for the ground deformation analysis of the slope and for the detection of landslide critical sectors. The detailed cumulative displacement maps are represented with a colour scale visualization: (i) stable areas are shown in light green; (ii) areas characterized by displacement toward the sensor L.O.S. feature colours from yellow to red and purple (maximum cumulative displacement); and (iii) areas characterized by displacements away from the sensor L.O.S. feature colours from dark green to deep blue (Fig. 6).

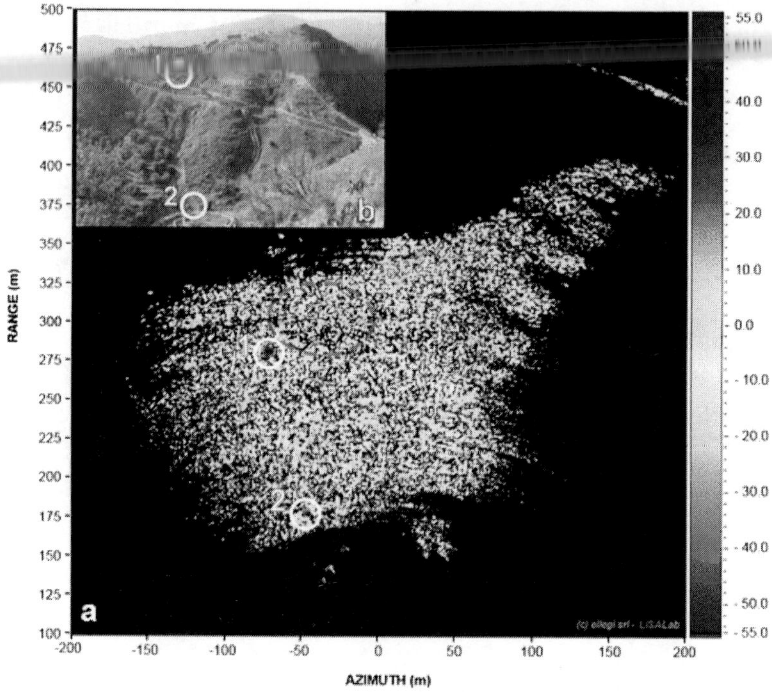

Figure 6. **a** Cumulative displacement map obtained by 1 h average interferograms acquired between 11 November 2015 17:59 h GMT + 1 and 21 November 2015 23:00 h GMT + 1 (the *red oval* represents sectors characterized by cumulative displacement values of up to 24 mm, whereas the *white circles* indicate sectors characterized by cumulative peak displacements of up to 55 mm); **b** corresponding optical image of the landslide scenario acquired on 17 November 2015

With the aim of analysing the landslide kinematics, displacement time series were extracted from 8 monitoring points in the displacement maps. These points were selected on the basis of high coherent values of the radar signal and high representativeness of the landslide behaviour (P1–P8) (Figs. 7 and 8). Specifically, the first six points (P1–P6) are located within the landslide body, and their cumulative displacements range from 4.5 mm (P5) up to 83 mm (P1). These points are representative of the areas that showed the highest displacement. The last two points (P7 and P8) were selected in order to be representative of stable areas: they are located outside the landslide body and correspond to a rocky outcrop and a sector of artificial drainage, respectively.

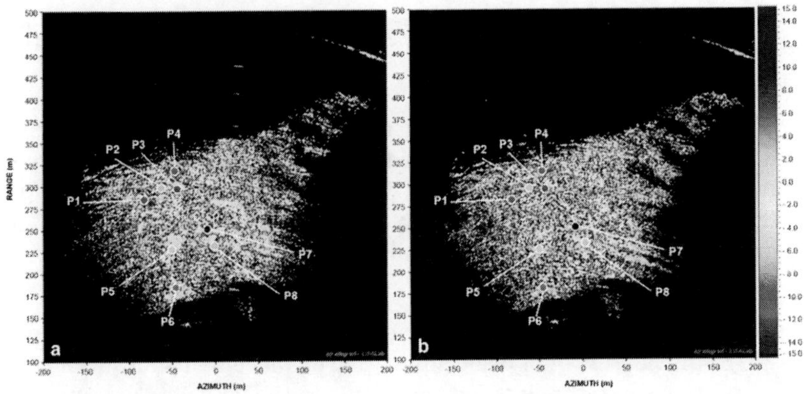

Figure 7. Cumulative displacement maps obtained based on 1 h averages acquired between **a** 25 November 2015 05:30 h GMT + 1 and 27 November 2015 17:33 h GMT + 1 and **b**20/12/2015 04:59 h GMT + 1 and 22 December 2015 20:12 h GMT + 1. P1-P8 represent the GB-InSAR monitoring points

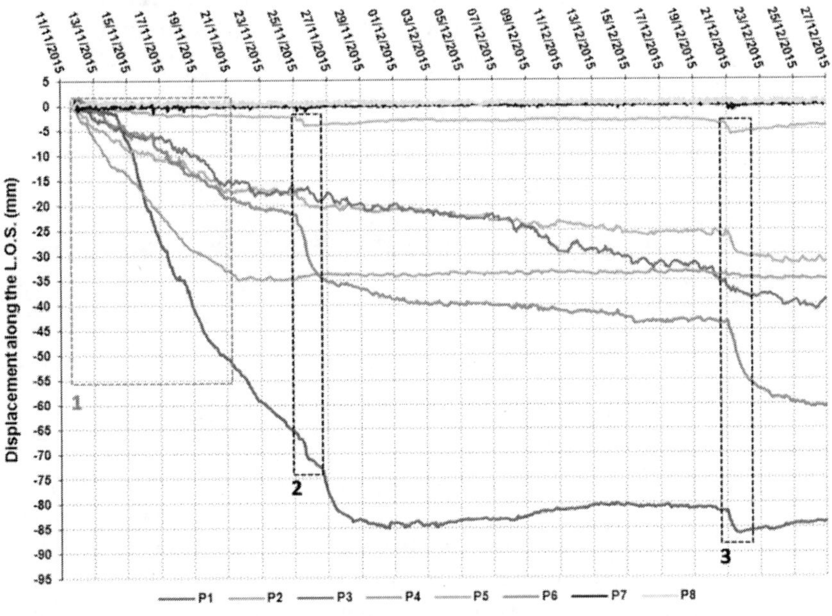

Figure 8. Time series of GB-InSAR monitoring points (*dotted red rectangle* highlights the detected displacement phase: 1 = first 10 days of monitoring; *dotted black rectangles* denote the phases after the two acceleration events: 2 = first period, 3 = second period).

By means of the radar data analysis, a first relevant displacement phase and two further phases after two acceleration events were detected and analysed (Table 1):

- The first 10 days of monitoring activity (11–21 November 2015) show a widespread sector characterized by cumulative displacement values of up to 24 mm (red oval in Fig. 6) corresponding to the upper landslide crown area. Peak cumulative displacements of up to 55 mm are located in small areas in the landslide top-middle and lowermost sectors (white circles in Fig. 6).
- The first phase (after the first acceleration event; 25–26 November 2015) displays 15-mm peak cumulative displacements localized in two restricted areas: the first corresponding to the landslide upper portion (around control point P4) and the second corresponding to the top-middle one (around control point P1) (Fig. 7a).
- The second phase (after the second acceleration event; 22–22 December 2015) exhibits a 15-mm peak cumulative displacement distributed in a widespread area of the landslide top-middle sector (around control points P1–P4) and along the right flank of the slope (Fig. 7b).

Table 1. Analysed cumulative displacement (in mm) and acceleration events (mean velocity in mm/day) of the GB-InSAR control points

Time period	Elapsed hours	Point 1 displacement (velocity)	Point 5 displacement (velocity)	Point 6 displacement (velocity)
11 Nov. 2015 17:59–21 Nov. 2015 23:03	244	54.3 (0.22)	2.1 (0.01)	19.6 (0.08)
25 Nov. 2015 02:25–26 Nov. 2015 17:00	37	7.6 (0.19)	1.6 (0.04)	12.8 (0.35)
20 Dec. 2015 23:43–22 Dec. 2015 20:12	45	3.7 (0.08)	0.9 (0.02)	13.7 (0.31)

To visualize the detected L.O.S. displacements directly on the 3D representation and therefore better localize the most critical areas of the landslide with respect to the work activities (landslide sectors, pipeline by-pass and bulldozer trails), the GB-InSAR data were merged with the TLS 3D model to obtain a 3D GB-InSAR cumulative displacement map (Fig. 9). Figure 9displays the selected control points and shows that the maximum cumulative displacements recorded during the monitoring period was approximately 100 mm and occurred near the control point P1.

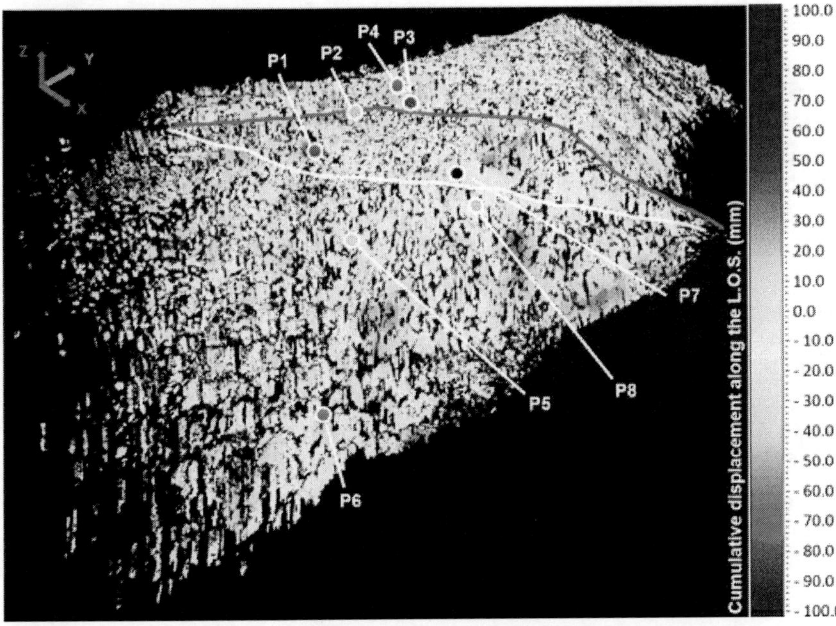

Figure 9. Cumulative 3D displacement map for the entire monitoring period (from 11 November to 31 December 2015). GB-InSAR monitoring points (P1-P8) are highlighted together with the pipeline by-pass trail (*blue line*) and the bulldozer trail (*white line*)

An early warning procedure was also implemented through the adoption of displacement and velocity thresholds calibrated during the first days of monitoring, when the movements were greatest. A simplified early warning system was developed, based on three different warning levels: ordinary, pre-alarm and alarm level. To ensure the safety of the workers involved in the restoration works, hourly displacement thresholds were adopted. A change in the level occurred if the following thresholds are surpassed: between 0.5 and 1.0 mm/h for the pre-alarm and >1.0 mm/h for the alarm level. To prevent possible damage to the pipeline by-pass (here representing the CI) and because this slow-moving landslide did not reach high deformation rates, daily displacement thresholds were adopted (between 12 and 24 mm/day for the pre-alarm and >24 mm/day for the alarm level). Communication, which is a fundamental issue of every early warning system (Intrieri et al., 2015), was achieved through the dispatch of monitoring bulletins every 12 h and whenever the warning thresholds were exceeded.

Remote 3D mapping

The high-resolution point cloud and the variable laser return intensity values also allowed us to add information regarding the main geomorphologic features of the landslide. In fact, the raw laser scanner product (Fig. 10) shows high intensity values in association with bare soils and rock outcrops. In contrast, the erosional channel thalwegs of the landslide are characterized by low intensity values, probably due to the moisture influencing the backscattered signal through absorption of the near-infrared wavelength beam (Franceschi et al. 2009). However, it is important to highlight that the difference in intensity values depends on several factors, such as the acquisition geometry (e.g. scanner-object distance and angle of incidence; Kaasalainen et al. 2011), the surface roughness and the presence of vegetation. The TLS point cloud was also used to validate the location of the selected GB-InSAR control points with respect to the landslide geomorphological features: right flank erosional channel topmost sector (P1), by-pass pipes trail (P2, P3), upper crown sector (P4), left flank erosional channel source area (P5), lowermost sector (P6), rock outcrop (P7) and upper sector of the artificial drainage (P8).

DISCUSSION

The main outcomes of this work suggest the effectiveness of the integrated monitoring system adopted to manage the Calatabiano landslide post-emergency phase. The peculiarity of this landslide event was its severe damage to CI, here represented by the disrupted local aqueduct, which produced a water shortage crisis in the city of Messina that lasted for several days and caused problems and discomfort for the local population. The adopted approach gave priority to a rapid system installation and near-real-time monitoring, in order to rapidly detect landslide displacements, ensure the safety of the workers involved in both the water pipeline by-pass construction and the landslide stabilization works and avoid further possible damage to the CI. Thanks to the versatility and portability of the devices, the GB-InSAR system was installed in a few hours on 11 November 2015, and the first displacement data became available while the TLS survey was performed. Furthermore, the employed remote sensing technique integration proved to be effective in surveying and monitoring the whole slope, overcoming the limitations of traditional single-point measurements.

The first 10 days of monitoring (11–21 November 2015; Figs. 6 and 8, Table 1) recorded the highest deformation rates (phase 1), which were related to the residual displacement of the landslide following the initial failure. As shown in Fig. 11, the landslide trigger was clearly related to the rainfall events that preceded on 24 October. In particular, more than 195 mm of cumulative precipitation fell during the period between 21 and 24 October (of which 182 mm occurred solely on 22 October).

Figure 10. TLS point cloud shaded by intensity values and GB-InSAR control points. *Yellow dashed polygons* indicate high intensity values related to bare soils and rock outcrops, corresponding to the following slope sectors, from top to bottom: the landslide crown, the by-pass pipes trail, the left bank erosional channel source area and the lowermost sector. The *red lines* mark the thalwegs of the erosion channels. P1–P8 represent the GB-InSAR monitoring points

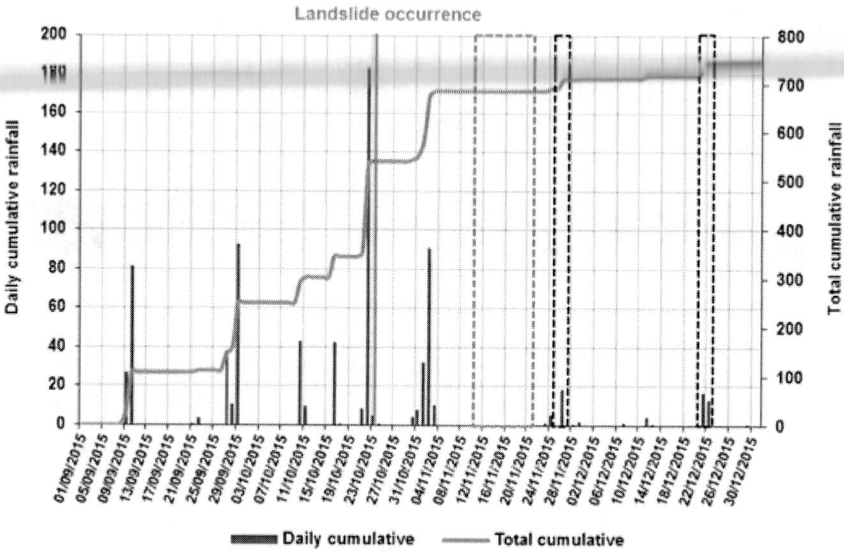

Figure 11. Rainfall data from the Alcantara rain gauge station (courtesy of Osservatorio delle Acque—Regione Siciliana). *Dashed rectangles* indicate the first 10 days of monitoring (in *red*) and the phases after two further acceleration events (in *black*)

Another rainfall event occurred between 30 October and 3 November (a total of 130 mm of cumulative precipitation was recorded, with a daily peak of 90 mm on 2 November; Fig. 11). This event may have contributed to the deformational trend. Two additional, albeit less intense, displacement periods (phases 2 and 3) were recorded during 25–26 November 2015 and 20–22 December 2015 (Figs. 7 and 8, Table 1), coinciding with two less intense rainfall events (approximately 25 and 30 mm of cumulative recorded precipitation, respectively) (Fig. 11). While the first displacement phase (particularly evident in the P1 and P4 control points time series) shows the features of post-landslide residual deformation, the second and third phases (particularly evident in the P1, P5 and P6 control points series; Fig. 8) were associated with impulsive acceleration events that were very similar in terms of magnitude and elapsed time (Table 1).

The interpretation of the remote sensing data collected from 11 November to 31 December 2015 was supported by means of the remote 3D products (Figs. 9 and 10). Furthermore, these 3D products were integrated to generate an accurate 3D geomorphological map of the landslide (Fig. 12), with the aims of accurately locating the recorded displacements with respect to the landslide geomorphological features and the excavation activities and detecting critical sectors on the monitored slope. The following areas were determined to be critical sectors:

- The uppermost area of the right flank erosional channel, which during the monitored period was affected by high and constant displacements (Figs. 6, 7a and 9), as confirmed by the control point P1 time series (Fig. 8)
- The left flank erosional channel source (control point P5 area) and lowermost sectors (control point P6 area) (minor scarp in Fig. 12). The latter sector proved to be one of the most active in the first 10 days of monitoring (Fig. 6), as displayed by the P6 cumulative displacement (approximately 62 mm, the second highest value within the control points). Corresponding to the erosional channels, the time series of control points P1, P5 and P6 clearly show the presence of displacement, probably induced by the high ground moisture content, as confirmed by the rainfall data (Fig. 11) and field inspections
- The landslide upper crown area (control point P4 area), which was largely involved in the residual deformation observed during the first 10 days of monitoring (Fig. 6)
- The by-pass pipes (located in the area of control points P2 and P3), displaying 32 and 39 mm of cumulative displacement, respectively (Figs. 8 and 9)
- Monitored slope areas located outside the landslide body, corresponding to both the earth movement associated with the excavation of the by-pass and bulldozer trails and the landslide slope re-profiling (Figs. 7 and 13).

The main limitations of the GB-InSAR displacement data interpretation included noise effects and temporal decorrelation, due to the widespread vegetation cover of the monitored slope and the influence of man-made excavations and earth movement related to the by-pass construction and landslide restoration works. Therefore, the remotely sensed data were validated by means of periodic field inspections (Fig. 13), which proved to be crucial in detecting and interpreting the evolution phases of the construction during the monitoring period, especially for the bulldozer trails and slope reprofiling area (Figs. 7b and 13d–f).

The 3D landslide geomorphological map was fundamental in understanding the kinematic mechanisms associated with the landslide mass and in assessing the areal distribution of the landslide (a total of approximately 4900 m^2) and its volume (on the order of approximately 25,000 m^3, considering a 5-m average thickness of the landslide body based on field surveys). Therefore, accurate mapping provided important information for the local authorities and technicians involved in the emergency management and in the restoration works

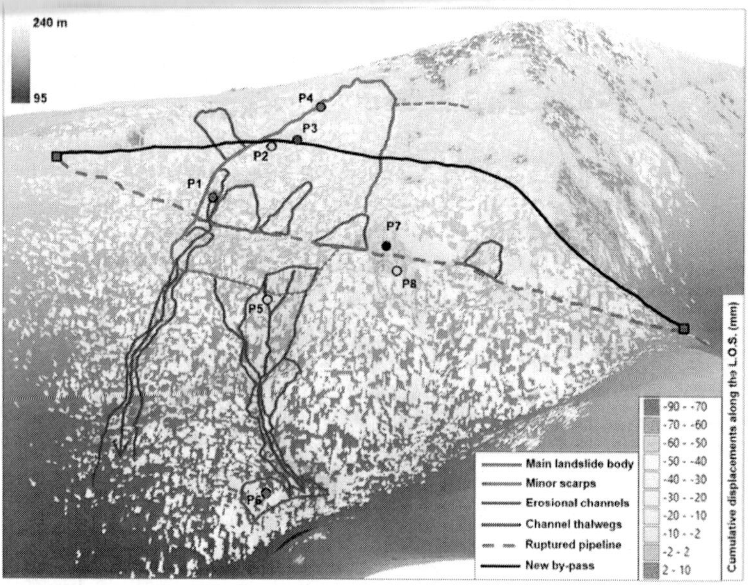

Figure 12. 3D geomorphological map of the Calatabiano landslide, projected on a high-resolution DEM, obtained by means of the TLS survey, and on the GB-InSAR cumulative displacement map. Control points, the pipeline by-pass trail (*black line*) and the ruptured pipeline (*dashed red line*) are highlighted. *P1–P8* represent the GB-InSAR monitoring points

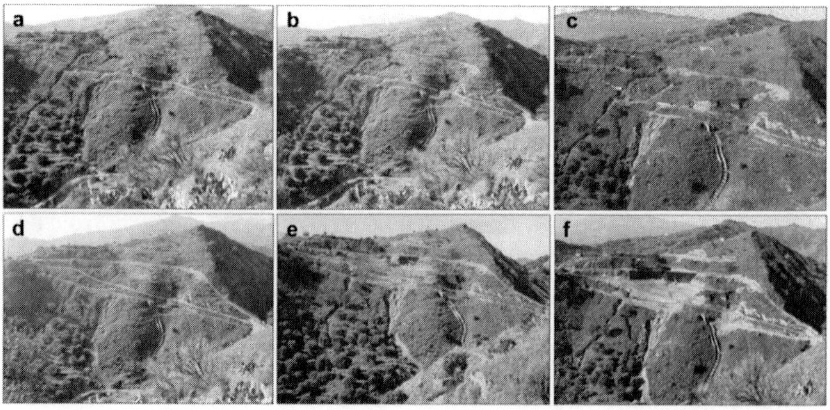

Figure 13. Evolution of the earthworks on the monitored slope: **a, b** by-pass trail construction (10–12 November); **c, d** beginning of the slope reprofiling and by-pass pipe emplacement (14–17 November); **e, f** geogrid positioning and intense slope reprofiling (21 November–27 December)

CONCLUSIONS

On 24 October 2015, intense rainfall triggered a landslide north of the town of Calatabiano (Sicily Island, Southern Italy). Although the portion of the slope affected by the failure was not exceptional in size or volume and no human casualties occurred, the event had an enormous social impact. The major pipeline providing water to the city of Messina was disrupted, and approximately 80 % of the city had limited access to this vital service for several days. The event demonstrated how the risk associated with landslides should not be assessed and managed only in terms of human safety but also in terms of the vulnerability of critical infrastructure. The remediation works required the creation of a by-pass trail around the damaged pipeline section and had to be performed directly on the unstable slope. Therefore, to assess the safety of the workers, a fully operational real-time monitoring network was deployed. The integrated use of GB-InSAR and TLS techniques allowed for a complete landslide characterization and a continuous assessment of the ongoing deformation rates on the slope. Two displacement thresholds and two velocity thresholds were defined in agreement with the emergency plan, which was based on a scale with three different warning levels. The triggering of the landslide was related to the rainfall events that preceded 24 October 2015, and the two accelerations recorded during the monitoring period (from 10 November to 31 December 2015) were correlated with the amount of precipitation in the area.

ACKNOWLEDGMENTS

The authors would like to thank the following groups for their support during the post-landslide emergency phase: the National Department of Civil Protection (DPCN), the Sicilian Regional Department of Civil Protection (DPCR) and the appointed Commissioner-Delegate, as well as the Osservatorio delle Acque—Regione Siciliana for providing rainfall data.

The GB-InSAR apparatus used in this application was designed and produced by Ellegi s.r.l. and was based on the proprietary technology GB-InSAR LiSALAB derived from the evolution and improvement of LiSA technology licenced by the Ispra Joint Research Centre of the European Commission.

REFERENCES

1. Abellan A, Vilaplana JM, Martinez J (2006) Application of a long-range terrestrial laser scanner to a detailed rockfall study at Vall de Nuria (eastern Pyrenees, Spain). Eng Geol 88:136–148
2. Bardi F, Frodella W, Ciampalini A, Bianchini S, Del Ventisette C, Gigli G, Fanti R, Moretti S, Basile G, Casagli N (2014) Integration between ground based and satellite SAR data in landslide mapping: the san Fratello case study. Geomorphology 223:45–60
3. Barla G, Antolini F, Barla M, Mensi E, Piovano G (2010) Monitoring of the Beauregard landslide (Aosta Valley, Italy) using advanced and conventional techniques. Eng Geol 116:218–235
4. Bozzano F, Cipriani I, Mazzanti P, Prestininzi A (2011) Displacement patterns of a landslide affected by human activities: insights from ground-based InSAR monitoring. Nat Hazards 59:1377–1396
5. Canuti P, Casagli N, Ermini L, Fanti R, Farina P (2004) Landslide activity as a geoindicator in Italy: significance and new perspectives from remote sensing. Environ Geol 45(907):919
6. Cassola P, Giammarino S, Puglisi D, Villa G (1991) Nuovi dati sedimentologico-petrografici e biostratigrafici sulla formazione di Piedimonte (Sicilia nord-orientale). Mem Soc Geol Ital 47:213–223
7. Catalano S, Corsaro RA, Marino M, Branca S, Cirrincione R, De Guidi G, Di Stefano A, Mazzoleni P (2010) Note illustrative della Carta Geologica d'Italia alla scala 1:50000, Foglio 613, Taormina. Ispra, S. EL. Ca, Firenze
8. Corsini A, Berti M, Monni A, Pizziolo M, Bonacini F, Cervi F, Ciccarese G, Ronchetti F, Bertacchini E, Capra A, Gallucci A, Generali M, Gozza G, Pancioli V, Pignone S, Truffelli G (2013) Rapid assessment of landslide activity in Emilia Romagna using GB-InSAR short surveys. In: Landslide science and practice. Springer, Berlin Heidelberg, pp. 391–399
9. Cruden DM, Varnes DJ (1996) Landslides types and processes. In: Turner AK, Schuster RL (eds) Landslides: investigation and mitigation. Transportation Research Board special report 247. National Academy Press, WA, pp. 36–75
10. Di Traglia F, Nolesini T, Intrieri E, Mugnai F, Leva D, Rosi M, Casagli N (2014) Review of ten years of volcano deformations recorded by the ground-based InSAR monitoring system at Stromboli Volcano: a tool to mitigate volcano flank dynamics and intense volcanic activity. Earth-Sci Rev 139:317–335
11. Fanti R, Gigli G, Lombardi L, Tapete D, Canuti P (2012) Terrestrial laser scanning for rockfall stability in the cultural heritage site of Pitigliano (Italy). Landslides 5:1–1
12. Franceschi M, Teza G, Preto N, Pesci A, Galgaro A, Girardi S (2009) Discrimination between marls and limestones using intensity data from terrestrial laser scanner. ISPRS J Photogramm 64:522–52
13. Frodella W, Morelli S, Fidolini F, Pazzi V, Fanti R (2014) Geomorphology of the Rotolon landslide (Veneto region, Italy). Journal of Maps 10(3):394–401
14. Geertsema M, Schwab JW, Blais-Stevens A, Sakals ME (2009) Landslides impacting linear infrastructure in west Central British Columbia. Nat Hazards 48:59–72
15. Gigli G, Intrieri E, Lombardi L, Nocentini M, Frodella W, Balducci M, Venanti LD, Casagli N (2014) Event scenario analysis for the design of rockslide countermeasures. J Mt Sci 11(6):1521–1530
16. Guzzetti F, Mondini AC, Cardinali M, Fiorucci M, Santangelo M, Chang KT (2012) Landslide inventory maps: new tools for an old problem. Earth-Sci Rev 112:1–25
17. Herrera G, Fernández-Merodo JA, Mulas J, Pastor M, Luzi G, Monserrat O (2009) A landslide forecasting model using ground based SAR data: the Portalet case study. Eng Geol 105:220–230

18. Intrieri E, Gigli G, Nocentini M, Lombardi L, Mugnai F, Fidolini F, Casagli N (2015) Sinkhole monitoring and early warning: an experimental and successful GB-InSAR application. Geomorphology 241:304–314
19. Jaboyedoff M, Oppikofer T, Locat A, Locat J, Turmel D, Robitaille D, Demers D, Locat P (2009) Use of ground-based LIDAR for the analysis of retrogressive landslides in sensitive clay and of rotational landslides in river banks. Can Geotech J 46:1379–1390
20. Kaasalainen S, Jaakkola A, Kaasalainen M, Krooks A, Kukko A (2011) Analysis of incidence angle and distance effects on terrestrial laser scanner intensity: search for correction methods. Remote Sens 3:2207–2221
21. Kadri F, Birregah B, Châtelet E (2014) The impact of natural disasters on critical infrastructures: a domino effect-based study. J Homel Secur Emerg 11:217–241
22. Kjekstad O, Highland L (2009) Economic and social impacts of landslides. In: Landslides–disaster risk reduction. Springer, Berlin Heidelberg, pp. 573–587
23. Luzi G (2010) Ground based SAR interferometry: a novel tool for geoscience. INTECH Open Access Publisher
24. Luzi G, Pieraccini M, Mecatti D, Noferini L, Guidi G, Moia F, Atzeni C (2004) Ground-based radar interferometry for landslides monitoring: atmospheric and instrumental decorrelation sources on experimental data. IEEE T Geosci Remote 42:2454–2466
25. Monserrat O, Crosetto M, Luzi G (2014) A review of ground-based SAR interferometry for deformation measurement. ISPRS J Photogramm 93:40–48
26. Murray AT, Grubesic TH (2007) Overview of reliability and vulnerability in critical infrastructure. In: Critical Infrastructure. Springer, Berlin Heidelberg, pp. 1–8
27. Nolesini T, Di Traglia F, Del Ventisette C, Moretti S, Casagli N (2013) Deformations and slope instability on Stromboli Volcano: integration of GBInSAR data and analog modeling. Geomorphology 180:242–254
28. Oppikofer T, Jaboyedoff M, Blikra LH, Derron MH (2008) Characterization and monitoring of the Aknes rockslide using terrestrial laser scanning. Proceedings of the 4th Canadian conference on geohazards: from causes to management, pp. 211–218
29. Pesci A, Teza G (2008) Effects of surface irregularities on intensity data from laser scanning an experimental approach. Ann Geophys-Italy 51:839–848
30. Petley D (2012) Global patterns of loss of life from landslides. Geology 40:927–930
31. RIEGL (2010) Data sheet of long range & high accuracy 3D terrestrial laser scanner LMS-Z420i. http://www.riegl.com/uploads/tx_pxpriegldownloads/10_DataSheet_Z420i_03-05-2010.pdf
32. Schulz WH, Coe JA, Shurtleff BL, Panosky J, Farina P, Ricci PP, Barsacchi G (2012) Kinematics of the Slumgullion landslide revealed by ground-based InSAR surveys. In Proc. Landslides and Engineered Slopes: Protecting Society through Improved Understanding—the 11th International and 2nd North American Symposium on Landslides and Engineered Slopes, Banff, Canada, pp. 3–8
33. Tarchi D, Casagli N, Fanti R, Leva D, Luzi G, Pasuto A, Pieraccini M, Silvano S (2003) Landslide monitoring by using ground-based SAR interferometry: an example of application to the Tessina landslide in Italy. Eng Geol 68:15–30
34. Voegtle T, Schwab I, Landes T (2008) Influences of different materials on the measurements of a Terrestrial Laser Scanner (TLS). The International Archives of the Photogrammetry, Remote Sensing and Spatial Information Sciences. Vol. XXXVIII. Part B5. Beijing

CHAPTER 12

The Impact of Climate Change on Landslides In Southeastern of High-Latitude Permafrost Regions of China

Wei Shan*, Zhaoguang Hu, Ying Guo, Chengcheng Zhang, Chuanjiao Wang, Hua Jiang, Yao Liu and Jitao Xiao

Institute of Cold Regions Science and Engineering, Northeast Forestry University, Harbin, China

ABSTRACT

Climate warming leads to permafrost degradation and permafrost melting phase transition, resulting in an increasing number of landslides. This study uses the road segments and road area at the intersection between Bei'an-Heihe Highway and the northwest section of the Lesser Khingan Range in north China as the study area. By means of geological survey combined with meteorological data, we analyzed the impact of climate change on landslide movement in the permafrost zone. Over a 60 year period, the average annual temperature of the study area has increased by 3.2°C, and permafrost degradation is severe. Loose soil on the hillside surface provides appropriate conditions for the infiltration of atmospheric precipitation and snowmelt, and seepage from thawing permafrost. As it infiltrates downwards, water is blocked by the underlying permafrost or dense soil, and infiltrates along this barrier layer toward lower positions, forming a potential sliding zone. The representative Landslide in the study area was examined in detail. Displacement monitoring points were set up on the surface of the landslide mass, and at the trailing edge of the landslide mass. The data collected were used to investigate the relationship between landslide movement and pore water pressure at the tailing edge as well as the ground temperature. The results show that the landslide movement process changes with the season, showing a notable annual cyclical characteristic and seasonal activity. Landslide movement is characterized by low angles and intermittence. The time of slide occurrence and the slip rate show a corresponding relationship with the pore water pressure at the tailing edge of the landslide mass. The seepage of water from thawing into the landslide mass will influence the pore water pressure at the tailing edge of the landslide mass, and is the main cause of landslide movement.

Keywords: climate change, high-latitude permafrost, permafrost degeneration, landslide movement, pore water pressure

INTRODUCTION

In northeast China, the area north of 47°N contains widespread permafrost. This is China's only high-latitude permafrost region, and is also China's second largest permafrost region (Guo et al., 1981; Zhou et al., 1996; Sun et al., 2007). In recent years, due to the effects of climate change, the southern boundary of China's northeast high-latitude permafrost region gradually moved northward. Permafrost near the southern boundary shows discontinuous island-like distribution and accelerated degradation (Jin et al., 2000; He et al., 2009a; Wei et al., 2010).

Climate is the long-term average state of atmospheric physical characteristics. Climate change refers to the change in the mean state of climate over time. In the last century, the average global surface temperature has been continuously increasing, which shows a consistent warming trend worldwide. The rate of warming during the past 50 years is almost twice that of the past 100 years [IPCC (Intergovernmental Panel on Climate Change), 2007]. Since the early 1980s, in most of the permafrost regions in the world, the temperature has increased. In some regions in northern Alaska, the observed temperature increase has reached 3°C, in the northern European region and Russia, the increase has reached 2°C. In the latter region, during the 1975–2005 period substantial reduction in the thickness and range of the permafrost layer was observed (IPCC, 2013). Climate change and its impact are an important research issue receiving extensive global attention.

Numerous studies have shown that the climate change in China exhibits the same trend as the global climate change (Ding et al., 2006). In the last 54 years, the rate of temperature increase in China was about 0.25°C/10a, far higher than the global or hemispheric average rate of warming. Northeast China is one of the areas showing the most significant warming and permafrost degradation in the country (Shi et al., 2014).

Forest and accumulated snow have a very important influence on the temperature change and the thawing process of the underlying seasonal frozen ground and permafrost (Chang et al., 2011). In the Great and Lesser Khingan Ranges of northeast China and the Outer Baikal region of Russia, temperature shifts caused by snow, vegetation, water, topography, atmospheric inversion, and other local factors are substantial, forming the "Khingan–Baikal type" that is permafrost distinctly from polar and high-altitude permafrost distribution (Zhou and Guo, 1982; Chang et al., 2013). Affected by the occurrence conditions, the process and form of high-latitude permafrost degradation in northeast China are also different from that of polar and high-altitude permafrost (Jin et al., 2009).

In recent years, landslide incidents triggered by climate change and extreme weather have increased (Blunden and Arndt, 2011), and have gradually received attention from national governments and relevant international academic organizations (Eu-Fp7, 2008; ICL, 2014). In particular, landslides in cold areas are becoming a hot issue in landslide research [ICL (International Consortium on Landslides), 2012; Guo et al., 2013].

Landslides are a natural geological phenomenon in mountainous areas. Their mechanism and evolution are closely related to the geological conditions and environmental factors, and are controlled not only by geological forces, lithologic structure, and other crustal internal factors; but also by topography, land cover, precipitation, changes in human activities, and environmental conditions. The spatial and temporal distribution of landslides has characteristics of uncertainty, sporadic nature, continuation, and irreversibility. Landslides are the result of geological and environmental changes and in turn also drastically change the geological environment (Shan et al., 2014b). Scientists have applied different methods in order to analyze the relationship between climate change and landslide mechanisms and evolution in cold areas (Shan et al., 2014a). The impact of landslides induced by glacier and permafrost degradation in cold regions; on the topography, geological environment, water resources, and biodiversity has been examined; and the role of climate, a main factor influencing the landslide movement, in the evolution of landslides in cold areas has been discussed (Fischer et al., 2013; Haeberli, 2013; Kliem et al., 2013; Starnberger et al., 2013; Ballantyne et al., 2014; Grab and Linde, 2014; Nussbaumer et al., 2014). However, due to lack of monitoring data, the majority of these studies were large-scale (Stoffel et al., 2014). Currently, there has been no report on the mechanisms, movement characteristics, and patterns of landslides induced by the combined effect of permafrost thawing and extreme weather events due to climate change and geological conditions.

In the present study, the road area where the Bei'an to Heihe Highway crosses the northwest section of the Lesser Khingan Range is used as the study area. We performed a geological survey, an engineering survey, and field measurements which were plotted on the topographic map of the landslide area and a relevant geological cross section. Using the meteorological data of Sunwu County (30 km from the study area), which was released by the China Meteorological Data Sharing Service System (http://cdc.cma.gov.cn/home.do), we analyzed the impact of climate change on permafrost thaw, geological environment, and landslide mechanisms in the study area. Using Landslide K178 + 530 in the landslide area as an example, we used monitoring data of point displacement on the landslide mass, ground temperature at the trailing edge, and pore water pressure to perform a comprehensive analysis of the impact of climate change on the pore water pressure at the measuring points and the landslide movement process.

BACKGROUND

China's Bei'an-Heihe Highway intersects the northwest section of the Lesser Khingan Range at the junction between Sunwu County and the Aihui District. The study area lies between east longitude 127°17′31″-127°21′24″ and north latitude 49°30′57″-49°41′50″ (Figure 1). The area is located on the southern fringe of China's high-latitude permafrost region and has typical periglacial landforms. The island-like permafrost in this region is the result of to residual paleo-glacial deposition and is currently in the degradation stage. The geological conditions are extremely unstable.

Topography and Geological Structure

The study area is located in the northwest section of the Lesser Khingan Range. It has a hilly landscape with undulating terrain. The slope of the ground is generally 10–20°, and the upper part of the slopes are relatively steep, about 25–30°, but locally can be up to 40°. The area can be further divided into the valley bottomland and hills. The altitude of the entire area ranges from 210–330 m.

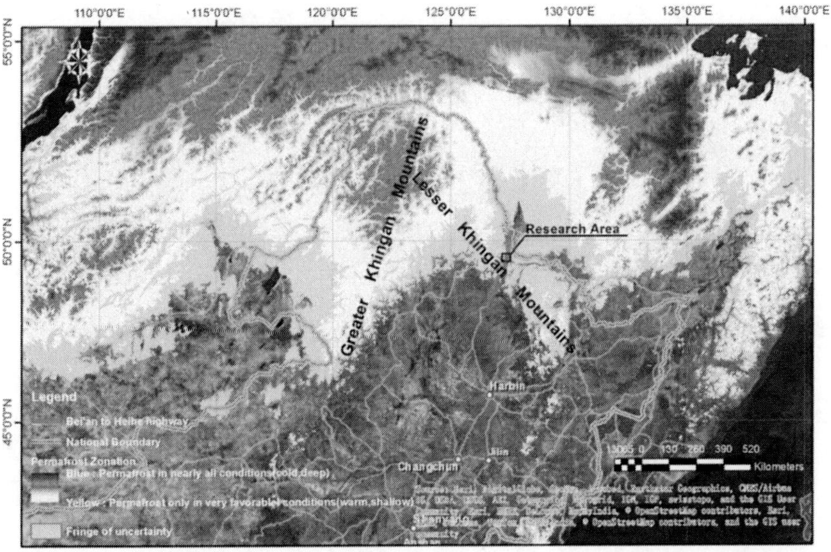

FIGURE 1. High-latitude permafrost distribution in northeast china (data from zurich university, switzerland), and the location of study area.

The geological structure of the study area belongs to the Khingan-Haixi fold belt. From the bottom up, the stratigraphy is composed by Cretaceous mudstone, Tertiary pebbly sandstone, silty mudstone, and powdery sandstone. From the late Tertiary to the early Quaternary, the Lesser Khingan Range experienced block uplift. Due to long-term erosion and leveling, loose sediments on the summit and the slope of the hills have gradually thinned, and the thickness of the current residual layer is generally only 1–2 m. The loose deposits accumulate mainly in the basin and valley areas between mountains, with a thickness of about 10 m. The soil is mainly composed of clayey silt, mild clay, and gravelly sand, and the surface is covered with a relatively thick layer of grass peat and turf. The surface vegetation is that of grassland and woodland, and there are inverted trees in the woodlands. Figure 2 shows the geological map of the landside road area of the study area plotted from the field survey conducted in June 2010.

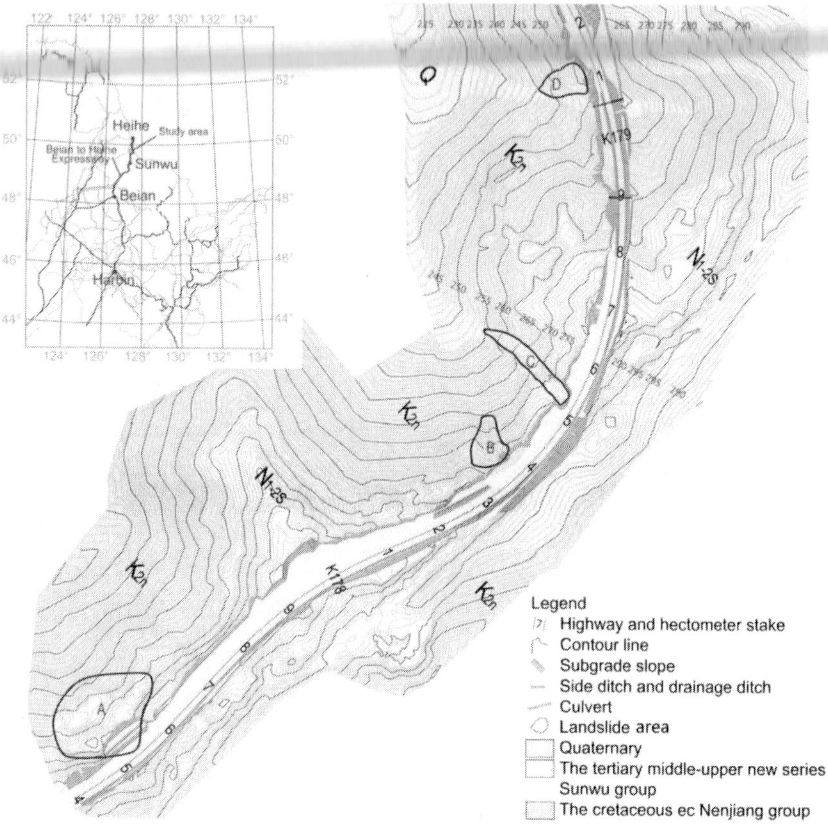

FIGURE 2. Geological map of k177+400-k179+200 section of bei'an-heihe highway.

Climate Conditions

The study area is located in the transition zone between the north part of the middle-temperate zone and the cold-temperate zone. The area is affected by alternating influences of high and low pressures from inland and the sea as well as by monsoons. Overall the climate is characterized by long, dry and cold winters and short, hot and humid summers. In other words, it belongs to the continental monsoon climate zone.

The average annual temperature in this area ranges −2–1°C, with an extreme maximum of 38.6°C, and an extreme minimum of −48.1°C. The annual average wind speed is 2.7–4.0 m/s. The average annual precipitation is 530–552 mm, with a maximum of 800 mm. Precipitation is mostly concentrated in the summer, from July to September, accounting for 61–67% of the total annual precipitation. The first snow fall is typically in mid-October, and the final snow fall is often in late March or early April of the following year. The maximum annual evaporation is greater than 1000 mm, whereas the minimum is 850 mm. The average annual sunshine is 2500 h, with a maximum of 2800 h and a minimum of 2200 h. The total amount of annual radiation averages 1148 kcal/cm^2, with a maximum of 1229 kcal/cm^2, and a minimum of 1039 kcal/cm^2.

Permafrost Distribution

The climate in Northeast China is controlled by the Siberia—Mongolia high-pressure control. The inversion layer is widely distributed in this region and has an important impact on the development process and regional distribution of permafrost. Meanwhile, the northeast region of China is China's main forest distribution area, where the ecological environments of wetlands, grasslands, and forests coexist with permafrost mutually restraining and influencing each other. Forests, bushes, mosses, and other ground covers reduce solar radiation, hindering the increase of air and ground temperatures. The low-lying river terraces, valleys, wetlands, and shady slopes provide good conditions for the development and preservation of island-like permafrost and slow permafrost degradation.

The seasonal ground freezing in the study area reaches maximum depth at the end of May, and the observed maximum depth is 2.26–2.67 m. In the mountains, the maximum seasonal freezing depth exceeds this value. April to September is the thawing period of seasonal frost; in dry areas all the seasonal frost thaws in early July, whereas in swamped thick peat and humus zones, seasonal frosts do not completely thaw until the end of October. On shady slopes and in valley areas, island-like permafrost is distributed.

Because there is no long-term meteorological data for the study area obtained from field measurements, meteorological data from Sunwu County (30 km from the study area), published by the China Meteorological Data Sharing Service System, was used for the analysis of climate change. Figure 3 plots the monthly average temperatures, soil temperatures in 40 cm depth, atmospheric precipitations, and maximum permafrost thicknesses from 1971–2000 using meteorological data of Sunwu County.

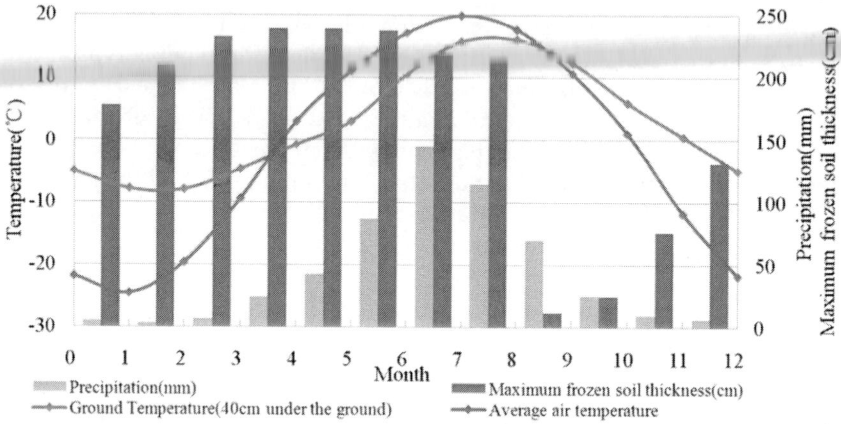

FIGURE 3. Monthly average temperature/soil temperature in 40 cm depth/permafrost thickness/precipitation in sunwu county (1971–2000, data: china meteorological data sharing service system).

Climate Change of the Study Area

Affected by global climate change, the study area has become one of the regions in northeast China with the largest temperature increase in the past 50 years. Meteorological data of Sunwu County show that in the past 60 years, among all climate indicators, change in temperature was the most significant. Figure 4shows the annual average maximum temperatures, average temperatures, average minimum temperatures, and average precipitation from 1954–2013 using meteorological data of Sunwu County. By applying linear regression, we found that, over the 60 year period from 1954 to 2013, the annual average temperature in the study area increased by 3.2°C; the average annual maximum temperature increased by 1.5°C, or 21.99%; the average annual minimum temperature increased by 5.2°C, or 69.04%; and the average annual precipitation decreased by 4.85 cm, or 8.93%. The increase in the average annual minimum temperature was 3.45 times that of the average annual maximum temperature.

Climate Change and Landslides

Climate change leads to permafrost degradation in the study area, and human engineering projects further accelerate the process of permafrost degradation (He et al., 2009b). In 1999, a survey conducted for constructing the secondary road from Bei'an to Heihe showed that there were 17 permafrost road segments along the entire road length (Wang et al., 2001; Zhang et al., 2001). Yet in 2009, a survey conducted right before the construction of the Bei'an-Heihe Highway showed that there were only six permafrost road segments remaining; the other 11 segments had been completely degraded.

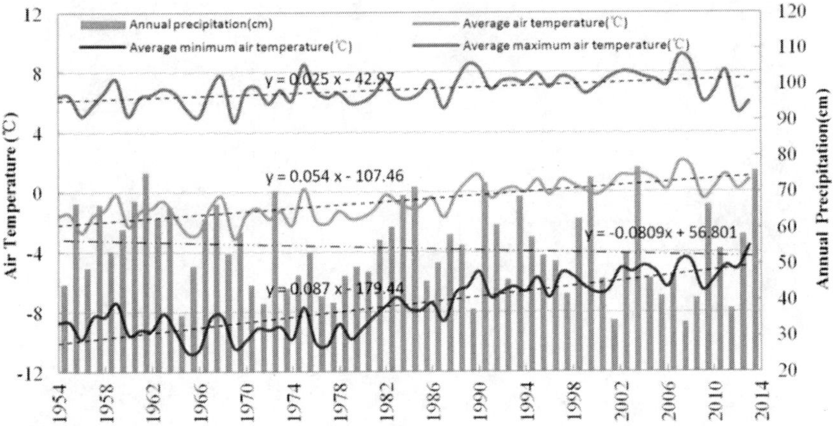

FIGURE 4. Annual average maximum temperatures, average temperatures, average minimum temperatures and average precipitation in sunwu county (1954–2013).

During the thawing of permafrost, the phase change of water has a severe impact on the mechanical properties of the soil (Wang et al., 2014b). Permafrost degradation causes a lot of geological engineering problems in the construction of roads. The construction of the Bei'an-Heihe secondary road started in 1999, and in August 2000 on the K176 + 900 – 178 + 200 section a landslide was induced by thawing of permafrost in the roadbed. This caused instability of the entire roadbed in this section, which had to be completely abandoned. The road was then redirected toward the left along the ridge, as shown in Figure 5A.

The survey conducted before the 2009 project of widening the Bei'an-Heihe road and constructing the Bei'an-Heihe Highway also showed that, in the K177 + 400 - K179 + 200 section, within 10 m of the left of the roadbed there were four landslides with a volume of over 20000 m^3, as illustrated in Figure 2. In addition, a number of landslides were found within 3 km of this road area (Figure 6), thus forming the landslide group at the intersection where the Bei'an-Heihe Highway crosses the northwest section of the Lesser Khingan Range. When compared with the permafrost distribution map obtained from the Enhanced Thematic Mapper (ETM+) imaging data collected from satellite Landsat7 in 2009 (Figure 7) (Wang et al., 2014a), it can be seen that the locations of the landslides fit the permafrost distribution very well. Thus, we deduced that these landslides were induced by the thawing of permafrost.

FIGURE 5. Satellite photos of k176 + 500 - k179 + 900 section of bei'an-heihe highway (google earth). **(A)** Landslide in K178+530 section (in 2000). **(B)** Satellite photos of K178 + 530 (2004.6). **(C)** Satellite photos of K178+530 (2010.9).

FIGURE 6. Full view of k175 + 500 - k180 + 200 section of bei'an-heihe highway (blue line is the boundary of the landslides, the orange area is corn field).

FIGURE 7. Permafrost distribution map in k176 + 500 - k180 + 000 section of bei'an-heihe highway.

Monitoring of Landslide Movement

In June 2010, on the basis of geological survey and topographic measurements of the study area, comprehensive examination and dynamic monitoring of the slip movement were performed on a representative landslide that may threaten the roadbed in order to investigate the impact of permafrost thaw on landslide movement.

Landslide K178 + 530 on the Bei'an-Heihe Highway (Figure 2C) is one of the landslides that is closest to the roadbed and has the fastest slip rate. Because survey and monitoring data prior to 2009 are not available, the time of the initial slip could not be determined. From the satellite photo taken on June 15, 2004, it can be seen that at that time the straight-line distance between the leading edge and the tailing edge of the landslide mass was 101.26 m (Figure 5B). The satellite photo taken on September 12, 2010 showed that the straight-line distance between the leading and tailing edges of the landslide mass had increased to 145.05 m (Figure 5C). Over this 6 year period, the leading edge of the landslide moved forward 43.79 m, whereas the position of the trailing edge basically did not change.

Figure 8 shows the distribution of the drillings, the geophysical measurement lines and points for monitoring displacement, ground temperature, and pore water pressure for Landslide K178 + 530. The points

A, B, C, D, E, and F in the figure denote the positions of the drilling, and lines G and H denote the high-density resistivity measurement line and the ground-penetrating radar (GPR) measurement line, respectively. During the drilling in holes E and F, permafrost was found at a depth 2.2 m below the ground surface, and the permafrost thicknesses at points E and F were 3.7 m and 2.4 m, respectively. The area of permafrost distribution in Figure 8 is inferred according to Figure 7.

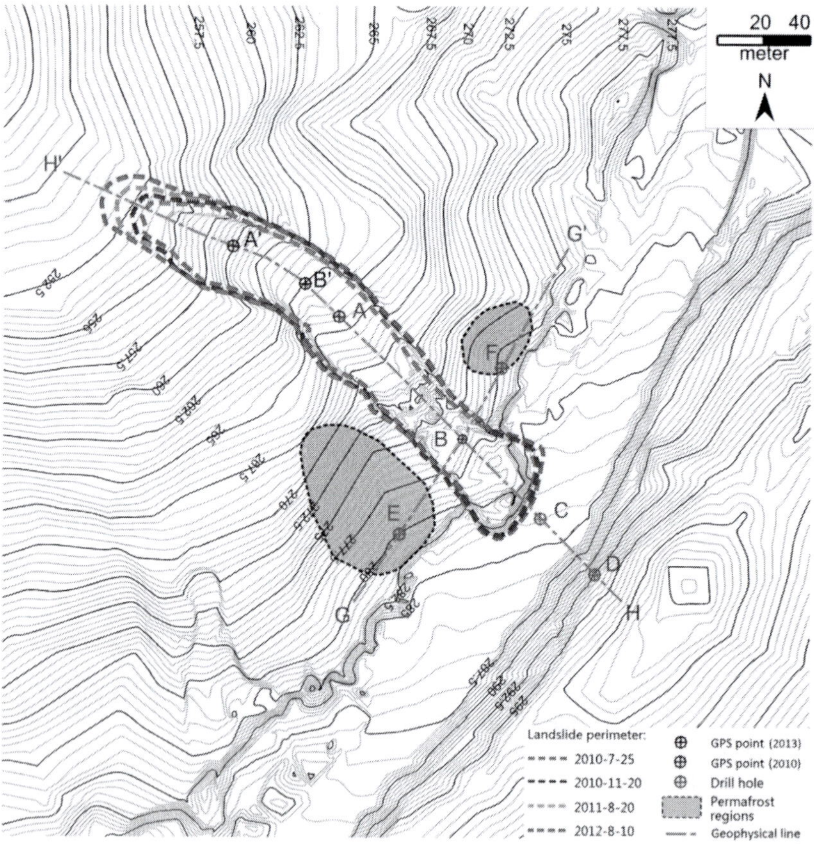

FIGURE 8. Distribution of the drillings, the geophysical measurement lines and points for monitoring displacement, ground temperature, and pore water pressure in landslide k178 + 530.

Through drilling, GPR, and high-density resistivity prospecting; the profiles of Landslide K178 + 530 in cross section H along the sliding direction and in cross section G perpendicular to the slide direction were obtained (Figures 9, 10).

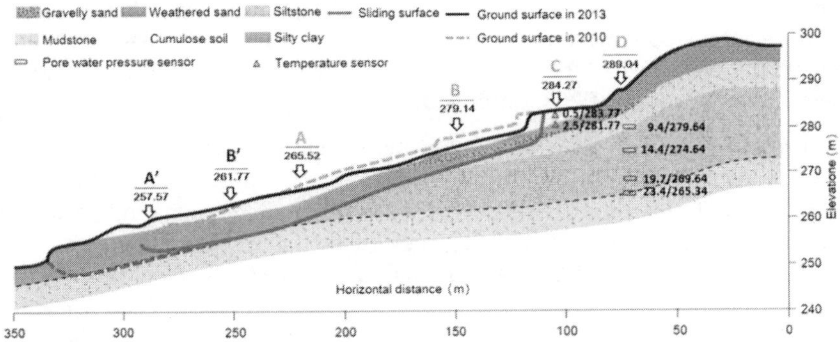

FIGURE 9. The profiles of landslide k178 + 530 along the sliding direction.

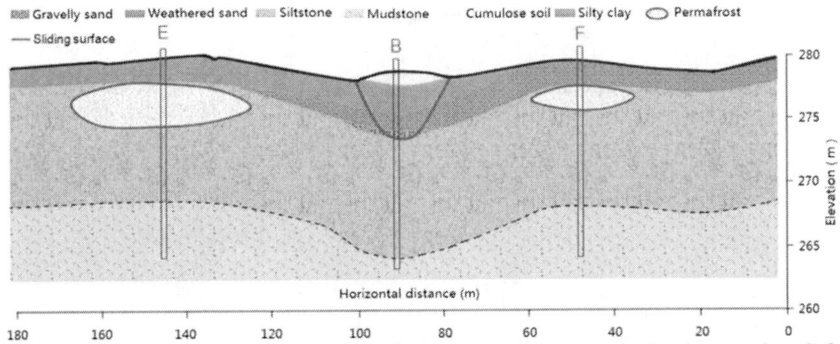

FIGURE 10. The profiles of landslide k178 + 530 perpendicular to the slide direction.

In boreholes A and B shown in Figure 8, plastic displacement measurement piles with a length of 400 cm and an above-ground height of 30 cm were buried; the top of the pile was used to monitor the slope displacement of the landslide mass at points A and B. C served as the monitoring point of the ground temperature; in the borehole, thermistor soil temperature sensors were set up at every 0.5 m from 0.5 to 2.5 m below the ground surface. D served as the monitoring point of pore water pressure; in the borehole, steel-wire soil pore water pressure sensors were set up at 3.8, 9.4, 19.4, and 23.7 m below the ground surface. Regular monitoring of relevant parameters started in July 2010.

The real-time dynamic deformation monitoring system, a Real Time Kinematic -Global Positioning System (RTK-GPS), was used to monitor the displacements of the measuring points on the landslide surface. From July 2010 to August 2014, point A slid horizontally by 69.03 m, and its elevation dropped by 7.95 m; point B slid horizontally by 113.97 m, and its elevation dropped by 17.37 m. Based on these values, it can be deduced that the angle between the slope and the horizontal plane decreased from 8.07° to 6.02°. Points A′ and B′ in Figures 8, 9 show the positions of points A and B respectively, in August 2014 after sliding.

Figure 11 shows the average daily ground temperatures and atmospheric precipitations over time at point C on Landslide K178 + 530 at depths of 0.5 and 2.0 m. The ground temperature data is on-site monitoring data; the average daily atmospheric temperatures and precipitations are meteorological data collected at Sunwu County weather station.

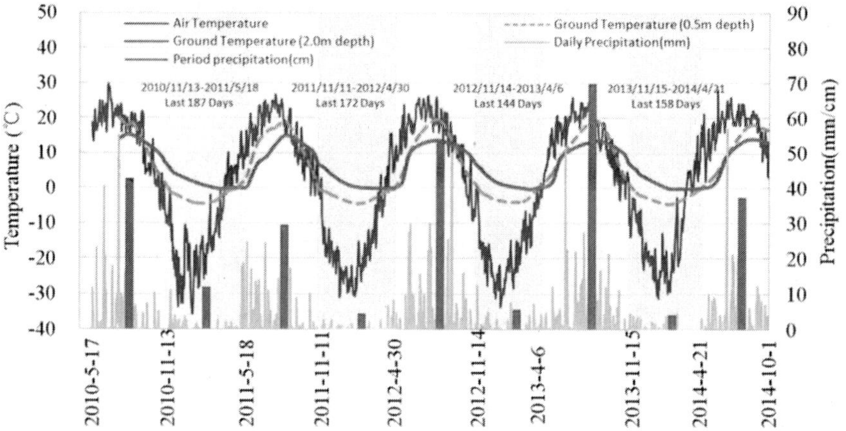

FIGURE 11. Daily average air temperature/precipitation/soil temperature in the depth of 50 cm, 200 cm in k178 + 530 section of bei'an-heihe highway.

To facilitate data analysis, the soil freeze period was defined as the period from the time when the thermistor sensor at 0.5 m below ground surface detected temperature below 0°C in the autumn to the time when the thermistor sensor at 2.0 m below ground surface detected temperature changing from below to above 0°C in the spring. This period is displayed in light blue in Figure 11. The sums of atmospheric precipitation over the soil freeze period and non-freeze period were calculated, and are denoted by the dark blue bars in Figure 11. During the soil freeze period, atmospheric precipitation and snowmelt water are blocked by the seasonal frozen soil layer and cannot normally infiltrate downward to deep soil.

Figure 12 shows the pore water pressure measured at point D, 19.4 m below the ground surface. At point D at the time of drilling, water was first observed at a depth of 3.8 m below the ground surface, but the stable water level measured during the monitoring period ranged from 15.8 to 16.6 m.

Soil at 3.8 and 9.4 m below the ground surface was found to be unsaturated during the monitoring period. Readings from the pore water pressure sensor placed at 23.7 m below the ground surface were unstable during the late stage of monitoring, and thus data collected from the sensor placed at 19.4 m below the ground surface are used to represent pore water pressure at point D.

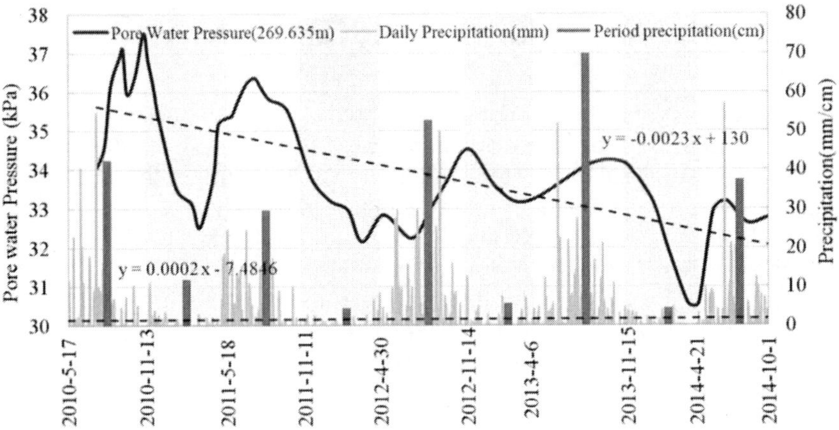

FIGURE 12. Daily precipitation and pore water pressure in the depth of 19.4 cm of d point in k178 + 530 section of bei'an-heihe highway.

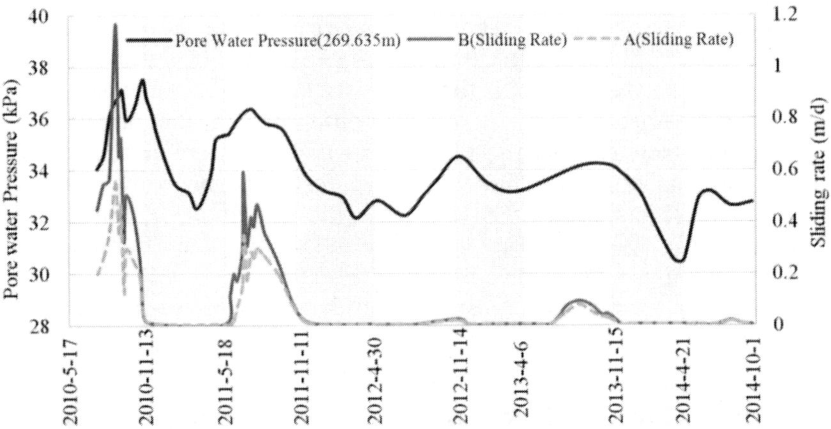

FIGURE 13. Horizontal displacement rates at point a and point b on the landslide mass and pore water pressure at point d (19.4 m depth) in k178 + 530 section of bei'an-heihe highway.

Figure 13 illustrates changes in the displacement rates at point A and point B on the landslide mass, and in pore water pressure at point D over time. To obtain the displacement rate, displacement data collected using the RTK-GPS real-time dynamic deformation monitoring system were divided by the monitoring interval.

RESULTS, ANALYSIS, AND DISCUSSION

The permafrost distribution determined by the present study is consistent with previous investigations (Wang et al., 2014a, 2015). In the current study, meteorological data of Sunwu County, 30 km south of the study area, was analyzed to examine the climate change in the study area. The causes of landslide in the study area were evaluated preliminarily by means of permafrost distribution, topography, geological survey of the study area, and investigation on the specific case of roadbed instability occurring in section K176 + 900 – 178 + 200 on a Bei'an-Heihe secondary road in August 2000.

The Sunwu County meteorological data show that, over the 60 year period from 1954 to 2013, the variation in atmospheric precipitation was 8.93%, and that this had little long-term effect on the thermal state of the soil in the study area. As shown in Figure 4, the increase in the average annual minimum temperature was substantially higher than the increase in the average annual maximum temperature. This suggests that the main cause of the temperature rising was not the increase in the amount of direct radiation, but rather the decrease in the heat demand of the soil from the atmosphere and the increase in the ability of the soil to radiate heat into the atmosphere (Chen et al., 2006). The difference in the amplitude of temperature increase between the average annual maximum temperature and the average annual minimum temperature was caused by the latent heat of permafrost thaw. The fact that the amplitude of temperature increase in the average annual maximum temperature and in the average annual minimum temperature differed by 47.05% over a 60 year period demonstrates that the permafrost in the study area is decreasing in amount, going through rapid degeneration.

In Figure 7, permafrost is not found in section K176 + 900 – 178 + 200 (where overall roadbed instability occurred) of the previous Bei'an-Heihe secondary road. This is because the ground temperature data used to plot Figure 7 are the ETM+ data collected by satellite Landsat7 in 2009, whereas the roadbed instability occurred in August 2000. After 9 years, the permafrost in this road section had degraded entirely. In Figure 7, along the road direction on both sides of point B, which is to the right of the road section K178 + 530, there are permafrost distributions. Yet in Figure 8E, in borehole B no permafrost was found; in addition, in the exploration of line G using a high-density resistivity method and GPR, no data abnormality was detected below point B within 15 m along line G, demonstrating that there was no permafrost distribution. This may be because permafrost in this location was at a stage of severe degradation; and although the soil temperature was low, the phase transition of soil moisture had already taken place and had transitioned to warm permafrost or melting permafrost. It may also be related to the resolution of the Landsat7 satellite ETM+ data.

From Figure 11, it can be seen that each ear the initial soil freezing in the study area occurred in a relatively concentrated time period, between 11 November and 15 November. But for the same measurement point, the time needed for seasonal frost to complete thawing varied rather largely from year to year. This may be related to the amount of autumn precipitation and winter snowfall in the previous year, as well as the thermal condition of soil water near the maximum seasonal frost depth, but needs to be further investigated.

As shown in Figure 12, at point D the pore water pressure started to rise in the spring of each year, reached the peak value in the summer, and then fell, thus showing annual periodic changes. But on an interannual time scale, pore water pressure showed an overall downward trend. This change was not correlated with atmospheric precipitation. Linear analysis showed that, over the monitoring period, the atmospheric precipitation in the study area increased by 0.318 mm, whereas the pore water pressure at point D decreased by 3.675 kPa. In other words, the groundwater level dropped by 36.75 cm, and the amount of groundwater reduction was 1155 times the increase in the atmospheric precipitation over the same period.

From Figure 13, it can be seen that the slip rates at measurement points A and B on the landslide mass correspond well to pore water pressure at point D. During the monitoring period, with the snowmelt and thawing of seasonal frost each spring, the pore water pressure in the hillside soil gradually increased, and the landslide mass started to slip. During the concentrated rainfall in the summer, the pore water pressure reached the peak for the year, and the slip rate of the landslide reached the peak value simultaneously. Along with the landslide movement, the soil pore water pressure gradually decreased, and the slope sliding gradually stagnated. The landslide movement had notable characteristics of annual periodicity and seasonal activity.

Figures 9, 13 reveal that changes in the slip rates of points A and B on the landslide mass were consistent. Over the monitoring period, the displacement of point B was 1.65 times that of point A. We deduced that the movement of point A might be occurring due to the force coming from the rear part of the landslide mass. From Figure 13, it can also be seen that the annual maximum pore water pressures at point D from 2010 to 2014 were 37.52, 36.37, 34.53, 34.14, and 33.19 kPa, respectively, declining year by year, and the decline in amplitude was relatively large in 2012. The maximum slip rates of point B from 2010 to 2014 were 115.7, 59, 2.55, 9.63, and 2.2 cm/d, respectively. Rapid sliding mainly occurred in 2010 and the summer of 2011. The slip rate decreased year by year, and the largest decline occurred in 2012. But in the summer of 2013, there was a slight increase in the slip rate. The reason behind these phenomena might be related to the fact that at points E and F (Figure 8), which are located outside the lateral edge of the slope at point B, permafrost thawed and supplemented water to the landslide mass. In addition, this might be related to the amount of summer precipitation. As shown in Figure 12, in 2013 the summer atmospheric precipitation reached the highest value over the entire monitoring period, 672 mm; this was 14.33 mm higher than the second highest summer precipitation (occurring in 2012) and 1.46 times the mean summer atmospheric precipitation over the monitoring period. We thus concluded that, in the monitoring period, the slip rate of Landslide K178 + 530 was mainly controlled by pore water pressure at the tailing edge of the landslide mass, and meanwhile it was also affected by the amount of summer precipitation in the study area. The pore water pressure at the tailing edge of the landslide mass was related to the thawing of permafrost at the lateral edge of the slope. As permafrost thawed year by year, its volume was reduced, which gradually reduced the pore water pressure at the tailing edge of the slope. As a result, the slip rate of the landslide mass was gradually reduced year by year.

CONCLUSION

In this study, we performed a comprehensive analysis of climate change data in the study area and data on the displacement, ground temperature, and pore water pressure at measurement points on Landslide K178 + 530 in Bei'an-Heihe Highway, and from this analysis we obtained the following main conclusions.

First, affected by the global climate change, the northwest section of the Lesser Khingan Range located at the southern boundary of the high-latitude permafrost region in northeast China experiences temperature increase, and the amplitude of the increase in average annual minimum temperature is 3.45 times that in average annual maximum temperature. Permafrost degradation is severe. Permafrost degradation will lead to further temperature rise in this region, and the northward movement of the southern boundary of the permafrost region will also accelerate.

Second, permafrost thaw causes the number of landslides in the northwest section of China's Lesser Khingan Range to increase, which results in topographical changes in this region. Human construction activity accelerates the thawing of permafrost.

Third, water seepage from thawing permafrost and infiltration of concentrated summer precipitation together increase the local moisture content in hillside soil. This is the main cause of landslides in the northwest section of the Lesser Khingan Range in China.

Fourth, in the northwest section of China's Lesser Khingan Range, landslide movement begins in the summer season with concentrated precipitation each year, and gradually stops in the autumn when the soil freezes. The landslide movement clearly has characteristics of seasonal and annual periodicity.

Fifth, the slip rate and movement process of landslides in the northwest section of China's Lesser Khingan Range are controlled by the thawing process of permafrost on and near the landslide mass. As the permafrost thaws year by year, its volume is reduced, and its ability to supplement water to the landslide mass is also reduced. This causes the slip rate of the landslides to decrease year by year until it reaches zero, and then the landslides tend to become stable.

CONFLICT OF INTEREST STATEMENT

The authors declare that the research was conducted in the absence of any commercial or financial relationships that could be construed as a potential conflict of interest.

ACKNOWLEDGMENTS

We thank the Science and Technology Project of the Chinese Ministry of Transport (2011318223630) and the International Landslide Research Program (IPL-167) for funding support. We are also grateful to an anonymous referee who helped improve the manuscript.

REFERENCES

1. Ballantyne, C. K., Sandeman, G. F., Stone, J. O., and Wilson, P. (2014). Rock-slope failure following Late Pleistocene deglaciation on tectonically stable mountainous terrain. *Q. Sci. Rev.* 86, 144–157.
2. Blunden, J., and Arndt, D. S. (2011). *State of the Climate in 2011*. American Meteorological Society. Available online at: http://www.ncdc.noaa.gov/bams-state-of-the-climate/
3. Chang, X. L., Jin, H. J., and He, R. X. (2013). Review of permafrost monitoring in the northern Da Hinggan Mountains, Northeast China. *J. Glaciol. Geocryol.* 35, 93–100.
4. Chang, X. L., Jin, H. J., Yu, S. P., Sun, H. B., He, R. X., Luo, D. L., et al. (2011). Influence of vegetation on frozen ground temperatures the forested area in the Da Xing'anling Mountains, Northeastern China. *Acta Ecol. Sin.* 31, 5138–5147.
5. Chen, J., Sheng, Y., and Cheng, G. D. (2006). Discussion on protection measures of permafrost under the action of engineering from the point of earth surface energy balance equation in Qinghai Tibetan Plateau. *J. Glaciol. Geocryol.* 28, 223–228.
6. Ding, Y. H., Ren, G. Y., and Shi, G. (2006). National assessment report of climate change (I): climate change in China and its future trend. *Adv. Clim. Change Res.* 2, 3–8.
7. Eu-Fp7. (2008). *ACQWA: Assessing Climate Impacts on the Quantity and Quality of Water A Large Integrating Project Under EU Framework Programme 7 (FP7), A Summary for Policymakers.* Available online at: www.acqwa.ch
8. Fischer, L., Hugge, C., Kääb, A., and Haeberli, W. (2013). Slope failures and erosion rates on a glacierized high-mountain face under climatic changes Earth Surf. *Process. Landforms* 38, 836–846.
9. Grab, S. W., and Linde, J. H. (2014). Mapping exposure to snow in a developing African context: implications for human and livestock vulnerability in Lesotho. *Nat. Haz.* 71, 1537–1560.
10. Guo, D. X., Wang, S. L., Lu, G., Dai, J. B., and Li, E. Y. (1981). Regionalization of permafrost in the Da and XiaoXing'anling Mountains in northeastern China. *J. Glaciol. Geocryol.* 3, 1–9.
11. Guo, Y., Canuti, P., Strom, A., Hideaki, M., and Shan, W. (2013). The First Meeting of ICL landslides in cold regions network, harbin, 2012. *Landslides* 10, 99–102.
12. Haeberli, W. (2013). Mountain permafrost - research frontiers and a special long-term challenge. *Cold Reg. Sci. Technol.* 96, 71–76.
13. He, R. X., Jin, H. J., Chang, X. L., Lv, L. Z., Yu, S. P., Yang, S. Z., et al. (2009b). Degradation of permafrost in the northern part of northeastern China: present state and causal analysis. *J. Glaciol. Geocryol.* 31, 829–834.
14. He, R. X., Jin, H. J., Lv, L. Z., Yu, S. P., Chang, X. L., Yang, S. Z., et al. (2009a). Recent changes of permafrost and cold regions environments in the northern part of northeastern China. *J. Glaciol. Geocryol.* 31, 525–531.
15. ICL. (2014). *"The 2014 Beijing Declaration Landslide Risk Mitigation: Toward a Safer Geo-environment."* Available online at: http://iplhq.org/category/home/
16. ICL (International Consortium on Landslides). (2012). *"International Consortium on Landslides Strategic Plan 2012-2021- To create a safer geo-environment."* Available online at: http://iplhq.org/category/home/
17. IPCC. (2013). *Summary for Policymakers. Working Group I Contribution to the IPCC Fifth Assessment Report Climate Change 2013: The Physical Science Basis.* Cambridge, UK: Cambridge University Press.
18. IPCC (Intergovernmental Panel on Climate Change). (2007). *Summary for Policymakers. Climate Change 2007: The Physical Science Basis. Contribution of Working Group I to the Fourth Assessment Report of the Intergovernmental Panel on Climate Change.* Cambridge: Cambridge University Press.

19. Jin, H. J., Li, S. X., Wang, S. L., and Zhao, L. (2000). Impacts of climatic change on permafrost and cold regions environments in China. *Acta Geogr. Sin.* 55, 161–173.
20. Jin, H. J., Wang, S. L., and Lv, L. Z. (2009). Features of permafrost degradation in Hinggan Mountains, northeastern China. *Sci. Geogr. Sin.* 2, 223–228.
21. Kliem, P., Buylaert, J. P., Hahn, A., Mayrd, C., Murray, A. S., Ohlendorf, C., et al. (2013). Magnitude, geomorphologic response and climate links of lake level oscillations at Laguna Potrok Aike, Patagonian steppe. *Q. Sci. Rev.* 71, 131–146.
22. Nussbaumer, S., Schaub, Y., Huggel, C., and Nat, A. W. (2014). Risk estimation for future glacier lake outburst floods based on local land-use changes. *Hazard. Earth Sys. Sci.* 14, 1611–1624.
23. Shan, W., Guo, Y., Wang, F., Marui, H., and Strom, A. (2014a). *Landslides in Cold Regions in the Context of Climate Change.* New York, NY: Springer, Environmental Science and Engineering.
24. Shan, W., Guo, Y., Zhang, C., Hu, Z., Jiang, H., and Wang, C. (2014b). Climate-change impacts on embankments and slope stability in permafrost regions of bei'an-heihe highway. *Landslide Sci. Safer Geoenviron.* 1, 155–160.
25. Shi, P. J., Sun, S., and Wang, M. (2014). Climate change regionalization in China (1961-2010). *Sci. China: Earth Sci.* 44, 2294–2306.
26. Starnberger, R., Drescher, S. R., Reitner, J. M., Rodnight, H., Reimer, P. J., and Spötl, C. (2013). Late Pleistocene climate change and landscape dynamics in the Eastern Alps: the inner-alpine Unterangerberg record (Austria). *Q. Sci. Rev.* 68, 17–42.
27. Stoffel, M., Tiranti, D., and Huggel, C. (2014). Climate change impacts on mass movements - case studies from the European Alps. *Sci. Total Environ.* 2014, 1255–1266.
28. Sun, G. Y., Yu, S. P., and Wang, H. X. (2007). Causes south borderline and subareas of permafrost in Da Hinggan Mountains and Xiao Hinggan Mountains. *Sci. Geogr. Sin.* 27, 68–74.
29. Wang, B., Sheng, Y., and Liu, J. P. (2001). Distribution and degradation of permafrost in Xiao Hinggan Mountains along the Heihe -Dalian Highway. *J. Glaciol. Geocryol.* 23, 302–306.
30. Wang, C., Shan, W., Guo, Y., Hu, Z., and Jiang, H. (2014a). Permafrost distribution study based on landsat ETM+ imagery of the northwest section of the Lesser Khingan Range. *Landslide Sci. Safer Geoenviron.* 3, 529–534.
31. Wang, C., Shan, W., Guo, Y., Hu, Z., and Jiang, H. (2015). Permafrost distribution research based on remote sensing technology in northwest section of lesser Khingan Range in China. *Eng. Geol. Soc. Territory* 1, 285–290.
32. Wang, S. H., Qi, J. L., Yin, Z. Y., Zhang, J. M., and Ma, W. (2014b). A simple rheological element based creep model for frozen soils. *Cold Reg. Sci. Technol.* 106, 47–54.
33. Wei, Z., Jin, H. J., and Zhang, J. M. (2010). Prediction of permafrost changes in Northeastern China under a changing climate. *Sci. China Earth Sci.* 41, 74–84.
34. Zhang, Y., Wu, Q. B., and Liu, J. P. (2001). Distribution characteristics of the permafrost in the section from Heihe to Bei' an in the Xiao Hinggan Mountains. *J. Glaciol. Geocryol.* 23, 312–317.
35. Zhou, Y. W., and Guo, D. X. (1982). Principal characteristics of permafrost in China. *J. Glaciol. Geocryol.* 4, 1–19.
36. Zhou, Y. W., Wang, Y. X., Gao, X. W., and Yue, H. S. (1996). Ground temperature, permafrost distribution and climate warming in northeastern China. *J. Glaclol. Gecryol.* 18, 139–147.

CHAPTER 13

Geophysical and Geotechnical Investigations of a Landslide in Kekem Area, Western Cameroon

Pouyon Dieudonné Epada[1,2*], Ganno Sylvestre[3], Tabod Charles Tabod[1]

[1]*Laboratory of Geophysics, University of Yaounde I, Yaounde, Cameroon*

[2]*National Civil Engineering Laboratory, Yaounde, Cameroon*

[3]*Department of Earth Sciences, University of Yaoundé I, Yaounde, Cameroon*

ABSTRACT

Geophysical and geotechnical surveys were conducted in the Western Cameroon (Kekem area) following a landslide on argillaceous material in order to understand the triggering processes and mechanisms of this landslide and to assess the stability of the slope. The geophysical soundings consisting of vertical electrical soundings with the Schlumberger electrode array configuration were carried out to monitor the behaviour of electrical resistivity in the landslide. Geoelectrical data showed a zone of low resistivity values identified as a clayey sand-filled aquifer. This aquifer played an important role in the triggering process of the landslide. Geotechnical soundings showed that the aquifer had a thickness of 7.0 m. The depth from the landslide crest level to the failure surface reached 3.0 m and 20.6 m. Laboratory tests were then carried out in order to evaluate the cohesion of the soil and the angle of internal friction, and to calculate the safety factor in view of making a stability analysis. The laboratory results exhibited a soil with low consistency, almost doughy. The mean value of the safety factor (1.4) been lower than the slope stability coefficient (1.5), revealed that the slope is unstable, likely to know at any moment a reactivation of the slide. This study showed that electrical soundings coupled with geotechnical surveys are useful tools for the characterization of landslides.

Keywords: Landslide; Resistivity; Argillaceous material; Basal undercutting; Kekem; Cameroon

1. INTRODUCTION

Geophysical methods are more and more used for landslide investigations. Soil modifications due to ground movement generally lead to the change of physical parameters of the sub soil. The measure of these parameters can then be used to image the movement of the land mass and monitor its evolution. Two main situations can potentially occur: 1) the geophysical contrast is related to a lithological change [1,2]; 2) the movement takes place within a homogeneous geological layer, but alters its physical characteristics. In this case, the geophysical contrast exists between the moving mass and the unaffected consolidated mass [3-6]. The geophysical contrast is then related to the cumulated or individual effects of the mechanical dislocation, the alteration and/or the increase of water content. Geophysical surveys applied to ground movements can be used to estimate the lateral extent of the landslide, the depth to the failure surface [7,8]; and also to detect water in the active mass [9].

Landslides are regular in Western Cameroon, and some of them often lead to human deaths; yet their causes are not sometimes well understood. In this paper, we analyze a landslide which occurred at Kekem (Western Cameroon) in 2007. The aim of this study is to understand the triggering processes and the mechanism of this landslide as well as examine the possibility of another landslide by seeking natural causes and anthropogenic factors. The methodology used is as follow:

- The study of the geological and geomorphological context of the site will provide useful information on the instability processes of the slope;
- The use of vertical electrical soundings with a Schlumberger array in order to monitor the behaviour of electrical resistivity in the sliding area and thus characterize the geological materials in the zone. Electrical resistivity is an interesting physical parameter to image

contrasts between different geological units, or the variations of the fracturing rate within the same geological structure [10,11].

- The use of geotechnical tests in order to define mechanical characteristics of the soils (cohesion, angle of internal friction, consistency index and plasticity index).

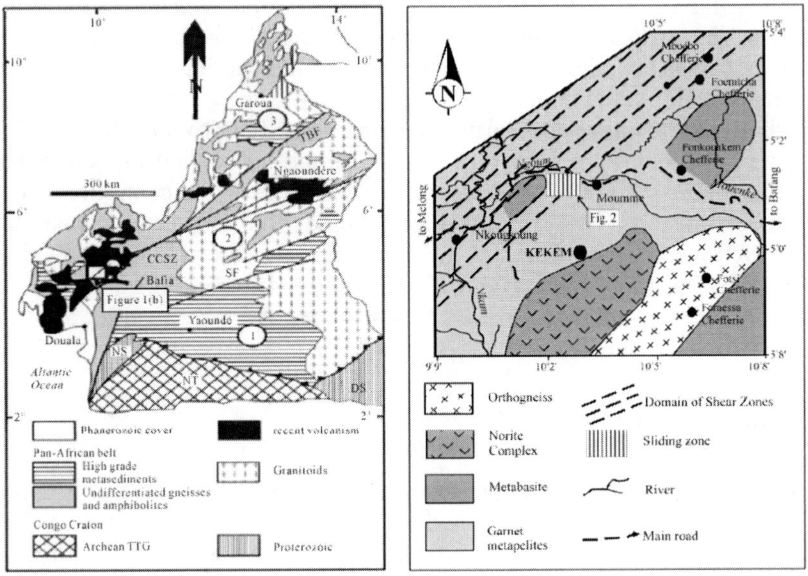

(a) (b)

Figure 1. (a) Geological map of Cameroon with the three main domains of the Pan-African orogenic belt, from [24,28]. (1) southern domain corresponding to the Yaounde series thrusted onto the Congo Craton; (2) Central domain; (3) Northern domain; CCSZ: Central Cameroon Shear Zone; DS: Dja series; NS: Nyong series; NT: Ntem complex; SF: Sanaga fault; TBF: Tibati-Banyo Fault. Location of Kekem area is marked by a square; (b) Geological map of the Kekem area, modified from [28]. Sliding zone is represented by box.

2. GEOLOGICAL SETTING

2.1. Topographic Context

The complex landslide which affected the hillslope of Jenbong (Kekem) occurred in a rough topographic context. This chaotic topography is marked by escarpments, cracks of regression upstream and dislocated blocks; a vegetation little dense and thinned out by places marked by trees, some dead, others tilted upstream. The hillslope is disturbed in its foot by a river (Ngoum) which draws meanders and carries on a strong erosive activity at the foot of the slope; it is thrown in the Nkam.

The crest of the slide which peaks in 2200 m roughly draws an arc of a circle. At its southeastern side, the slope has angle of 20%. The northwestern side is steeper and has a convex slope, which is of the order of 54%. The vertical distance between the crest and the foot of the hillslope is 57 m. Under the summit, the northwestern side of the slope locally forms a quasi-circular catchment area. This catchment area gives rise to torrents which affect the superficial formations and contribute to notch the hillside of several meters (approximately 20 m), forming a shredded landscape. The grounds of the slope are exploited, in particular the debris cones because of their fertility. At the top and all around the slope, are inhabited constructions; a clear growth of the population produces an intensification of the agro-pastoral activity and a widespread clearing. This strong agro-pastoral activity increases the risk of instability of the slope; until a recent past, ground movements of low magnitude were observed there.

2.2. Lithologic Context

In Cameroon, two main formations are distinguished: the Archean Congo Craton to the south and the Pan-African Belt to the north (**Figure 1**(a)). The Pan-African belt also known as Central African Fold Belt (CAFB) or Pan-African North Equatorial Fold Belt [12], is a Neoproterozoic orogen linked to the Trans-Saharan belt of western Africa and to the Braziliano Orogen of northeastern Brazil. The Neoproterozoic belt in Cameroon is subdivided into three main domains [13-17], from south to north (**Figure 1**(a)):

1) The southern domain comprises Neoproterozoic metasedimentary units, such as the Ntui-Betamba, Yaoundé and Mbalmayo units; the protoliths of these units were deposited in a passive margin environment at the northern edge of the Congo Craton and were metamorphosed under high P conditions at 616 Ma [12,15,18]. The rocks of this southern domain were thrusted onto the Archean Congo Craton towards the south [19]. The thrust continues towards the east, forming the Oubanguides Nappe in the Central African Republic;

2) The central domain is located between the Sanaga fault to the south and the Tibati-Banyo fault to the north. These large NE-striking transcurrent faults, as well as the Adamaoua fault inside the central domain, are regarded as possible prolongations of the major shear zones of NE Brazil in a pre-drift Gondwana reconstruction [20]. This central domain consists of

archean to paleoproterozoic high-grade gneisses intruded by widespread Neoproterozoic syntectonic plutonic rocks of high-K calc-alkaline affinities [13,14,17,21-26];

3) The northern domain consists of subordinate 830 Ma-old metavolcanics of tholeiitic and alkaline affinities associating with metasediments known as the Poli series. Widespread 630 - 660 Ma-old calc-alkaline granitoids, presently orthogneissified, result from a major episode of crustal accretion. A Paleoproterozoic crustal source in this region is attested by the presence of 2-Ga old inherited zircons in the granitoids [27].

The study area (Kekem) is located to the southwestern part of the central domain (**Figure 1**(b)). Recent works [28,29] distinguished two lithologic associations: a metamorphic unit intruded by magmatic complex. The metamorphic unit is composed of metabasites (mainly garnetpyroxene gneisses, amphibolites and pyroxenites) and metasediments (mainly garnet-biotite/sillimanite gneisses, garnet-biotite gneisses). All these rocks outcrop as inselberg, flagstones, lenses or boudins. Granoblastic textures prevail in all rock types, although flaser and mylonitic ones are also commonly observed. It records a first phase of deformation D_1 of tangential tectonics with pressure peak of a prograde metamorphism developed in the high temperature amphibolites facies conditions (720°C/12 Kbars - 750°C/7 Kbars; [29]). The magmatic unit comprises more or less orthogneissified granitoids, which crop out close to shear zones characterizing the second phase of deformation D_2 and were emplaced during D_2, and post D_2 granitoids (biotiteand amphibole-rich granites, two micas granites, monzodiorite, monzonite and norite). The second phase of deformation D_2 is of transcurrent tectonics and took place between 580 to 560 Ma (U-Th-Pb monazite age, [28]). Structures of this phase transpose early D_1 structures and are coeval with granite emplacement.

3. MORPHOLOGY AND CHARACTERISTICS OF THE SLIDING

On October 20, 2007, towards the end of the great rainy season, a landslide occurred at Kekem on the National road No. 5 connecting Bafang to Melong (Western, Cameroon). This sliding took place exactly at a village called Jenbong. The slipped mass of earth covered a volume of about 30,000 $m^3$of materials, sweeping off road over a distance of close to 100 m and about 4 m deep, destroying houses and plantations, and killing one woman. It occurs in geological formations particularly sensitive to this type of movement because of the heterogeneity of their structure and the relative impermeability of some layers. The mass of materials put in motion has a length of about 152 m and a width of more than 53 m at the level of the crest of the sliding. It widens in its median part. The scar of wrenching shows a rejection of 5 m of height. A series of streaks notched in the clays testify to the traction which affected the ground. The intense fracturing of the clayey benches causes the detachment of some pieces of rocks and provokes crumblings which continue to feed the body of the sliding. The body of the sliding is constituted by a mass of chaotic materials. Under a lateritic cover of 1 to 2 m, are greyish clays encrusted with blocks of granitic rock more or less altered of variable size, some being able to reach 2 to 3 m of diameter. From the head of the sliding to the median part, a series of secondary fractures, rejections less important than the main steep slope, gives to the body of the sliding a structure in steps. These steps have a surface weakly tilted towards the hillslope. Given the absence of channels drainage, the secondary fractures in counterslope allow the accumulation of rainwater, which gives to these steps a muddy environment during the rainy season. The affected area covers a surface of approximately 19.8 hectare. In September 2008, a mudslide was observed at the same place

4. METHODS

4.1. Geoelectrical Soundings

Three vertical electrical soundings S_1, S_2 and S_3 were carried out using a Schlumberger array (**Figure 2**). All profiles are almost parallel to the direction of the slide. Directions of expansion and the length of profiles were constrained by topography; there was only one direction in which electrodes could be taken a sufficient distance in a straight line. The geoelectrical apparatus used for this campaign was the resistivimeter TERRATEST TT 800. The data processing and inversion were made using GeoElect.Mod. This software uses the Gosh [30] linear digital filter.

4.2. Geotechnical Tests

1) In situ tests

Thirty four (34) tests using the heavy dynamic penetrometer were carried out. Ten (10) sampling manual wells were dug on the stable part of the slope in order to take undisturbed soil samples. The heavy dynamic penetrometer test enables to appreciate the soil resistance and is useful to determine the depth to the failure surface and the aquifer thickness. Undisturbed soils samples taken from the wells were used for physical and mechanical purposes in the laboratory.

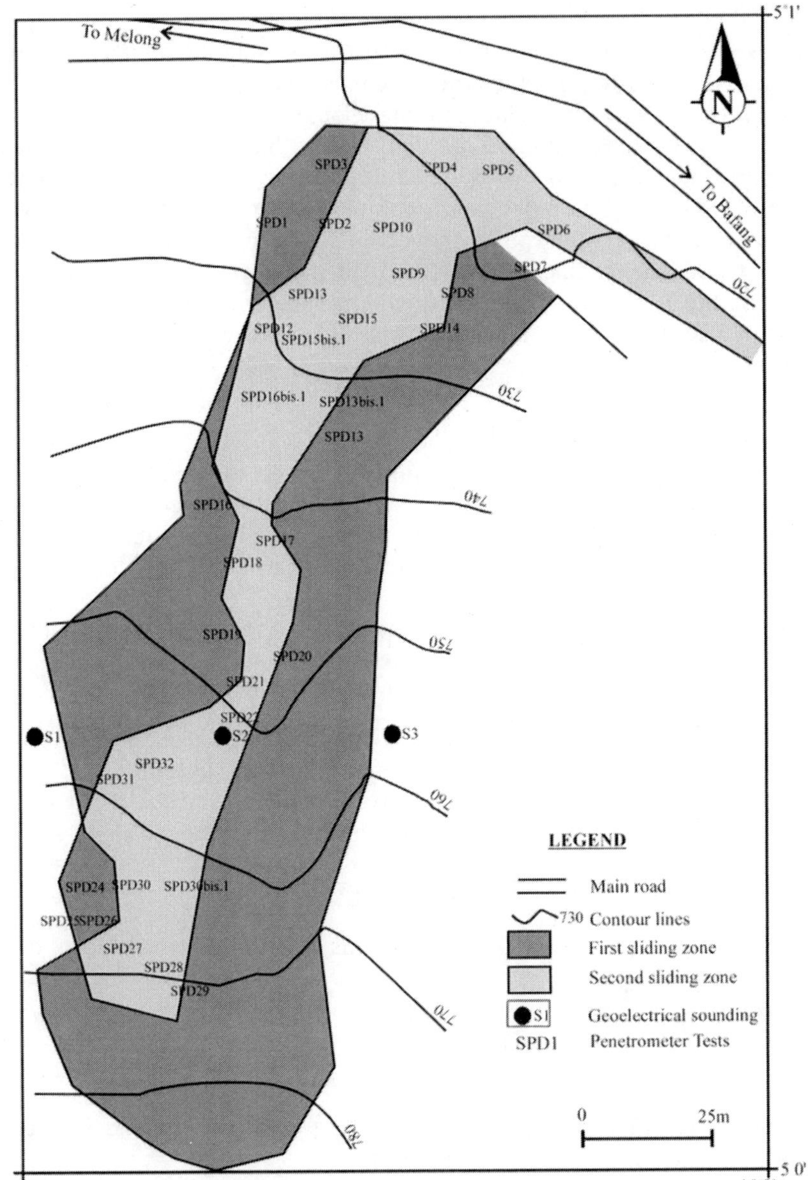

Figure 2. Map of the sliding zone and location of geoelectrical soundings and penetrometer tests.

2) Laboratory tests

Lab tests were carried out on the undisturbed samples taken from the manual wells in order to determine the cohesion of the soil and the angle of internal friction and to calculate the safety factor to evaluate the stability of the slope. The adopted method of calculation is that of Fellenius because it underestimates the value of the factor of safety and thus goes in the direction of safety. The hillslope of Jenbong (Kekem) having already been the subject of several ground movements, the minimum design slope stability coefficient was taken equal to 1.5 because of the strong susceptibility to landslides which this slope presents, and thus high risk that it induces by causing damage and loss of lives. The following tests were then carried out:

- Atterberg limits w_L and w_P the purpose of which is to define the states of moisture corresponding to the limits between the four states of the ground (solid, semi-solid, plastic and liquid). The consistency and behaviour in each state of the soil are different, and so are its mechanical properties. The boundary between each state can be defined based on a change in the soil's behaviour. The liquid limit w_L is the water content where a soil changes from plastic to liquid behaviour, plastic limit w_P is the water content where a soil starts to exhibit plastic behaviour. These characteristics are supplemented by the plasticity index $I_P = w_L - w_P$ which characterizes the state of plasticity of a soil. It is the size of the range of water content where the soil exhibits plastic properties [31]. Atterberg limits w_L and w_P are useful parameters for clayey soils.

- Consistency index I_c (this index informs about the consistency of a soil): $I_c = \dfrac{w_L - w_P}{I_P}$ where w represents the natural water content of a soil.

For the penetrometer tests (Spd) the sliding area was divided into two zones because of the topography of the site.

5. RESULTS AND DISCUSSION

5.1. Results

Geoelectrical soundings results are given in the tables 1-3. The resistivity curves (Figures 3(a)-(c)) obtained from the geoelectrical data gave a three layered earth model:

1) A conductive surface layer of about 90 cm of thickness (resistivity ρ varies from 80 Ω·m to 160 Ω·m).

2) An intermediate layer of lower resistivity (50 Ω·m to 180 Ω·m) and thickness varying from 2.0 m to 7.0 m. Its low resistivity indicates the presence of an aquifer. Geotechnical soundings and geological investigations identified this zone as a clayey sand-filled fractured reservoir. Hydrodynamic characteristics of this aquifer are the following: effective porosity: 14%; hydraulic conductivity: 2×10^{-8} m/s; average transmissivity: $13 \times 10^{-8\,m2}$/s.

3) A deep relatively resistant layer (250 Ω·m to 400 Ω·m) with a thickness varying from 2.5 m to 18.0 m lying on a granito-gneissic basement (ρ > 1000 Ω·m).

Table 1. (a) Earth configuration of P1; (b) Measured and computed data along P1.

(a)

Layer No	Resistivity (Ohm-m)	Thickness (m)
1	160	0.6
2	180	2.0
3	130	3.0
4	400	2.0
5	60	6.5
6	400	6.0
7	1000.00	Infinity

(b)

	Apparent resistivity	
AB/2 (m)	Measured (Ohm-m)	Computed (Ohm-m)
1.0	158.71	164.57
2.0	197.30	170.65
3.0	94.51	171.26
4.0	159.26	169.00
5.0	150.87	166.42
7.0	174.36	163.34
10.0	154.53	161.87
15.0	256.87	159.79
20.0	166.71	160.63
25.0	116.81	167.16
30.0	147.26	178.72
35.0	181.64	193.65
40.0	211.37	210.49
50.0	251.02	246.20
60.0	299.49	281.54
80.00	311.11	346.41
110.00	449.46	428.43

Geotechnical soundings showed that the depth to the failure surface varies between 3.0 m and 20.6 m. In addition, these tests confirmed the aquifer thickness of 7.0 m.

The analysis of the results of the laboratory tests carried out on the undisturbed samples taken on the site emphasizes that the remained

Table 2. (a) Earth configuration of P2; (b) Measured and computed data along P2.

(a)

Layer No	Resistivity (Ohm-m)	Thickness (m)
1	80	0.6
2	160	0.9
3	70	3.0
4	400	2.5
5	150	7.0
6	5000	infinity

(b)

	Apparent resistivity	
AB/2 (m)	**Measured** (Ohm-m)	**Computed** (Ohm-m)
1.0	81.63	93.37
2.0	142.20	108.60
3.0	114.12	110.44
4.0	126.68	108.83
5.0	140.82	108.72
7.0	169.13	115.34
10.0	143.23	134.17
15.0	194.84	169.28
20.0	293.02	204.46
25.0	203.10	240.78
30.0	260.97	278.41
35.0	302.47	316.92
40.0	316.12	355.82
50.0	415.01	433.48
60.0	424.49	509.78
80.00	588.49	656.54
110.00	768.30	862.11

movable ground from there after the sliding is a rather compressible deteriorated rock and whose mechanical characteristics in the short run (the angle of internal friction φ_u and the cohesion C_u), vary respectively from 24° to 32° and from 0.013 to 0.19 bar. These results suggest materials with low cohesion.

The measurement of the Atterberg limits yielded the following values: w_L : 79%; w_P : 42%, which made it possible to have a plasticity index I_P of 37%, the ground in place thus consists of a rather plastic clay. The value of the consistency index ($I_c = 0.35$) testifies to a not very consistent soil, almost doughy.

The calculation of the stability of this slope from the geomechanical characteristics of the soil led to a mean safety factor equal to 1.4, lower than the slope stability coefficient.

5.2. Discussion: The Triggering Systems of the Kekem landslide

Three major factors (meteorological conditions, geological factors and anthropogenic causes) triggered the Kekem landslide:

1) Meteorological conditions

This landslide occurred in the course of the rainy season, after intense precipitations. These precipitations decreased the cohesion of the soil materials by the wetting of the soil matrix [32]. Heavy rains that fell during the months of September and October caused deep imbibetions of clay materials of the slope and triggered in an instantaneous and brutal way the moving of a considerable quantity of materials.

2) Geological factors

Here lie a number of factors: the structure and the lithology of the slope, and the geotechnical nature of affected materials as well as the dynamics of the river Ngoum.

The soil in the landslide zone is mainly argillaceous. Clays deteriorate generally only on very low thicknesses because of their impermeability. On tilted slopes, they deteriorate all the more quickly as the factors determining the permeability are active [33]: low cohesion, porosity, the density of the slits of withdrawal due to the regular alternation of

Table 3. (a) Earth configuration of P3; (b) Measured and computed data along P3.

(a)

Layer No	Resistivity (Ohm-m)	Thickness (m)
1	130	4.5
2	50	2.0
3	120	5.0
4	250	18.0
5	7000	Infinity

(b)

	Apparent resistivity	
AB/2 (m)	**Measured** (Ohm-m)	**Computed** (Ohm-m)
1.0	81.63	93.37
2.0	142.20	108.60
3.0	114.12	110.44
4.0	126.68	108.83
5.0	140.82	108.72
7.0	169.13	115.34
10.0	143.23	134.17
15.0	194.84	169.28
20.0	293.02	204.46
25.0	203.10	240.78
30.0	260.97	278.41
35.0	302.47	316.92
40.0	316.12	355.82
50.0	415.01	433.48
60.0	424.49	509.78
80.00	588.49	656.54
110.00	768.30	862.11

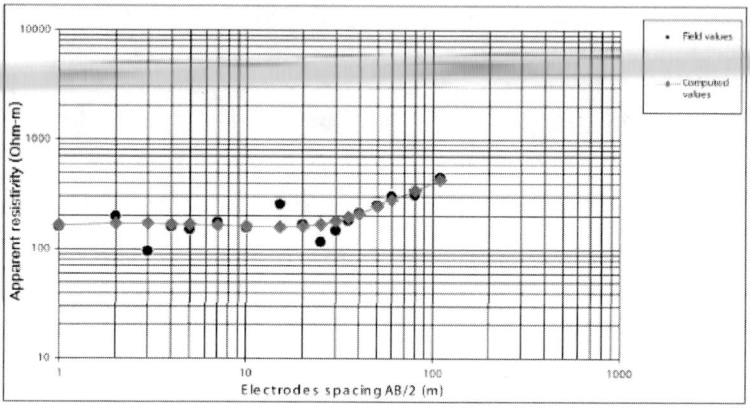

(a)

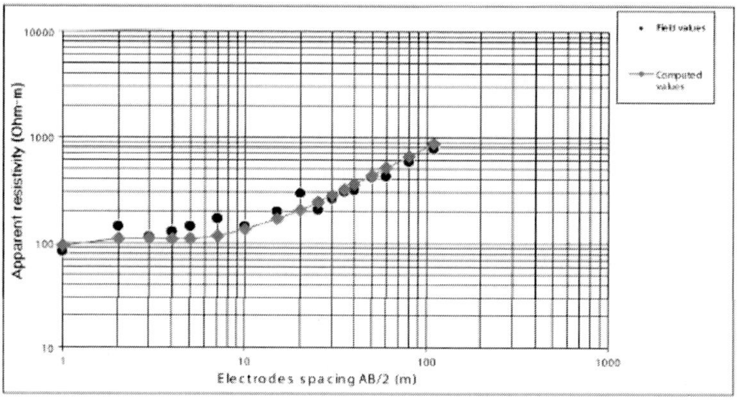

(b)

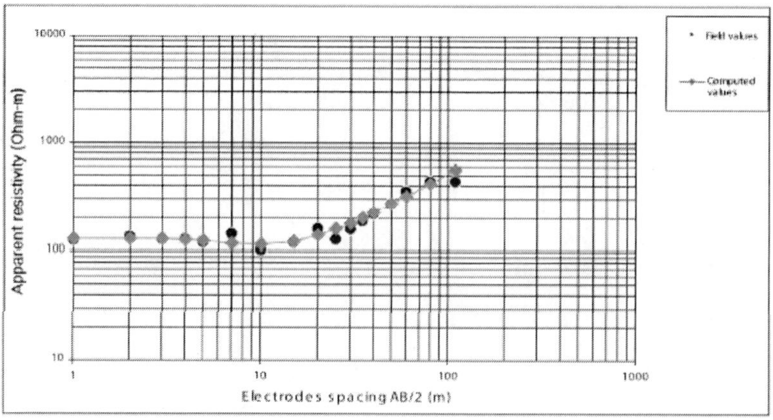

(c)

Figure 3. Resistivity curves of VES1 (a), VES 2 (b) and VES 3 (c).

moistening during the rainy season and deep desiccation during the dry season [34]. The material thus mobilized is subjected to erosion. After a short transport on the slope, a part accumulates at the bottom of the slope. The abundance of clays in materials of the slope has effects on stability according to their geotechnical and mineralogical characteristics. Indeed, as soon as the water content of clays exceeds the content of confined water and that the free water appears the resistance of materials falls. They pass from a solid state in a plastic state then to a liquid state. The density of the fracturing in a tilted gneiss material created weak zones which accelerated infiltration of water and the imbibitions of the clayey levels. These litho-structural conditions combined to a high slope angle of 54%, contributed to the destabilization of the slope and to the triggering of the landslide.

The intense undercutting of the foot of the slope by the Ngoum River contributed to the destabilization of the slope and so facilitated the landslide at Kekem. The hydrodynamic activity of this river was intensified by the important contributions of water, caused by the strong rains which fell in the area. This activity accentuated the stiffness of the slope to the foot of the slope. The call to the vacuum created by the undercutting of the base of the slope accelerated the slide.

3) Anthropogenic causes

The human activity contributed to the triggering of the landslide at Kekem in two ways: 1) the cultivating of these steep slopes renders the topmost layers of the earth loose and increase the rate of water infiltration; 2) the existence of a major carrying heavy traffic frequently provoked vibrations of the loose soil cover on the slope thus favoring the occurrence of the landslide.

6. CONCLUSIONS

The aim of this paper was to determine the triggering mechanisms of the Kekem landslide. The geoelectric study revealed zones of clayey sand-filled fractures and a spatial distribution of materials with argillaceous prevalence. Penetrometer surveys confirmed the presence of these fractures and the aquifer. These results made it possible to determine the depth to the failure surface; from the landslide crest level, it varies between 3.0 m and 20.6 m. The results of the laboratory tests showed a ground with low cohesion, not very consistent and rather compressible. The causes of the sliding of Kekem are varied: they are internal and external. The internal causes are the characteristics specific to the slope. It is about lithology, the structural disposition and the value of the angle of the slope. The external factors as for them destabilize by their actions the internal structures of the formations of the slope and cause their movement. It is of the climate, basal undercutting and the anthropogenic action. Water is the worsening factor of the landslides. It is the engine and the catalyst of the majority of the actions of the other factors. Under the action of water, the mechanical deformation of clays which form the dominant fraction of the geological units of the slope is the main cause. The geological formations rich in clays are generally impermeable with water, however, the structural disposition and the human intervention amplified the density of the zones of weakness which allowed an in-depth imbibitions. When all the factors are present, only gravity is required to provoke the sliding. This action of gravity was facilitated by basal undercutting by the river Ngoum at the foot of the hill.

For the stability study of this slope, the safety factor was calculated and found to have an average value of 1.4. This value, been smaller than the adopted value of the slope stability coefficient (1.5), revealed that the zone is unstable; likely to experience some disorders at any time. The disposition and the aspect of the body of the sliding support its reactivation. The crest of the slope favors the accumulation of rain waters. The infiltration and the percolation of water by the plans of fractures which have a strong hydraulic conductivity can play a part of lubricant for the handing-over moving and the reactivation of the sliding. It is a slower phase and less spectacular, sometimes unperceivable, but it is a latent instability which always constitutes a risk in the event of precipitations similar to those which triggered it. It is what explains the mudslide observed one year after the sliding.

This work shows that electrical soundings coupled with geotechnical surveys are useful tools for the characterization of landslides.

7. ACKNOWLEDGEMENTS

Pouyon Dieudonné Epada, Ganno Sylvestre, Tabod Charles Tabod (LABOGENIE) for providing the equipment used to carry out this study. The first author thanks Mr. NGAH OWONO Jules (Civil Engineer) for his contribution to the processing and the interpretation of geotechnical data.

REFERENCES

1. T. Glade, P. Stark and R. Dikau, "Determination of Potential Landslide Shear Plane Depth Using Seismic Refraction—A Case Study in Rheinhessen, Germany," Bulletin of Engineering Geology and the Environment, Vol. 64, No. 2, 2005, pp. 151-158.
2. V. Agnesi, M. Camardab, C. Conoscentia, A. Di Maggio, I. Dilibertoc, P. Madoniac and E. Rotiglianoa, "A Multidisciplinary Approach to the Evaluation of the Mechanism

That Triggered the Cerda Landslide (Sicily, Italy)," Geomorphology, Vol. 65, No. 1-2, 2005, pp. 101-116.

3. J. P. T. Caris and Th. W. J. Van Asch, "Geophysical, Geotechnical and Hydrological Investigations of a Small Landslide in the French Alps," Engineering Geology, Vol. 31, No. 3-4, 1991, pp. 249-276.
4. V. Lapenna, P. Lorenzo, A. Perrone, S. Piscitelli, E. Rizzo and F. Sdao, "2D Electrical Resistivity Imaging of Some Complex Landslides in Lucanian Apennine Chain, Southern Italy," Geophysics, Vol. 70, No. 3, 2005, pp. 11- 18.
5. T. Lebourg, S. Binet, E. Tric, H. Jomard and S. El Bedoui, "Geophysical Survey to Estimate the 3D Sliding Surface and the 4D Evolution of the Water Pressure on Part of a Deep Seated Landslide," Terra Nova, Vol. 17, No. 5, 2005, pp. 399-406.
6. V. Bogoslovsky and A. Ogilvy, "Geophysical Methods for the Investigation of Landslide," Geophysics, Vol. 42, No. 3, 1977, pp. 562-571.
7. D. Cummings and B. R. Clark, "Use of Seismic Refraction and Electrical Resistivity Surveys in Landslide Investigations," Bulletin of the Association of Engineering Geologists, Vol. 25, No. 4, 1998, pp. 459-464.
8. F. Bruno and F. Marillier, "Test of High-Resolution Seismic Reflection and Other Geophysical Techniques on the Boup Landslide in the Swiss Alps," Surveys in Geophysics, Vol. 21, No. 4, 2000, pp. 333-348.
9. H. J. Mauritsch, W. Seiberl, R. Arndt, A. Romer, K. Schneiderbauer and G. P. Sendlhofer, "Geophysical Investigations of Large Landslides in the Carnic Region of Southern Austria," Engineering Geology, Vol. 56, No. 3-4, 2000, pp. 373-388.
10. O. Meric, "Etude de Mouvements de Terrain par Méthodes Géophysiques," Ph.D. Thesis, Joseph Fourier University, Grenoble, 2006.
11. J. P. Nzenti, "Pétrogenèse des Migmatites de Yaoundé (Cameroun): Eléments Pour un Modèle Géodynamique de la Chaîne Pan-Africaine Nord-Equatoriale," Thèse Nouveau Doctorat, University of Nancy I, Lorraine, 1987.
12. J. P. Nzenti, P. Barbey, J. M. Bertrand and J. Macaudière, "La Chaîne Panafricaine au Cameroun: Cherchons Suture et Modèle," 15ème Réunion des Sciences de la Terre, Abstract, Nancy, 26-28 Avril 1994.
13. J. P. Nzenti, B. Kapajika, G. Wörner and R. T. Lubala, "Synkinematic Emplacement of Granitoids in a Pan-African Shear Zone in Central Cameroon," Journal of African Earth Sciences, Vol. 45, No. 1, 2006, pp. 74-86.
14. T. Ngnotué, J. P. Nzenti, P. Barbey and F. M. Tchoua, "The Ntui-Betamba High Grade Gneisses: A Northward Extension of the Pan-African Yaoundé Gneisses in Cameroon," Journal of African Earth Sciences, Vol. 31, No. 2, 2000, pp. 369-381.
15. V. Ngako, P. Affaton, J. M. Nnange and T. Njanko, "Pan-African Tectonic Evolution in Central and Southern Cameroon: Transpression and Transtension during Sinistral Shear Movements," Journal of African Earth Sciences, Vol. 36, 2003, pp. 207-214.
16. E. L. Tanko Njiosseu, J. P. Nzenti, T. Njanko, B. Kapajika and A. Nédelec, "New U-Pb Zircon Ages from Tonga (Cameroon): Coexisting Eburnean-Transamazonian (2.1 Ga) and Pan-African (0.6 Ga) Imprints," Comptes Rendus Géosciences, Vol. 337, No. 16, 2005, pp. 551- 562.
17. J. Penaye, S. F. Toteu, W. R. Van Schmus and J. P. Nzenti, "U-Pb and Sm-Nd Preliminary Geochronology Data on the Yaoundé Series, Cameroon: Re-Interpretation of the Granulitic Rocks as the Suture of Collision in the Centrafrican Belt," Comptes Rendus de l'Académie des Sciences de Paris, Vol. 317, No. 6, 1993, pp. 789-794.
18. J. P. Nzenti, P. Barbey, P. Jegouzo and C. Moreau, "Un Nouvel Exemple de Ceinture Granulitique Dans une Chaîne Protérozoïque de Transition: Les Migmatites de Yaoundé au Cameroun," Compte Rendu Académie des Sciences de Paris, Vol. 299, No. 17, 1984, pp. 1197-1199.
19. C. Castaing, J. L. Feybesse, D. Thieblemont, C. Triboulet and P. Chevremont, "Paleogeographical Reconstructions of the Pan-African/Brasiliano Orogen: Closure of an Oceanic Domain or Intracontinental Convergence between Major Blocks," Precambrian Research, Vol. 69, 1995, pp. 327-344.

20. C. Nzolang, H. Kagami, J. P. Nzenti and F. Holz, "Geochemistry and Preliminary Sr-Nd Isotopic Data on the Neoproterozoic Granitoids from the Bantoum Area, West Cameroon: Evidence for a Derivation from a Paleoproterozoic to Archean Crust," Polar Geoscience, Vol. 16, 2003, pp. 196-226.

21. M. L. Djouka-Fonkwe, B. Schulz, J. P. Tchouankoué and C. Nzolang, "Geochemistry of the Bafoussam Pan-African Iand S-type Granitoids in Western Cameroon," Journal of African Earth Sciences, Vol. 50, No. 2-4, 2008, pp. 148-167. H

22. A. Nzina Nchare, J. P. Nzenti, E. L. Tanko Njiosseu, S. Ganno and T. Ngnotué, "Synkinematic Ferro-Potassic Magmatism from the Mekwene-Njimafofire Foumban Massif, along the Foumban-Banyo Shear Zone in Central Domain of Cameroon Pan-African Fold Belt," Journal of Geology and Mining Research, Vol. 2, No. 6, 2010, pp. 142-158.

23. C. Chebeu, C. D. Ngo Nlend, J. P. Nzenti and S. Ganno, "Neoproterozoic High-K Calc-Alkaline Granitoids from Bapa-Batié, North Equatorial Fold Belt, Central Cameroon: Petrogenesis and Geodynamic Significance," The Open Geology Journal, Vol. 5, 2011, pp. 1-20.

24. A. A. Ganwa, W. Frisch, W. Siebel, G. E. Ekodeck, C. K. Shang and V. Ngako, "Archean Inheritances in the Pyroxene-Amphibole-Bearing Gneiss of the Méiganga Area (Central North Cameroon): Geochemical and Pb/Pb Age Imprints," Comptes Rendus Geosciences, Vol. 340, No. 4, 2008, pp. 211-222.

25. A. A. Ganwa, W. Siebel, C. K. Shang, N. Seguem and G. E. Ekodeck, "New Constraints from Pb-Evaporation Zircon Ages of the Méiganga Amphibole-Biotite Gneiss, Central Cameroon, on Proterozoic Crustal Evolution," International Journal of Geosciences, Vol. 2, 2011, pp. 138-147.

26. S. F. Toteu, A. Michard, J. M. Bertrand and G. Rocci, "U/Pb of Precambrian Rocks from Northern Cameroon, Orogenic Evolution and Chronology of the Pan-African Belt of Central Africa," Precambrian Research, Vol. 37, No. 1, 1987, pp. 71-87.

27. D. Tchaptchet Tchato, B. Schulz and J. P. Nzenti, "Electron Microprobe (EMP) Monazite Dating and P-T Data of the Neoproterozoic Metamorphic and Mylonitic Events in the Kekem Area, Cameroon North Equatorial Fold Belt," Neues Jahbur of Paleontology, Vol. 186, No. 1, 2009, pp. 95-109.

28. D. T. Tchato, J. P. Nzenti, E. L. Tanko Njiosseu, T. Ngnotué and S. Ganno, "Neoproterozoic Metamorphic Events in the Kekem Area (Central Domain of the Cameroon North Equatorial Fold Belt): P-T Data," Journal of Cameroon Academy of Sciences, Vol. 8, No. 2-3, 2009, pp. 91-106.

29. D. P. Gosh, "Inverse Filter Coefficients for the Computation of Apparent Resistivity Standard Curve for Horizontal Stratified Earth Geophysics," Prospectives, Vol. 19, 1971, pp. 769-775.

30. J. Costet and G. Sanglerat, "Cours Pratique de Mécanique des Sols (Deuxième Edition)," Tome I, Dunod, Paris, 1975.

31. Th. Van Asch, W. J. J. Buma and L. P. H. Van Beek, "A View on Some Hydrological Triggering Systems in Landslides," Geomorphology, Vol. 30, No. 1-2, 1999, pp. 25-32.

32. A. Faleh and A. Sadiki, "Glissement Rotationnel de Dhar El Harrag: Exemple d'Instabilité de Terrain Dans le Prérif Central (Maroc)," Bulletin de l'Institut Scientifique, Rabat, Section Sciences de la Terre, No. 24, 2002, pp. 41-48.

33. M. Borie, A. Marache, M. Frappa, J-F. Lataste, J. Riss and R. Fabre, "Application de la Méthode Electrique à l'Etude de la Stabilité de la Falaise Rocheuse des Viviers-Basques (64, Urrugne): Détermination de la Structure du Massif Rocheux et des Conditions Hydrauliques," Conference Proceeding, Journées AGAP, Lyon, 27-28 October 2004, pp. 34-37.

34. P. Mouroux, P. Margron and J. C. Pinte, "La Construction Economique sur Sols Gonflants," Edition BRGM, Paris, 1988.

CHAPTER 14

Landslide Investigation with Remote Sensing and Sensor Network: From Susceptibility Mapping and Scaled-down Simulation towards *in situ* Sensor Network Design

Gang Qiao [1], Ping Lu [1,*], Marco Scaioni [1], Shuying Xu [1], Xiaohua Tong [1], Tiantian Feng [1], Hangbin Wu [1], Wen Chen [1], Yixiang Tian [1], Weian Wang [1] and Rongxing Li [1,2]

[1]*Center for Spatial Information Science and Sustainable Development Applications, College of Surveying and Geo-Informatics, Tongji University, Shanghai 200092, China*

[2]*Mapping and GIS Lab, The Ohio State University, Columbus, OH 43210, USA*

ABSTRACT

This paper presents an integrated approach to landslide research based on remote sensing and sensor networks. This approach is composed of three important parts: (i) landslide susceptibility mapping using remote-sensing techniques for susceptible determination of landslide spots; (ii) scaled-down landslide simulation experiments for validation of sensor network for landslide monitoring, and (iii) *in situ* sensor network deployment for intensified landslide monitoring. The study site is the Taziping landslide located in Hongkou Town (Sichuan, China). The landslide features generated by landslides triggered by the 2008 Wenchuan Earthquake were first extracted by means of object oriented methods from the remote-sensing images before and after the landslides events. On the basis of correlations derived between spatial distribution of landslides and control factors, the landslide susceptibility mapping was carried out using the Artificial Neural Network (ANN) technique. Then the Taziping landslide, located in the above mentioned study area, was taken as an example to design and implement a scaled-down landslide simulation platform in Tongji University (Shanghai, China). The landslide monitoring sensors were carefully investigated and deployed for rainfall induced landslide simulation experiments. Finally, outcomes from the simulation experiments were adopted and employed to design the future *in situ* sensor network in Taziping landslide site where the sensor deployment is being implemented.

Keywords: landslide; sensor network; susceptibility mapping; remote sensing

1. INTRODUCTION

Landslides are major geo-hazards heavily impacting many regions of the world in terms of human lives and economic losses [1]. The large magnitude of natural forces that are involved in landslides makes actions of mitigation or prevention unfeasible, with exceptions for small occurrences or under particular conditions. According to [2], on the basis of methods employed landslide research can be classified into theoretical, analytical and numerical studies along with laboratory experiments, field investigations, monitoring and inventory mapping, as well as GIS/Remote Sensing application.

Landslide assessment, susceptibility mapping, monitoring (*in situ* and remotely) and early warning systems all play an important role in landslide investigations thereby being directly related to disaster reduction and hazard mitigation. Remote sensing, especially from high-resolution satellite imagery is gaining importance in landslide investigation due to its wide coverage [3]. With its increasing spatial and temporal resolution, remote sensing has been widely adopted in landslide mapping for rapid response and recovery after hazard occurrence by government agencies as well as research community [4]. When data are available, the comparison of images collected before and after the event gives relevant support to landslide recognition [5,6]. Apart from its use in emergency response, when incorporated with other vector data the landslide inventory data from landslide mapping results and historical records can be employed as a database for susceptibility mapping using GIS and statistical approaches [7–9].

Landslide susceptibility mapping is an attempt to derive spatial variation of area-based slope failure probability or instability at a regional scale. This is based on a number of factors categorized into (a) preparatory factors such as lithology and geomorphology; and (b) triggering factors such as seismicity, rainfall, land cover, and anthropogenic causes. This technique is a fundamental tool for planners, administrators, and emergency managers, who are responsible for identifying hazard zones [10,11].

As another important component in landslide research, landslide monitoring has usually been regarded as a point-based and site-specific slope stability analysis [12]. In a broad sense, the term "monitoring" refers to the observations over time of any change in the area of interest caused by the landslide process. This includes either qualitative properties (vegetation cover, urbanization, land use, surface processes, and the like) or geometric aspects (deformation monitoring) [13]. Monitoring is, on one hand, a fundamental tool for understanding and modeling all geo-processes related to landslide dynamics. On the other hand, the integration of models and observations can lead to a greater opportunity for predicting partial or full slope failures. Monitoring also plays a major role in early-warning systems, where it should provide timely and reliable information to evaluate the present risk. The closer is the relationship between monitoring and modeling, the smaller is the uncertainty of forecasted events. Indeed, both tasks can mutually benefit through the assimilation of observation data and empirical or physical models for the description of landslide behavior [14,15]. On one hand, dense and accurate observations can improve models. On the other, better models will help tailor the monitoring system to become more efficient in understanding geo-processes. This can be achieved, for example, by deployment of new sensors or by tuning the data acquisition rate according to the observed dynamics. In the latest generation of landslide monitoring systems, this capability to change the sensor parameters can be accomplished almost in an automatic way ("smart networks"). This, however, is a perspective of future development in this domain.

Landslides are complex geo-hazards requiring multiple observations to be investigated. For this reason, a large number of sensors could be installed as necessitated by the specific properties of each site. The term "sensor" is intended as in [16]. The logical connection between sensors is usually referred to as a "network", within which sensors are the "nodes". These nodes are implemented in a physical way using wired or wireless communication infrastructures [17] and different data transmission protocols [18]. In general, a Sensor Network (hereafter "SN") can comprehend the different categories of sensors that are in use for gathering information about the underground layers, about the topographic of the slope surface, and about meteorological conditions. Usually these data inputs need to be updated at high-to-medium level of frequency (*i.e.*, from minutes to within a few days). On the other hand, some low-frequency data can be useful for the completion of the observation budget. Major categories of sensors include [19]:

- Geotechnical sensors to gather observations from surface and sub-surface;
- Geophysical sensors focusing on the underground layers;
- Surveying techniques (including the Global Navigation Satellite System, or GNSS—see [20]) which provide point-wise deformation measurements;
- Remote-sensing techniques (photogrammetry, laser scanning, and Interferometric Synthetic Aperture Radar (InSAR)—see [21–24]) that can offer "area-based" observations from different platforms (spaceborne, airborne and ground-based); and
- Environmental sensors, which can cover local (e.g., whether stations), regional (e.g., meteorological radars), and wide-areas (*i.e.*, meteorological satellites).

Monitoring should be complemented by up-to-date information on triggering factors, such as earthquakes [25] or anthropogenic activities [13,26] other than rainfall. The definition of a general methodology [27] for designing of a SN for monitoring landslides is a challenging issue that depends strictly on the geological processes involved [28], on the extent of the area under investigation, on the specific features of each site, and on the available economic resources. A detailed study of the local geological properties and the inventory of past landslides in the same region, as well as the acquisition of available information on past slope failures would significantly improve the model. In addition, information concerning the surrounding region (human settlements, hydraulic networks, communication infrastructures) and important targets (such as river barrages, energy power-plants and the like) is fundamental for designing risk scenarios outlining the serious impacts that could be caused directly by the landslide or by other domino effect such as the Na-Tech disasters [29,30].

2. BACKGROUND OF THE RESEARCH

This research proposes a methodology for designing a SN, and its application to a landslide located in Sichuan province in southwestern China. Two important earthquakes ("Wenchuan" earthquake on 12 May 2008 [25] and "Ya'an" earthquake on 20 April 2013) have occurred in this area in the last five years, affecting different mountain locations. Both events were characterized by the triggering of several slope instability factors that, in the case of the 2008 earthquake, also coincided with intense rainfall resulting in floods and debris flows. In addition to the landslide runouts that occurred immediately after these earthquakes, many other slopes became unstable and represented a significant risk to man and property [31]. Considering the large area involved and the similarity between numerous unstable slopes (geometric properties, lithology, slope, soil, and ground coverage), the development of a unified, sustainable approach to landslide susceptibility mapping is quite important.

The key points of emphasis are: (i) the use of remote-sensing data together with some geo-referenced information from archives to determine the susceptible areas (see Section 5); and (ii) the construction of a scaled-down model of the slope to artificially reproduce the landslide process using a rainfall trigger (see Section 6). It is considered that future rainfall might be the highest probable triggering factor of new slope failures in the area [32]. The scaled-down model has two purposes. For one, it can be used to test the SN before it is deployed on the real slope [33]. In addition, it can provide observation results to be used in the process of numerical modeling of landslides.

3. RESEARCH AREA AND DATASET

The research area is located in Hongkou Town (Sichuan, China) within the transition zone of the Chengdu Plain and the Western Sichuan Plateau. The approximate extent of the area is 30 km^2 and is characterized by relatively high and rugged mountains with elevations ranging from 700 to 1,700 m above sea level as well as deeply incised valleys [34] (Figure 1). This area is situated in the Yingxiu-Beichuan fault zone, where earthquakes and heavy precipitation lead to geological hazards such as landslides and debris flows, for example after shock events of 2008 Wenchuan Earthquake [31]. The major mountain range in this area is the Longmenshan Mountains, mainly composed of conglomerate, carbonat, tuffaceous, arkose, limestone, and granite rock (*cf.*, National Geological Archives of China, NGAC). The 2008 Wenchuan Earthquake triggered a large number of landslides, rock avalanches, and debris flows. Some of the landslides formed natural dams in the rivers, causing the potential secondary hazard of subsequent flooding. One third of the estimated 88,000 casualties of the earthquake were thought to be caused by landslides alone [35].

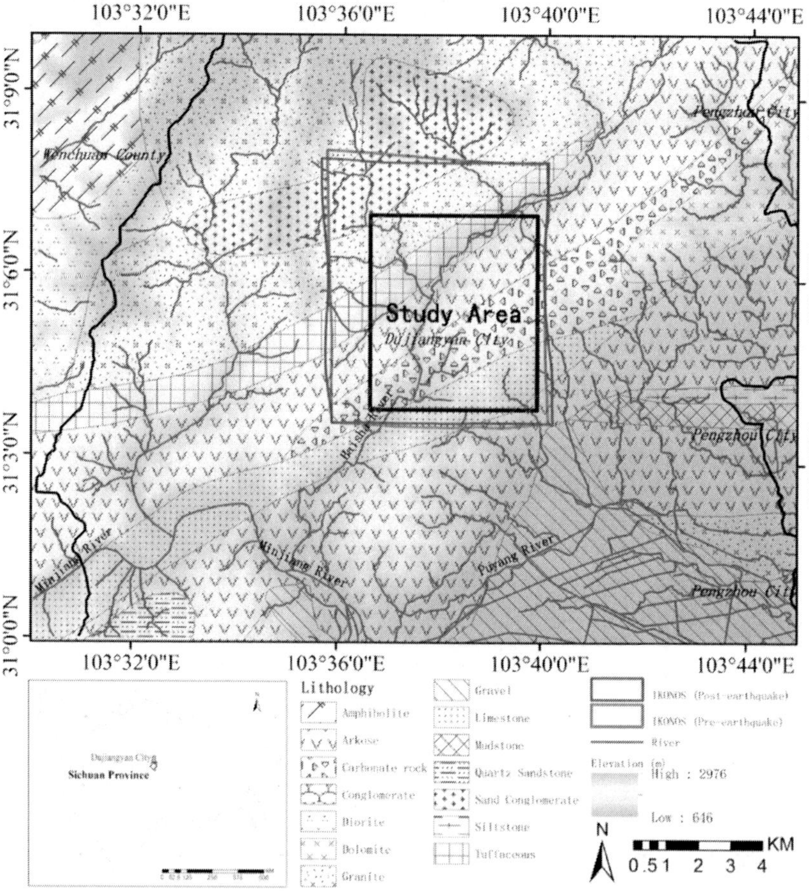

Figure 1. Study area and profile of the data coverage.

To map the landslides induced by the earthquake, two scenes of IKONOS imagery with almost the same coverage were used (Figure 1).The pre-event image was acquired on 14 September 2007, and the post-event one was taken one and a half months after the event on 28 June 2008.

In addition to this imagery, a DEM covering the research area having a grid resolution of 25 m × 25 m was provided by the National Geomatics Center of China (NGCC). Updated in 2011, this DEM is suitable to generate basic topographic information for landslide control factor analysis after the 2008 Wenchuan Earthquake. Updates of the road network and river data were obtained in 2011 from the Digital Line Graph (DLG) data offered by the NGCC. The lithology data and 2008-Wenchuan-Earthquake fault information were generated by the NGAC and the China Earthquake Networks Center (CENC), respectively.

4. METHODOLOGY

As shown in the flowchart (Figure 2), this study is composed of three parts: landslide susceptibility mapping for susceptible spot determination, a scaled-down landslide simulation experiment for SN test, and the *in situ* SN deployment of sensors for intensified observation of landslide activity. In the first part, landslides were extracted by means of an object-oriented method using the pre- and post-event IKONOS images. Then landslide control factors (including aspect, elevation, slope, lithology, distance to nearest river, distance to nearest road, and seismic intensity) were derived from the DEM and other vector data as listed in Table 1. The relationships between the distribution of the extracted landslides and the control factors were analyzed, and a landslide susceptibility map was generated based on Artificial Neural Network (ANN) to determine any susceptible spots (see Section 5 for details). In the second part, the Taziping landslide was selected as an example to design and implement a scaled-down landslide simulation platform at Tongji University (Shanghai, China). The landslide monitoring sensors were carefully investigated and deployed for this landslide simulation experiment triggered by rainfall, and conclusions concerning SN deployment were drawn. In the final part, current landslide prevention infrastructure and monitoring sensors were first introduced and discussed, and finally with the considerations from the simulation experiment, suggestions for future *in situ* SN deployment in Taziping were given.

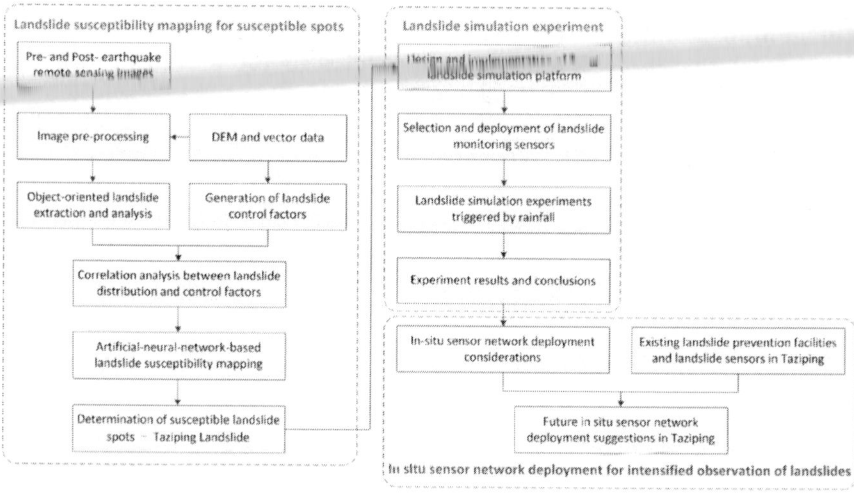

Figure 2. Flowchart of the research approach.

Table 1. Selected information on the IKONOS images and vector data used in the study.

	IKONOS Images		Vector Data				
Item	Pre-earthquake	Post-earthquake	DEM	Road	River	Lithology	Fault
Date	2007.9.14 12:12 pm	2008.6.28 12:02 pm	2011	2011	2011	2008	2008
Scale	NA	NA	1:50,000	1:50,000	1:50,000	1:500,000	NA
Resolution	1.0 m	1.0 m	25.0 m	NA	NA	NA	NA
Source	GeoEye Company (now DigitalGlobe)			NGCC		NGAC	CENC

5. LANDSLIDE SUSCEPTIBILITY MAPPING

5.1. Landslide Extraction Based on Pre- and Post-Earthquake IKONOS Images

The IKONOS images were first enhanced, rectified and geo-referenced [36] based on the corresponding points obtained from vector data such as road networks, rivers, and land-use maps. Then an object-oriented approach [37,38] was employed to extract the earthquake-induced landslides. Since such geo-hazards usually occur on higher slopes and thus destroy the vegetation-cover, for most of the rugged mountains in this area any changes in vegetation-cover between the pre and post-event images could be viewed as landslide indicators [39,40]. A

Vegetation Damage Index (VDI) is hence defined here and used for landslide recognition:

$$VDI = NDVI_{Pre} - NDVI_{Post} \tag{1}$$

where $NDVI_{Pre}$ is the Normalized Difference Vegetation Index (NDVI) [41] in pre-event image, and $NDVI_{Post}$ is the corresponding parameter in the post-event image.

A multi-resolution segmentation algorithm [37,42,43] was applied to the post-event IKONOS image to identify the homogeneous regions (namely "image objects") and the $NDVI_{Post}$ value for each object was computed. Then the image objects were overlaid with the pre-event IKONOS image to obtain the $NDVI_{Pre}$ values. An empirical threshold for the VDI of 0.36 was obtained after several trials based on visual inspection of resulting VDI map compared with observed landslides in the post-event IKONOS image in the research area. This value was used for the extraction of landslides from both images. Results were visually checked, and any errors (e.g., features such as roads sheltered by trees in the pre-event image and revealed in the post-event image, and some of the landslides with smaller VDI due to lower contrast) were edited to ensure the extraction accuracy by minimizing the number and area of the declassified landslides.

Figure 3a shows the extracted landslides overlaid on the post-event image as well as the river layer. In general, the landslide area shows dark brown contrast where it is compared with the nearby surroundings in remotely sensed images. The path of debris movement could be discerned easily, as the landslide resulted in newly denuded vegetation in comparison with the pre-event image [44]. The landslides were mainly situated on the hanging wall of the precipitous mountain ridges and on into the steep valley sides of the Baisha River and its tributaries. Most of the landslides took place in the western part of the study area with some very large ones in the southwestern section while only a few occurred in the eastern part where the terrain relief is relatively flat (see also Figure 4c). Figure 3b illustrates a detailed landslide pattern by the side of a village area with a background of the pre- and post- event images to demonstrate the apparent changes caused by debris movement. Here it can easily be observed that some buildings were damaged, and a major trunk road was buried and blocked by a large landslide.

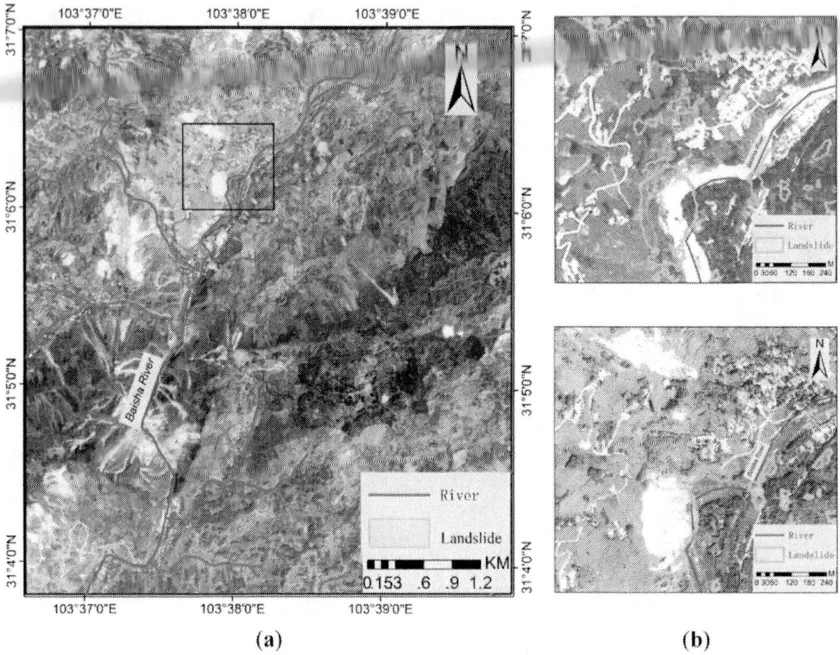

Figure 3. The landslides extracted from the two IKONOS images: (**a**) Landslide distribution overlaid on the post-event IKONOS and the river maps; and (**b**) A detailed landslide pattern (the outlined area in (a)) with a background of pre-event (upper) and post-event (lower) IKONOS images.

Table 2 shows the landslide numbers and area extracted from the remote-sensing images. For the IKONOS images having a 1.0 m spatial resolution, landslides with an area of less than 10 m^2 could sometimes be identified, but errors could also be introduced in this case, so they were removed from the statistics. In total, the extracted landslides cover a surface area of about 3.25 km^2, more than 10% of the research area, indicating severe damage by the landslides. The dimensions of individual landslide ranged from about 10 m^2 to more than 10,000 m^2. Most of the selected 740 landslides are with an area of 100–500 m^2, making almost half of the total number (41.62%), while a relatively small number (40) of large landslides (more than 10,000 m^2 in area) make the highest percentage (82.05%).

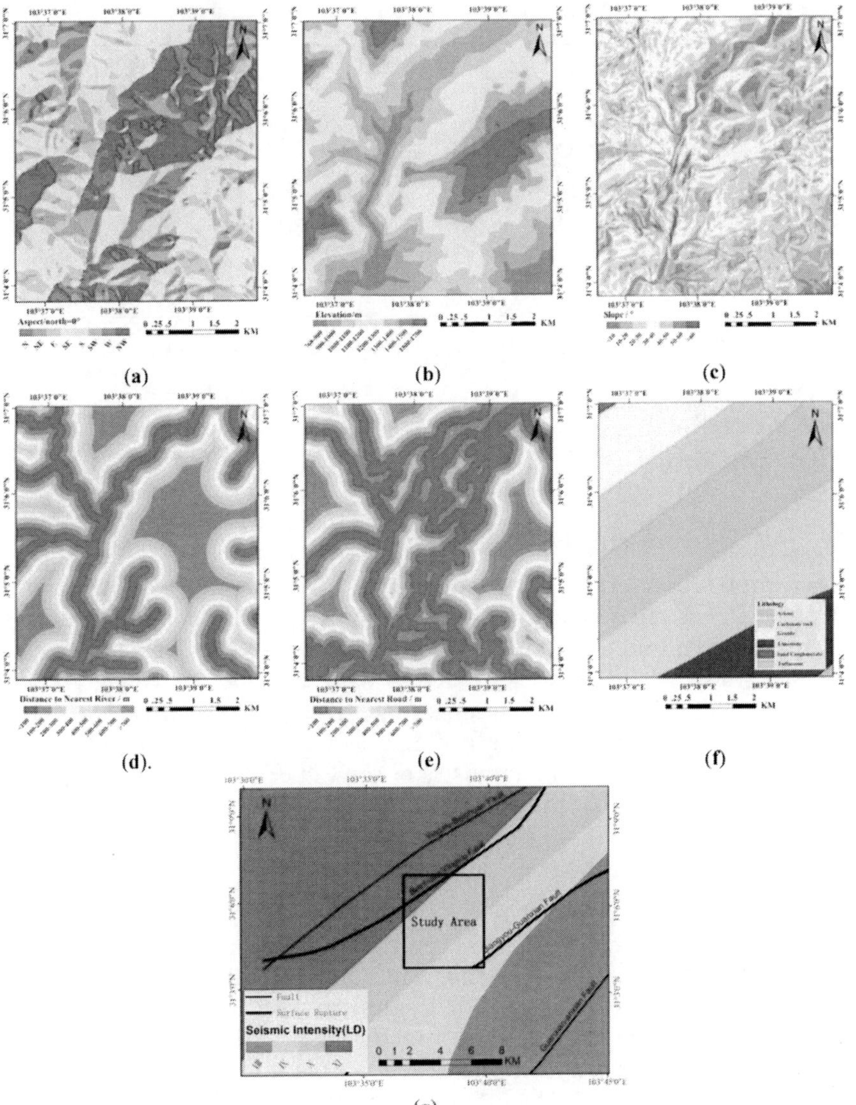

Figure 4. The topographical (**a–e**), geological (**f**), and seismic (**g**) factors for landslides in the study area. (a) Aspect; (b) Elevation; (c) Slope; (d) Distance to nearest river; (e) Distance to nearest road; (f) Lithology; (g) Seismic intensity.

Table 2. Extracted landslide statistics in the research area.

Landslide Dimension (m^2)	Landslide Area		Landslide Number	
	Area (m^2)	Percentage (%)	Number	Percentage (%)
10–50	1,718.76	0.05	58	7.84
50–100	6,659.45	0.20	88	11.89
100–500	78,931.20	2.43	308	41.62
500–1,000	62,232.23	1.91	90	12.16
1,000–5,000	271,108.12	8.34	133	17.97
5,000–10,000	163,055.36	5.01	23	3.11
>10,000	2,667,869.77	82.05	40	5.41
Total	3,251,574.89	100.00	740	100.00

5.2. Correlation between Control Factors and Landslide Distribution

5.2.1. Generation of Control Factors

The control factors mentioned above are generally divided into topographical and geological factors [45,46]. In addition, seismic factors also are taken into account for earthquake-triggered landslides [47,48]. In this study, all three types of factors were considered. Topographical factors included aspect, elevation, slope, distance to nearest road, and distance to nearest river. The geological factor adopted was lithology. The seismic factor was the intensity of the 2008 Wenchuan Earthquake. It should be noted that the distance from the epicenter (approx. 30 km) was not used for this analysis because of the relatively small extent (approx. 30 km^2) of the study area. This parameter would give a similar contribution in the whole region.

Figure 4 shows the landslide control factors in the study area: topographical (Figure 4a–e), geological (Figure 4f, lithology), and seismic (Figure 4g, seismic intensity). For the topographical factors, the aspect, elevation, and slope were derived from the DEM, and both the distance to nearest river and distance to nearest road were derived from the DLG data provided by NGCC. The lithology was generated from the 1:500,000 geological maps provided by the NGAC, and the 2008 Wenchuan Earthquake seismic intensity map was provided by the CENC.

5.2.2. Correlation Analysis between the Control Factors and Landslide Spatial Distribution

In order to analyze the correlation between control factors and landslide spatial distribution, the former were first classified into different grades according to their specific characteristics and data ranges (see Table 3).

The aspect factor was divided based on different direction classes with an amplitude of 45°. The elevation was classified based on 100-m intervals within the height range in the study area, while for the slope factor, classes of 10° were selected. For the distance-to-nearest-river/road factors, a 100-m interval was applied within the 700 m distance to obtain the grades. The lithology was classified on the basis of main rocks in the study area. The IX, X and XI seismic intensities (Richter scale) of the 2008 Wenchuan Earthquake were within the research area for further analysis.

Table 3. Detailed grades of the control factors used for correlation analysis.

Grade	Aspect (°)	Elevation (m)	Slope (°)	Distance to Nearest River (m)	Distance to Nearest Road (m)	Lithology	Seismic Intensity
1	N	768–900	<10	<100	<100	Conglomerate	IX
2	NE	900–1,000	10–20	100–200	100–200	Carbonate Rock	X
3	E	1,000–1,100	20–30	200–300	200–300	Tuffaceous	XI
4	SE	1,100–1,200	30–40	300–400	300–400	Arkose	
5	S	1,200–1,300	40–50	400–500	400–500	Limestone	
6	SW	1,300–1,400	50–60	500–600	500–600	Granite	
7	W	1,400–1,500	60–68	600–700	600–700		
8	NW	1,500–1,706		>700	>700		

On the basis of this dataset, two parameters were determined for analysis of the correlation between landslides and control factors: (1) Landslide Occurrence Rate in All landslides (LORA) and (2) Landslide Occurrence Rate in a certain landslide Grade (LORG) [49]. The former can be defined as the ratio of the landslide area within a certain grade with respect to the total landslide area extracted from satellite images of the area. The latter is the ratio of the landslide area within a certain grade to the area of this grade in the research region. LORA represents the percentage of the landslide area in a certain grade, while LORG denotes the possibility of landslide occurrence for a certain grade within the study area.

Figure 5 shows the correlation between landslides and the control factors with the blue line representing the LORA and the red the LORG. For the aspect factor, the values of both LORA and LORG are similar in the research region, and the two indexes are higher in the NE, E, SE and SW than in other directions, revealing that there are more landslides occurring in these directions. For the elevation factor, the LORA value (landslide area) is larger in the elevation range of 900–1,400 m, but the LORG is relatively smooth for all the grades, indicating no obvious correlation in elevation factor. For the slope factor, the LORA first increases and then decreases with the summit at 30°–40°, indicating the larger landslide area within the middle of the slope angle range, while the LORG ascends along with the rising slope angle, showing clear dependency between the landslide possibility and the slope angle and that higher slope areas are more unstable for the occurrence of the earthquake-triggered landslides. For the distance-to-nearest-river/road factors, a similar trend could be viewed for the LORA lines, and both values decrease progressively as the distance increases, showing that more landslides occurred near the rivers or roads. For the LORG lines, the distance-to-nearest-river [50] value has a similar degression tendency, indicating that the landslides occurred easily in valleys with flowing rivers. The distance-to-nearest-road value shows an increasing and then decreasing curve with the maximum in a range of 200-to-500 m, which is most probably due to the fact that roads are built mainly within this distance to mountains having a slope of 30°–40°. For the lithology factor, the landslides occurred mostly in regions of arkose and carbonate rock, while granite and tuffaceous are rocks most vulnerable to earthquake-triggered landslides. As to the seismic intensity factor, the X degree takes most of the landslide area, while possibility of landslides increases with the seismic intensity from IX, X to XI. This is consistent with common knowledge.

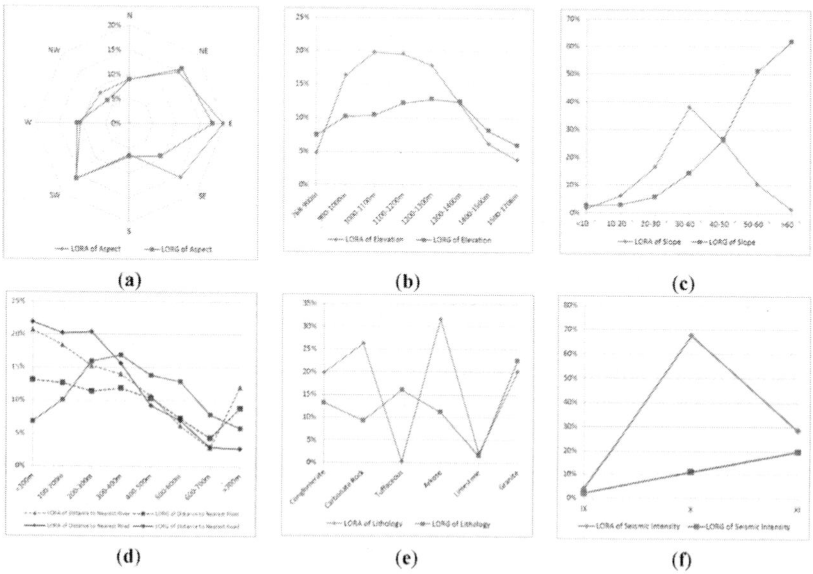

Figure 5. Relationships between landslides and the control factors: aspect, elevation, slope, distance to nearest river/road, lithology, and seismic intensity. (**a**) Aspect; (**b**) Elevation; (**c**) Slope; (**d**) Distance to nearest river/road; (**e**) Lithology; (**f**) Seismic intensity.

5.3. Landslide Susceptibility Mapping Based on an Artificial Neural Network

Landslide susceptibility mapping will provide an effective means for determining the susceptible regions for landslide monitoring. Landslide susceptibility mapping is commonly used in landslide research and prediction [51–53]. As a widely used non-linear method of prediction, the Artificial Neural Network (ANN), which is a generalization of mathematical models to simulate the human cognition and neural biology and thus to predict outputs from inputs [54], was employed in this research to derive the landslide susceptibility map. ANN has been recently used in landslide susceptibility mapping, and was proved to be an effective means for landslide prediction due to its capability to establish the complex and non-linear relationships between landslide distribution and the control factors [11].

In a typical landslide susceptibility mapping application, the ANN is usually composed of three steps: training, validation, and susceptibility mapping [55,56]. A back-propagation learning algorithm, consisting of input layer, hidden layers, and output layer was used to train the ANN in this research. The back-propagation algorithm is a procedure that minimizes the Sum of Squares for Error (SSE) in output layer by modifying the weights of ANN parameters to approximate the input sample and output sample. This multi-layer back-propagation algorithm was selected in this paper because it is one of the most commonly used ANN models that has the advantages of non-linear mapping, self-learning, and fault tolerance with respect to other models [57].

The landslides extracted from the IKONOS images were randomly divided into two groups, one used for training the ANN, and the other for accuracy assessment. The training sites were selected based on analysis of the topographical (aspect, elevation, slope, and distance to river/road), geology (lithology), and seismic (seismic intensity) landslide control factors. Based on the analysis described in the previous section, the LORG index of the control factors was used as the standard inputs for the network training due to its more reasonable evaluation compared with LORA, and the outputs of the training were the spatial distributed landslides extracted from satellite images. We tested different training methods, including the improved back-propagation algorithms and the numerical optimization based techniques, and found that the Levenberg-Marquardt approach featured the best performance in terms of Mean Square Error (MSE) [58,59], So it was selected in the back-propagation algorithm to calculate weights between the input layer and the hidden layer, as well as between the hidden layer and the output layer, by modifying the number of hidden layers and the learning rate [60] (see Table 4). The activation function in the hidden layer was set to tansig, and that in the output layer was set to purelin [50]. The input node (neuron number) was set to seven for the corresponding control factors and the output node (neuron number) was set to 1 for the susceptible value. The neuron number in the hidden layer was determined to be nine by evaluation of the training time, iterations and the resulting SSE (see Figure 6). 80% of the landslides were randomly selected as training data for the ANN, and the remainder was used as check points to validate the accuracy of the susceptibility mapping. Finally, with 58 iterations, an MSE (Mean Square Error) of 0.308 was achieved for the Levenberg-Marquardt method, and an accuracy of 95% could be reached for the assessment.

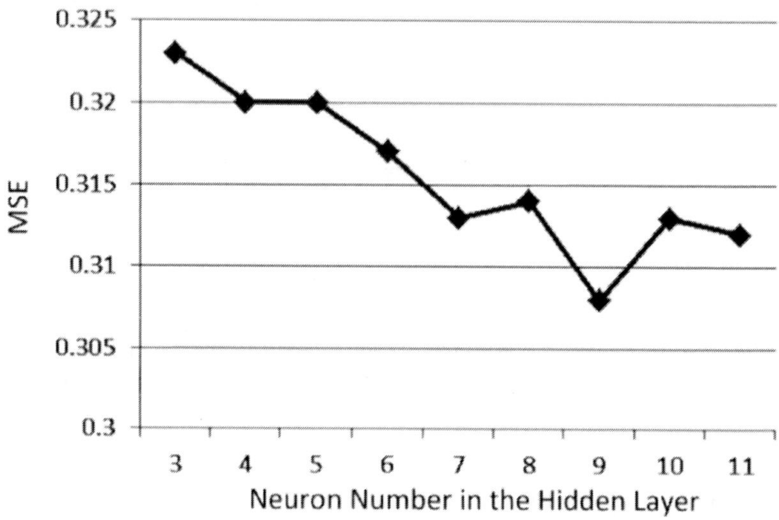

Figure 6. Mean Square Error (MSE) *vs.* neuron number in the hidden layer (Levenberg-Marquardt method).

Table 4. The Weights of the Artificial Neural Network (ANN) Parameters (Levenberg-Marquardt method).

Network Parameters	Weights
Learning Rate	0.05
Threshold of Residue	9.5238×10^{-4}, or SSE = 0.02
Error Increase Factor	1.04
Minimum Gradient	1×10^{-10}
Number of Maximum Training	2,000

The landslide susceptibility map derived from the ANN is shown in Figure 7, where the level of susceptibility was divided into five categories according to the output value r: very low ($r < 0$), low ($0 \leq r < 0.25$), moderate ($0.25 \leq r < 0.5$), high ($0.5 \leq r < 0.75$), and very high ($r \geq 0.75$). The overlay rate between the extracted landslides and the ones predicted by the susceptibility map is 89%, indicating the high reliability of the ANN approach.

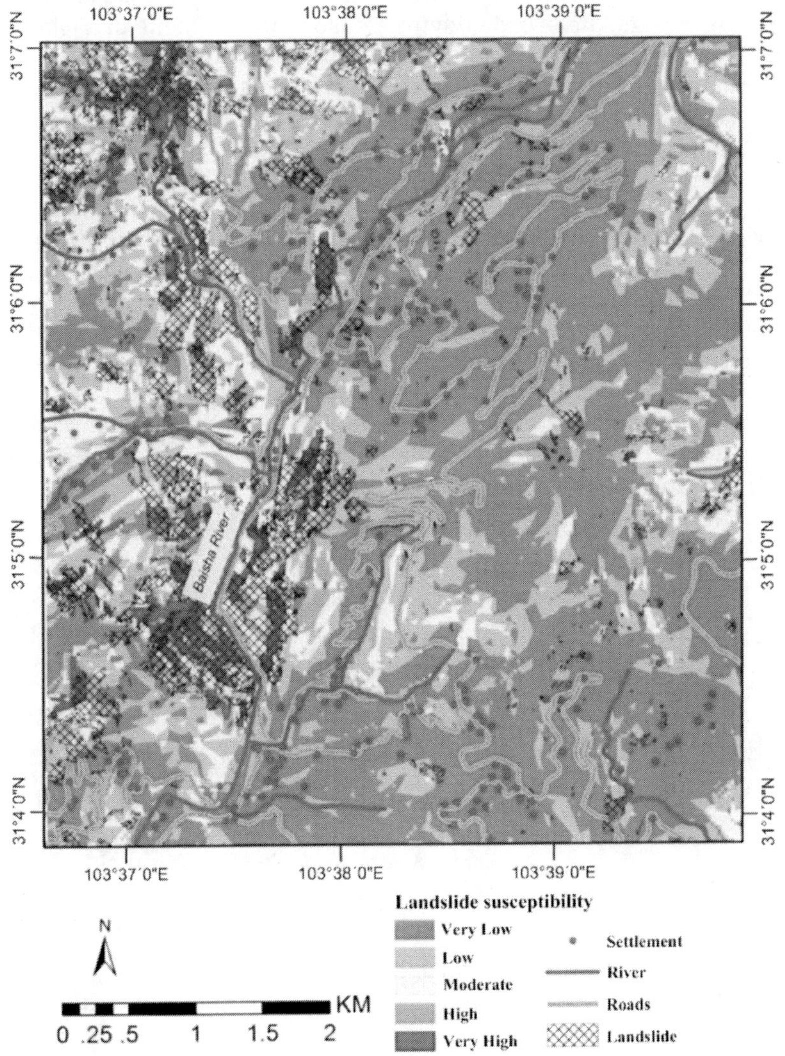

Figure 7. Landslide susceptibility map derived from Artificial Neural Network (ANN) (Levenberg-Marquardt method).

Figure 7 shows that the more landslide-prone regions are located mainly in the rugged mountainous areas having high slopes. The western and the northwestern areas are those most commonly susceptible to landslides, and there are relatively fewer settlements in these regions. One of the most susceptible spots, Taziping, was selected for landslide monitoring based on an *in situ* geo-sensor network because there is a river flowing nearby, and because there are relatively densely populated settlements in the area (Figure 7) which could be severely damaged if a landslide occurs. Before deploying sensors in the field of Sichuan, a scaled-down landslide simulation and monitoring system was constructed on the campus of Tongji University to test the usability and deployment strategy of these *in situ* sensors. The simulation experiment is detailed in the next section.

6. THE SCALED-DOWN LANDSLIDE SIMULATION

A simulation platform can be used as a first step in defining the set of *in situ* sensors to be adopted in the field. In addition, it can be used to understand the data interpretation and processing approach. For these purposes, a scaled-down landslide simulation platform (Figure 8) has been set up on the campus of Tongji University. The main purpose of this facility is to test the potential of various sensors in detecting early-warning signals.

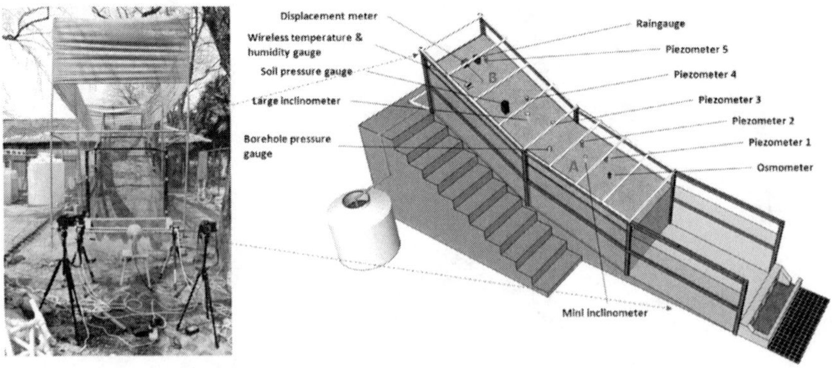

(a) (b)

Figure 8. The scaled-down landslide simulation platform on the campus of Tongji University (**a**) and the spatial distribution of *in situ* sensors for the experiment (**b**).

This scaled-down landslide simulation experiment system is composed of a simulation platform of mass movement, a spatial SN, a data collection and communication sub-system, a data storage server, and a visualization system [33]. Firstly, the simulation platform was designed to simulate the movement type, the geological configurations, and the triggering factors of intensive rainfall for the real-scene landslide that occurred in Taziping, Sichuan Province. Secondly, the SN includes a set of contact and non-contact sensors that were installed for observing and measuring different types of parameters that are related to landslide initialization, reactivation, acceleration, and, finally, failure of the slope. Thirdly, the data collection and communication sub-system was composed of the following components: (1) a data taker (dataTaker®, DT80) to collect data from all the contact sensors and then wirelessly transfer the signals through a General Packet Radio Service (GPRS) network; (2) a serial-port sever to gather some of the non-contact sensors and then transmit to the database server through 3G service, and (3) a self-organizing and self-healing mesh network to collect and transfer the video stream in real time through an iMesh device developed by the OTEC Communication Technology Co. (Guangzhou, China) [61]. Fourthly, the data storage server receives data from the wireless network and then interprets and stores this data for different sensors into different categories of the database. Finally, a visualization system composed of nine screens (46 inches each) was installed in the monitoring headquarters. This can instantly visualize the sensor positions, real-time video stream of mass movement, 2D spatial distribution of piezometers, temporal readings of sensors and the status of different sensors.

To simulate the Taziping landslide, the platform dimensions were designed to be 1.5 m in width and 6 m in length. The platform was divided into three equal sections having slope angles of 5°, 15°, and 30°, respectively. Tempered glass windows framed in steel in both sides of the platform were fixed to hold the landslide mass, and for the visual observation of the internal changes. To simulate the rainfall-triggered landslide, an artificial precipitation system was installed at the top of the landslide body. This system is composed of a water tank of 3,000 L in volume for the water supply, a pump to lift water for rainfall, and five evenly distributed water sprays that can be controlled and manipulated for precipitation duration and intensity. In addition, a steel frame was constructed to support a water-tight canvas enclosure to keep the sprayed water within the landslide body.

In this experiment, a total of 13 spatial *in situ* sensors were deployed (Figure 8). These sensors were used mainly to monitor landslide characteristics and triggering factors such as surface displacements on the landslide body, slope inclination, pore water pressure of groundwater flow, rainfall intensity, soil pressure, acceleration of slope movement, etc. For example, five piezometers were aligned along the sliding direction at the bottom of the landslide body with a depth of 1.5 m so as to interpret the changes in pore water pressure before and after slope failure. However, due to the limitations of scale, some sensors that will be used in the field, such as the GNSS, are not meaningful in a scaled-down model, so these kinds of sensors were not tested for the experiments. In addition to *in situ* sensors, in order to observe the surface displacement and landslide failure process and to evaluate the landslide magnitude, some stereo-camera systems and 3D imaging sensors were integrated into the simulation platform. Technological aspects that should be tackled on the real-scene landslides require a more in-depth analysis before moving from the simulation platform to the field.

Two technological issues need to be better focused before illustrating the results from the simulation experiments. The first concerns the use of wired connections between each contact sensor and the corresponding data acquisition unit (chiefly the data Taker®), which seems to contradict the use of a Wireless Sensor Network (WSN). For real-time field applications, the use of clusters of sensors has been devised to be the optimal solution to coping with the following problems. First, a solar power supply (consisting of a solar panel and its control and storage unit) can be shared among a large number of sensors; based on this hypothesis, power cables would be required for energy supply, and thus can be used for data communication as well. A local acquisition unit can be located close to the power source and operate for the entire sensor cluster; the successive communication from the local control unit to a central control unit outside the field then can be based on wireless technology. From this point of view, the simulation platform reproduces a sensor cluster. Second, the deployment of several sensors in a neighborhood allows for the cross-validation of spatially correlated observations. The second technological concern is related to the use of oversized communication channels. This selection is motivated by the required scalability of the SN, which should be able to accommodate a variable number of sensors under the same network structure.

The significant number of sensor observations was derived from the rainfall induced-landslide simulation experiment conducted on 18 November 2011. This experiment started at 14:08 and lasted about 2 h and 40 min. Due to a wireless communication technical problem, there was a period of data loss lasting for 25 min (14:37–15:02). The experiment was performed under the simulation of an intensive rainfall of approximate 280 mm within 3 h (as recorded by the rain gauge), with two periods of extremely heavy rainfall (14:25–14:40, 15:40–15:50) (Figure 9).

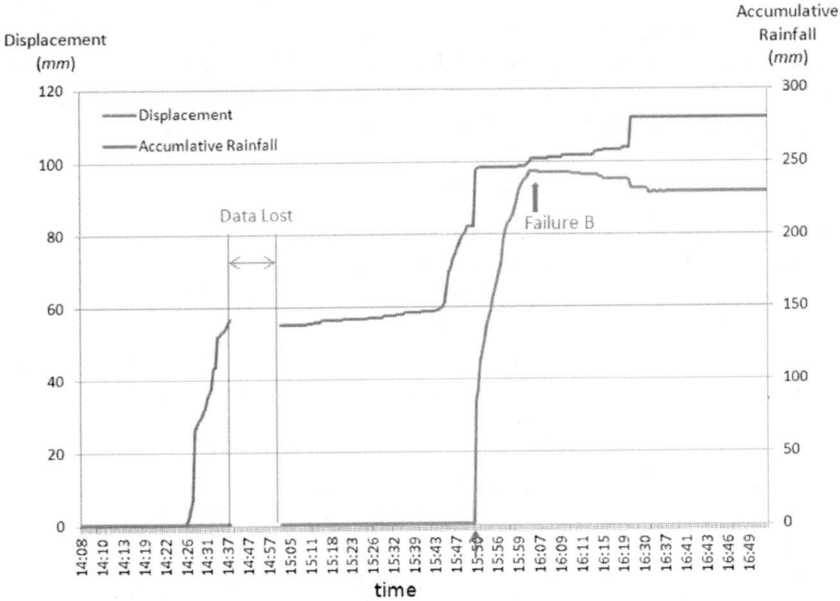

Figure 9. Monitoring data from the rain gauge and the displacement meter.

The flume test for a simulated landslide always renders the research concern on the rainfall condition, which necessarily needs a consideration of the field site and the amount of rainfall to trigger the slope failure in the experiment. As a result, it is necessary to set rainfall conditions that can evaluate the relationship between actual rainfall amount in the field and slope failure during the flume test. The regional area where the Taziping landslide is located is along the seismic zones that were strongly affected by the 2008 Wenchuan Earthquake. In this area, the amount of rainfall necessary to trigger landslides and debris flows nevertheless has not been well studied by previous researchers. [32] reported an accumulative rainfall of approx. 330 mm in 40 h (23–24 September 2008) which induced 969 new landslides in this seismic region. In our simulation, a rainfall accumulation of 280 mm over 3 h was used, with higher hourly rainfall intensity than the actual amounts in the field. The reason for using higher rainfall intensity during the experiment was to include the consideration of landslide scale. The simulated landslide is at a much smaller scale than the real-scene landslide. To match the scale, the rainfall simulation facility has been designed to produce decreased sizes of raindrops and increased density of raindrops. This is also the reason why natural rainfall is not used in the experiment considering its larger size of raindrop. In terms of the applied tipping bucket rain gauge with the same diameter (203 mm) of cylindrical container, it records higher rainfall intensity for dense, small-sized raindrops. This does not indicate that the measurements from the rain gauge are not correct, but it reflects the fact that in order to simulate the same effect of rainfall as in the field, a scaled rainfall simulation system with higher rainfall intensity is needed.

The first slope failure started from the bottom of the landslide body (Figure 8) at 15:50. This was observed by the borehole pressure gauge mounted at the bottom of the landslide with a depth of 1.5 m (Figure 10) and the mini inclinometer (Figure 11) set up in the lower part of the slope with a depth of 0.5 m. These gauges detected significantly increasing signals starting 5 min and 20 min, respectively, before the failure in section A. A subsequent slope failure occurred in the upper part of the slope at 16:05. This was recorded by the soil pressure gauge installed just below the slope surface (Figure 10), the displacement meter put on the slope surface (Figure 9), and the two-axis large inclinometer covering the whole depth of the landslide (Figure 11). These gauges noted a sudden signal change 25 min, 15 min and 20 min, respectively, before the failure in section B. In addition, the mass movement can be noticed 20 min before the initial failure by changes in the pore water pressure (Figure 12) that were measured by an osmometer and a set of piezometers affixed to the bottom the landslide body with a depth of 1.5 m.

While the high-resolution cameras provide images throughout the entire duration of a simulation experiment, high-speed cameras are used mainly to analyze the final slope failure, which might feature very rapid movements. Both camera systems gather sequences of stereo images to be used for photogrammetric analysis (volume computation, surface point tracking); at the moment, these processing tasks are applied after the end of the experiment, in a manner similar to how satellite remote-sensing observations will be used in field applications. On the other hand, the high-speed camera system could be used for both tasks after selection of the most suitable time span of the images to process for each part of the experiment.

In recent years, several 3D imaging sensors have been introduced that can provide the direct acquisition of a 3D surface by means of phase-shift analysis (range cameras) or by using triangulation principles [62]. Although the resolution and precision are lower than the ones obtained from the above-mentioned photogrammetric systems, they would allow for real-time processing.

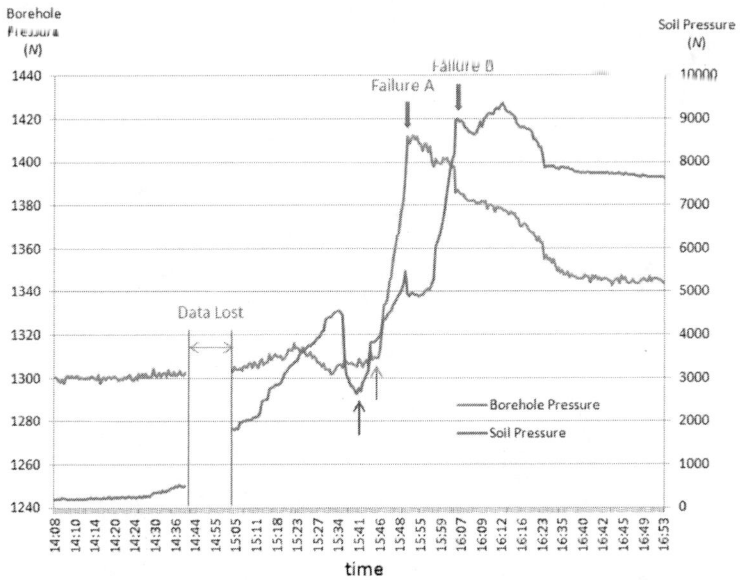

Figure 10. Monitoring result of the borehole pressure gauge and the soil pressure gauge.

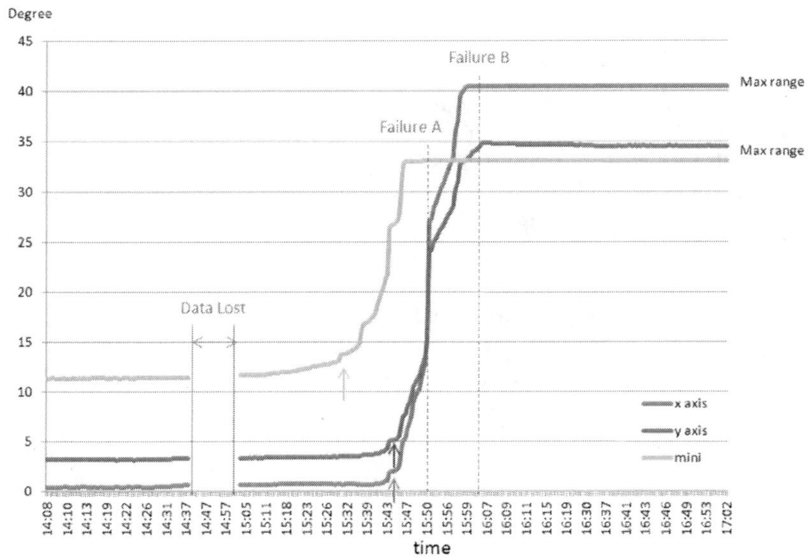

Figure 11. Monitoring data from the two-axis large and one-axis mini inclinometers.

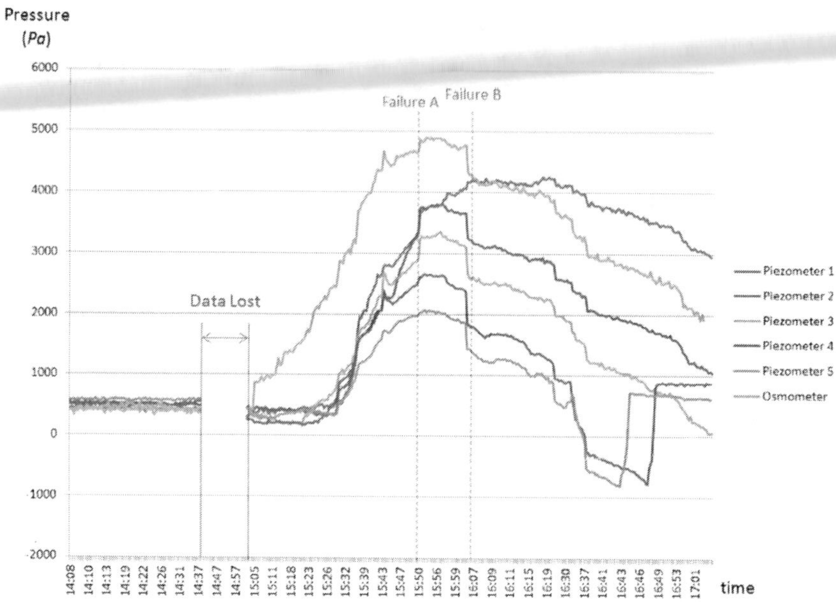

Figure 12. Monitoring data collected by the osmometer and the piezometers.

In a nutshell, the above-mentioned landslide simulation experiment confirms the usefulness of *in situ* sensors (such as inclinometers, piezometers, borehole, and soil pressure gauges) in real-time landslide monitoring and detection of early-warning signals. These sensors can be further employed reliably in real-scene landslide monitoring in Taziping, Sichuan Province, China. In addition, the experiment shows the possibility of transferring real-time video streams from the infrared video camera through a wireless network. As new approaches for landslide monitoring, high-speed stereo cameras and 3D imaging sensors can provide alternative means for landslide stability evaluation through remote-sensing approaches. However, due to scale differences and field conditions, they need to be further tested and evaluated in the field for real-scene landslide monitoring.

7. *IN SITU* SENSOR NETWORK DEPLOYMENT IN TAZIPING LANDSLIDE

7.1. The Existing Landslide Prevention Facilities and Monitoring Sensors

The Taziping landslide (Figure 13) is located above Hongkou Town, which was strongly affected by the 2008 Wenchuan Earthquake. The elevations of the source and the toe areas are about 1,370 m and 1,007 m, respectively, with a relative elevation difference of about 363 m. The slope ranges from 25° to 40°, with relatively steep landslide source areas and a relatively flat toe of mass accumulation. After the event, the landslide was frequently accelerated by intensive rainfall and subsequent water infiltrations, especially with the accumulation of loose debris and the fragile soil characteristics triggered by the earthquake [32]. Due to the extensive amounts of loose debris carried from the source areas, this landslide has the potential to turn into debris flows with intensive surface water adjunction. In addition, frequent human activities such as road building, forest cutting and fruit cultivation can often further decrease the stability of the slope and accordingly accelerate the sliding processes. The Taziping landslide represents a major threat to the town at the foot of the mountain not only in terms of human lives, but also in terms of infrastructure and local economic development.

Due to the high vulnerability of the slope, since 2010 the Taziping landslide has been continuously reinforced by the constructions of mitigation infrastructure. Efforts include the positioning of anchors, mitigation walls, shallow drainage trenches and a system for surface water run-off (Figures 13 and 14). In order to monitor the effectiveness of these mitigation efforts, three types of preliminary sensors have been installed on the landslide. First, a MEMS (Micro-Electro-Mechanical Systems) inclinometer with a depth of 15 m has been set up in the middle of the slope to measure the sliding surface and evaluate the activeness of the landslide. Second, three surface clinometers were established close to the mitigation wall in order to measure the surface displacement of the slope and evaluate the effectiveness of the mitigation program. Third, for the purpose of measuring ground water, three soil moisture sensors were mounted close to the surface drainage system. Although these sensors provide a useful approach to monitor the status of the slope, the system does not provide the possibility for real-time transfer of acquired data and it requires frequently manual readings of data collected by *in situ*data collection units. As a result, a more sophisticated real-time monitoring system is needed.

Figure 13. An overview of the Taziping landslide, the mitigation infrastructure and the currently available sensors for monitoring.

7.2. Design Considerations for Future Sensor Network Deployment at the Taziping Landslide

The SN tested in the landslide simulation experiment was successful in the following aspects. First, it enabled a rapid data transferring from the landslide to the monitoring headquarters in real time, which satisfied emergency management requirements when the slope is approaching failure. Second, a combination of three communication approaches (GPRS, 3G and broadband) separated different types of data and thus avoided transfer congestion at a single data node. Third, the experiment showed that a network-based monitoring system is more reliable than a point-based strategy because sensors can work collaboratively and cross-validation of data reliability is made possible. Fourth, the networked sensors ensure the transmission efficiency and accordingly avoid great delay of data transferring and storage. Finally, the experiment encompasses a complete and sophisticated monitoring system that includes *in situ* SN, data collection units, diverse communication approaches, data processing and storage units, and a visualization system.

Figure 14. Some existing facilities and sensors at the Taziping landslide: (**a**) Installation for soil moisture sensor; (**b**) Inclinometer; (**c**) Mitigation wall; and (**d**) Drainage trench.

Such a system will be transferred in the future to real-scene landslide monitoring in Taziping. A preliminary design of the improved network is sketched in Figure 15. The plan of this network is to use five communication relay stations on the slope of the landslide, each of which serves as a gateway node, namely working as a local data collection unit and as an intermediate point for data distribution and communication. These five stations are designed to form a connected SN, provided that the data communication between them can be maintained. If there are significant obstacles (e.g., forests) which block communication routes, more stations will be established, with each installation maintaining a certain distance above the ground. Each station will apply a multi-hop device for fast roaming communication that allows for a fast shift (within 0.01 s).

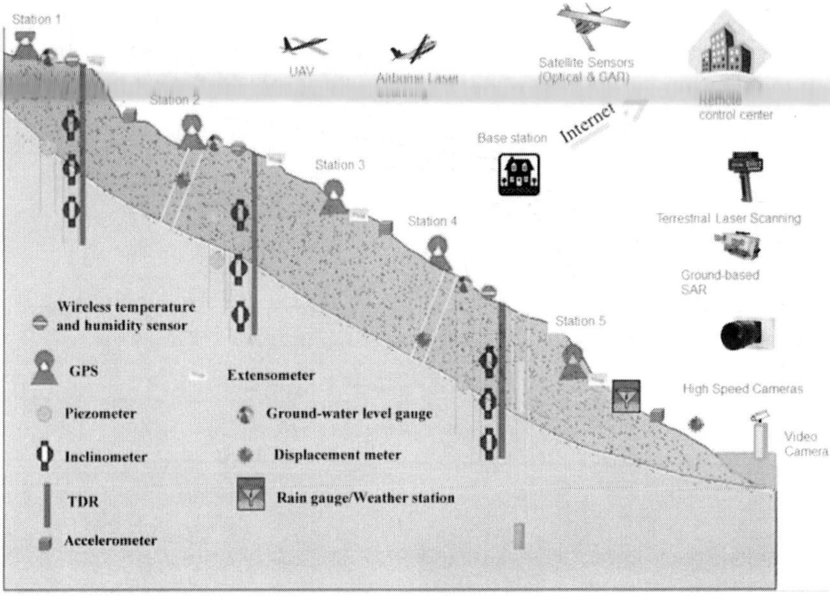

Figure 15. Design of the spatial sensor network to be deployed at the Taziping landslide.

For each sensor node of the station, a series of sensors will be connected for local data storage and distribution. Inside the drilled borehole, piezometers will be installed to monitor changes of pore water pressure in the ground water, especially for real-time monitoring of water infiltration after an intensive rainfall event. Also, inclinometers will be assembled and then installed inside the slope so as to measure slope movement both vertically and horizontally. Each gateway node also will include an extensometer that measures surface displacement. In terms of a rainfall triggering factor, a rain gauge will be used to monitor accumulative rainfall in real time. By employing a weather station that takes advantage of the National Oceanic and Atmospheric Administration (NOAA) weather forecast model [63], it is possible to forecast rainfall levels within 12 h, thus allocating the monitoring priority to the rain gauge and the weather station. These sensors were well tested during the landslide simulation experiment and can reach expected results for landslide monitoring.

In addition, the following sensors have been considered for more comprehensive observations. Time Domain Reflectometry (TDR) is to be installed to detect internal fractures of the slope. In the meanwhile, it can measure the level of ground water, providing comparisons for the observations from the piezometers. In addition, plans call for mounting high-frequency (80 Hz) accelerometers to be used for detecting surface vibration caused by the mass movement and, subsequently, for evaluation of the amount of acceleration. If the satellite signals can be received, GNSS then can be added as a ground reference for measuring surface displacement. Besides *in situ* sensors, certain remote-sensing platforms are scheduled to be installed. A ground-based InSAR [64,65] will be installed at the toe of the landslide so as to scan the entire slope and measure surface displacement along the Line-Of-Sight (LOS). Moreover, a pair of stereo high-speed cameras has already been tested and will be employed to capture the moment of slope failure so as to further analyze the landslide mechanism especially during the critical moment of slope failure. Since topographic information concerning a slope is fundamental for further analysis of landslide process, a long-range terrestrial laser scanning system [66] is to be put at the foot of the mountain so as to obtain a high-resolution DTM of the slope and, with multi-temporal acquisitions, to measure the surface displacement from both cloud points and sequential DTMs. The point clouds can also be acquired from airborne laser scanning [67], which allows for a larger scanning extent useful for regional analysis. Owing to its flexibility and low cost, it is planned to employ a Unmanned Aerial Vehicle (UAV) [68], embedded with a high-resolution camera, to acquire high-resolution aerial photos, to detect small rock fissures of the slope, and to generate a DTM for comparison with that formed from laser scanning. Finally, based on the satellite platforms, the ground movement can also be well detected using InSAR [69], especially for shorter wavelength sensors such as TerraSAR-X and COSMO-SkyMed, which allows a detection precision of millimeters.

Nevertheless, for a successful transplantation of the SN from the simulation platform to the *in situ* environment, careful consideration still needs to be made, particularly in the following aspects. First, the power supply system using solar panels needs to be established and it is necessary to ensure all sensor and gateway nodes can be supplied with more than 20% residuals during rainy or cloudy days. Second, for remote control of the installed sensors, an XML file is needed to have sending and editing functions, aiming at an adjustment of sensor parameters through the client server. Third, in case of data lost, it is crucial to establish a VPN transferring route between the local data center in Taziping and the monitoring headquarters in Shanghai, to ensure the safety of data security and data transfer. Fourth, a synchronization approach using GNSS should be tested so as to ensure that the synchronization rate is within 0.1 ms. Finally, when there is network congestion, the priority of data transfer needs to be defined with a first consideration of critical sensors relating to monitoring of the mass movement.

8. CONCLUSIONS

In this study, a systematic approach to landslide investigation based on remote sensing and Sensor Network (SN) has been developed and assessed. This approach is composed of three parts, landslide susceptibility mapping for susceptible spots determination in Hongkou Town (Sichuan, China), scaled-down landslide simulation for SN prototype system test on the campus of Tongji University (Shanghai, China), and *in situ* SN design for field landslide observation at the Taziping landslide, Hongkou Town. First, the landslides triggered by the 2008 Wenchuan Earthquake were extracted by means of an object-oriented method from a pair of geo-referenced pre- and post- event IKONOS images (3.25 km^2 of landslide area compared with the 30 km^2 study site), followed by the correlation analysis between the landslide distribution and the control factors generated by DEM and vector data. The landslide susceptibility map was then produced by the Artificial Neural Network (ANN) algorithm to determine the susceptible landslide spots, and the Taziping landslide was selected for further investigation and SN deployment. Second, the SN was tested on a scaled-down landslide simulation platform before the *in situ* deployment to examine the usability of the sensors as well as the deployment strategy. Finally, the considerations and suggestions were given for the future *in situ* SN deployment in Taziping.

The research results support the following conclusions:

(1) The spatial distribution of earthquake-triggered landslides is found to correlate with some of the control factors, e.g., it has positive correlation with the slope and seismic intensity factors, and the landslides occurred more easily in valleys with flowing rivers as well as within 200-to-500 m distance to roads.

(2) The ANN is proved to be effective for generation of landslide susceptibility map to identify the zones of high risk. Correspondences of estimated susceptible areas compared with landslides extracted from the high-resolution IKONOS images were found in 89% of the cases.

(3) The landslide simulation experiment demonstrates the applicability of the *in situ* sensors and SN in real-time landslide monitoring triggered by rainfall, and the potential for detection of early-warning signals useful for landslide prediction before the slope failure (e.g., 10 to 20 min in advance).

(4) The integration of remote sensing approach and SN technique contributes to landslide research by bridging the gaps between large-scale regional landslide investigation and small-scale individual landslide monitoring, allowing the localization of slopes with higher hazard-risk probability for intensified and real-time landslide observation with deployment of SN.

Despite the achievements in this research, there are currently several limitations that need further improvement in the future. For example, the relatively short pre-alerting time requires the use of real-time data collection, communication and analysis to be exploited in the monitoring system for the deployment of SN in Taziping. Due to scale differences and field conditions between the simulation platform and the slope in Taziping, further tests in the field are required for real-scene landslide monitoring. Finally, the proposed methodology suffers from approximations entailed in the scaled-down models. These approximations, which however would be involved in any other modeling techniques (e.g., in numerical or empirical modeling), call for a careful validation with respect to observations collected from the real landslide in the future.

ACKNOWLEDGMENTS

This study is supported by the National High Technology Research and Development Program of China (2012AA121302), the State Key Development Program for Basic Research of China (2012CB957701, 2013CB733204 and 2013CB733203), the National Science Foundation of China (41201425, 41201424, and 41101443), the Key Laboratory of Advanced Engineering Surveying of SBSM (TJES0905) and the Key Laboratory of Mapping from Space of SBSM (K201005), and the Key Laboratory for Land Environment and Disaster Monitoring of SBSM (LEDM2009B02). The authors would like to thank Leslie B. Smith from the Mapping and GIS Lab, OSU for her English editing of the manuscript, and the anonymous reviewers for their valuable comments and suggestions.

CONFLICTS OF INTEREST

The authors declare no conflict of interest.

REFERENCES

1. Schuster, R.L. Socioeconomic Significance of Landslides. In Landslides: Investigation and Mitigation; Turner, A.K., Schuster, R.L., Eds.; Transportation Research Board Special Report 247; National Academies Press: Washington, DC, USA, 1996; pp. 12–35.
2. Sassa, K.; Tsuchiya, S.; Ugai, K.; Wakai, A.; Uchimura, T. Landslides: A review of achievements in the first 5 years (2004–2009). Landslides 2009, 6, 275–286.
3. Delacourt, C.; Allemand, P.; Berthier, E.; Raucoules, D.; Casson, B.; Grandjean, P.; Pambrun, C.; Varel, E. Remote-sensing techniques for analysing landslide kinematics: A review. Bull. Soc. Géol. Fr 2007, 178, 89–100.
4. Metternicht, G.; Hurni, L.; Gogu, R. Remote sensing of landslides: An analysis of the potential contribution to geo-spatial systems for hazard assessment in mountainous environments. Remote Sens. Environ 2005, 98, 284–303.
5. Guzzetti, F.; Mondini, A.C.; Cardinali, M.; Fiorucci, F.; Santangelo, M.; Chang, K.-T. Landslide inventory maps: New tools for an old problem. Earth Sci. Rev 2012, 112, 42–66.
6. Tung, S.-H.; Shih, M.-H.; Sung, W.-P. Identification of the landslide using the satellite images and the digital image correlation method. Disa. Adv 2013, 6, 4–9.
7. Van Westen, C.J.; Castellanos, E.; Kuriakose, S.K. Spatial data for landslide susceptibility, hazard, and vulnerability assessment: An overview. Eng. Geol 2008, 102, 112–131.
8. Felicísimo, M.Á.; Cuartero, A.; Remondo, J.; Quirós, E. Mapping landslide susceptibility with logistic regression, multiple adaptive regression splines, classification and regression trees, and maximum entropy methods: A comparative study. Landslides 2013, 10, 175–189.
9. Strozzi, T.; Ambrosi, C.; Raetzo, H. Interpretation of aerial photographs and satellite SAR interferometry for the inventory of landslides. Remote Sens 2013, 5, 2554–2570.
10. Chauhan, S.; Sharma, M.; Arora, K.M. Landslide susceptibility zonation of the Chamoli region, Garhwal Himalayas, using logistic regression model. Landslides 2010, 7, 411–423.
11. Park, S.; Chio, C.; Kim, B.; Kim, J. Landslide susceptibility mapping using frequency ratio, analytic hierarchy process, logistic regression, and artificial neural network methods at the Inje area, Korea. Environ. Earth Sci 2013, 68, 1443–1464.
12. Tofani, V.; Raspini, F.; Catani, F.; Casagli, N. Persistent Scatterer Interferometry (PSI) technique for landslide characterization and monitoring. Remote Sens 2013, 5, 1045–1065.
13. Mantovani, F.; Soeters, R.; van Westen, C.J. Remote sensing techniques for landslide studies and hazard zonation in Europe. Geomorphology 1996, 15, 213–225.
14. Dunning, S.A.; Rosser, N.J.; Massey, C.I. The integration of terrestrial laser scanning and numerical modeling in landslide investigations. Q. J. Eng. Geol. Hydrogeol 2010, 43, 233–247.
15. Gigli, G.; Fanti, R.; Canuti, P.; Casagli, N. Integration of advanced monitoring and numerical modeling techniques for the complete risk scenario analysis of rockslides: The case of Mt. Beni (Florence, Italy). Eng. Geol 2011, 120, 48–59.
16. Li, R.; The CSISSD Research Team. Advanced Spatial Sensor Network Systems—Review, Status, and Applications. Proceedings of 2011 International Symposium on Image and Data Fusion, Tengchong, Yunnan, China, 9–11 August 2011.
17. Ramesh, M.V. Design, development, and deployment of a wireless sensor network for detection of landslides. Ad Hoc Netw. 2012.
18. Arattano, M.; Marchi, L. Systems and sensors for debris-flow monitoring and warning. Sensors 2008, 8, 2436–2452.

19. Angeli, M.; Pasuto, A.; Silvano, S. A critical review of landslide monitoring experiences. Eng. Geol 2000, 55, 133–147.
20. Akbarimehr, M.; Motagh, M.; Haghshenas-Haghighi, M. Slope stability assessment of the Sarcheshmeh Landslide, Northeast Iran, investigated using InSAR and GPS observations. Remote Sens 2013, 5, 3681–3700.
21. Jaboyedoff, M.; Oppikofer, T.; Abellán, A.; Derron, M.H.; Loye, A.; Metzger, R.; Pedrazzini, A. Use of LIDAR in landslide investigations: A review. Nat. Hazards 2012, 61, 1–24.
22. Lato, M.J.; Bevan, G.; Fergusson, M. Gigapixel imaging and photogrammetry: Development of a new long range remote imaging technique. Remote Sens 2012, 4, 3006–3021.
23. del Ventisette, C.; Ciampalini, A.; Manunta, M.; Calò, F.; Paglia, L.; Ardizzone, F.; Mondini, A.C.; Reichenbach, P.; Mateos, R.M.; Bianchini, S.; et al. Exploitation of large archives of ERS and ENVISAT C-Band SAR data to characterize ground deformations. Remote Sens 2013, 5, 3896–3917.
24. Ou, J.; Qiao, G.; Bao, F.; Wang, W.; Di, K.; Li, R. A new method for automatic large scale map updating using mobile mapping imagery. Photogramm. Rec 2013, 28, 240–260.
25. Sato, H.P.; Harp, E.L. Interpretation of earthquake-induced landslides triggered by the 12 May 2008, M7.9 Wenchuan earthquake in the Beichuan area, Sichuan Province, China, using satellite imagery and Google Earth. Landslides 2009, 6, 153–159.
26. Bozzano, F.; Cipriani, I.; Mazzanti, P.; Prestininzi, A. Displacement patterns of a landslide affected by human activities: Insights from ground-based InSAR monitoring. Nat. Hazards 2011, 59, 1377–1396.
27. Hong, Y.; Adler, R.F. Towards an early-warning system for global landslides triggered by rainfall and earthquake. Int. J. Remote. Sens 2007, 28, 3713–3719.
28. Cruden, D.M.; Varnes, D.J. Landslides Types and Processes. In Landslides: Investigation and Mitigation; Turner, A.K., Schuster, R.L., Eds.; Transportation Research Board Special Report 247,; National Academies Press: Washington, DC, USA, 1996; pp. 36–75.
29. Guzzetti, F.; Carrara, A.; Cardinali, M.; Reichenbach, P. Landslide hazard evaluation: A review of current techniques and their application in a multi-scale study, Central Italy. Geomorphology 1999, 31, 181–216.
30. Dai, F.C.; Lee, C.F.; Nhai, Y.Y. Landslide risk assessment and management: An overview. Eng. Geol 2002, 64, 65–87.
31. Dai, F.C.; Xu, C.; Yao, X.; Xu, L.; Tu, X.B.; Gong, Q.M. Spatial distribution of landslides triggered by the 2008 Ms 8.0 Wenchuan Earthquake, China. J. Asian Earth Sci 2011, 40, 883–895.
32. Tang, C.; Zhu, J.; Qi, X.; Ding, J. Landslides induced by the Wenchuan earthquake and the subsequent strong rainfall event: A case study in the Beichuan area of China. Eng. Geol 2011, 122, 22–33.
33. Scaioni, M.; Lu, P.; Chen, W.; Qiao, G.; Wu, H.; Feng, T.; Tong, X.; Wang, W.; Li, R. Analysis of spatial sensor network observations during landslide simulation experiments. Eur. J. Environ. Civil. Eng. 2013.
34. Qi, S.W.; Xu, Q.; Lan, H.X.; Zhang, B.; Liu, J.Y. Spatial distribution analysis of landslides triggered by 2008.5.12 Wenchuan Earthquake, China. Eng. Geol 2010, 116, 95–108.
35. Wang, F.W.; Cheng, Q.G.; Highland, L.; Miyajima, M.; Wang, H.B.; Yan, C.G. Preliminary investigation of some large landslides triggered by the 2008 Wenchuan earthquake, Sichuan Province, China. Landslides 2009, 6, 47–54.
36. Qiao, G.; Wang, W.; Wu, B.; Liu, C.; Li, R. Assessment of geo-positioning capability of high-resolution satellite imagery for densely populated high buildings in metropolitan areas. Photogramm. Eng. Remote. Sens 2010, 76, 923–934.
37. Lu, P.; Stumpf, A.; Kerle, N.; Casagli, N. Object-oriented change detection for landslide rapid mapping. IEEE Geosci. Remote Sens 2011, 8, 701–705.
38. Hölbling, D.; Füreder, P.; Antolini, F.; Cigna, F.; Casagli, N.; Lang, S. A semi-automated object-based approach for landslide detection validated by Persistent Scatterer

Interferometry measures and landslide inventories. Remote Sens2012, 4, 1310–1336.

39. Yang, C.J.L.; Ren, X.L.; Huang, H. The vegetation damage assessment of the Wenchuan earthquake of May 2008 using remote sensing and GIS. Nat. Hazards 2012, 62, [illegible]–55.

40. Lacroix, P.; Zavala, B.; Berthier, E.; Audin, L. Supervised method of landslide inventory using panchromatic SPOT5 images and application to the earthquake-triggered landslides of Pisco (Peru, 2007, Mw8.0). Remote Sens 2013, 5, 2590–2616.

41. Defries, S.R.; Townshend, J.R.G. NDVI-derived land cover classifications at a global scale. Int. J. Remote Sens 1994, 15, 3567–3586.

42. Marqués, F.; Gasull, A. Multiresolution Image Segmentation Based on Compound Random Fields: Application to Image Coding; Universitat Politècnica de Catalunya: Barcelona, Spain, 1992; pp. 86–114.

43. Tong, H.J.; Maxwell, T.; Zhang, Y.; Dey, V. A supervised and fuzzy-based approach to determine optimal multi-resolution image segmentation parameters. Photogramm. Eng. Remote. Sens 2012, 78, 1029–1044.

44. Gorum, T.; Fan, X.M.; van Westen, J.C.; Huang, R.Q.; Xu, Q.; Tang, C.; Wang, G.H. Distribution pattern of earthquake-induced landslides triggered by the 12 May 2008 Wenchuan earthquake. Geomorphology 2011, 133, 152–167.

45. Conoscenti, C.; Maggio, C.D.; Rotigliano, E. GIS analysis to assess landslide susceptibility in a fluvial basin of NW Sicily (Italy). Geomorphology 2008, 94, 325–339.

46. Neuhäuser, B.; Damm, B.; Terhorst, B. GIS-based assessment of landslide susceptibility on the base of the Weights-of-Evidence model. Landslides 2012, 9, 511–528.

47. Pareek, N.; Sharma, M.L.; Arora, K.M. Impact of seismic factors on landslide susceptibility zonation: A case study in part of Indian Himalayas. Landslides 2010, 7, 191–201.

48. Chen, H.; Lin, G.W.; Lu, M.H.; Shih, T.Y.; Horng, M.J.; Wu, S.J.; Chuang, B. Effects of topography, lithology, rainfall and earthquake on landslide and sediment discharge in mountain catchments of southeastern Taiwan. Geomorphology2011, 133, 132–142.

49. Akgun, A. A comparison of landslide susceptibility maps produced by logistic regression, multi-criteria decision, and likelihood ratio methods: A case study at İzmir, Turkey. Landslides 2012, 9, 93–106.

50. Othman, A.A.; Gloaguen, R. River courses affected by landslides and implications for hazard assessment: A high resolution remote sensing case study in NE Iraq–W Iran. Remote Sens 2013, 5, 1024–1044.

51. Moreiras, S.M. Landslide susceptibility zonation in the Rio Mendoza Valley, Argentina. Geomorphology 2005, 66, 345–357.

52. Akgun, A.; Dag, S.; Bulut, F. Landslide susceptibility mapping for a landslide-prone area (Findikli, NE of Turkey) by likelihood-frequency ratio and weighted linear combination models. Environ. Geol 2008, 54, 1127–1143.

53. Ray, R.L.; Jacobs, J.M.; Cosh, M.H. Landslide susceptibility mapping using downscaled AMSR-E soil moisture: A case study from Cleveland Corral, California, US. Remote Sens. Environ 2010, 114, 2624–2636.

54. Bartlett, E.B. A stochastic training algorithm for artificial neural networks. Neurocomputing 1994, 6, 31–43.

55. Li, Y.; Chen, G.; Tang, C.; Zhou, G.; Zheng, L. Rainfall and earthquake-induced landslide susceptibility assessment using GIS and Artificial Neural Network. Nat. Hazard Earth Syst. Sci 2012, 12, 2719–2729.

56. Liu, C.; Li, W.; Wu, H.; Lu, P.; Sang, K.; Sun, W.; Chen, W.; Hong, Y.; Li, R. Susceptibility evaluation and mapping of China's landslides based on multi-source data. Nat. Hazards 2013.

57. Holmstrom, L.; Koistinen, P. Using additive noise in back-propagation training. IEEE Trans. Neural Netw 1992, 3, 24–38.

58. Haykin, S. Neural Networks: A Comprehensive Foundation, 2nd ed; Prentice-Hall: Upper Saddle River, NJ, USA, 1999; p. 823.
59. Ma, C.F.; Jiang, L.H. Some research on Levenberg–Marquardt method for the nonlinear equations. Appl. Math. Comput 2007, 184, 1032–1040.
60. Lee, S.; Ryu, J.H.; Kim, L.S. Landslide susceptibility analysis and its verification using likelihood ratio, logistic regression, and artificial neural network models: Case study of Youngin, Korea. Landslides 2007, 4, 327–338.
61. OTEC Communication Technology Co. (Guangzhou, China). Available online http://www.cnotec.cn/1234/html/?18.html (assessed on 3 September 2013).
62. Sansoni, G.; Trebeschi, M.; Docchio, F. State-of-the-art and applications of 3D imaging sensors in industry, cultural heritage, medicine, and criminal investigation. Sensors 2009, 9, 568–601.
63. NOAA National Weather Service. Available online http://mag.ncep.noaa.gov/ (accessed on 3 September 2013).
64. Casagli, N.; Catani, F.; del Ventisette, C.; Luzi, G. Monitoring, prediction, and early warning using ground-based radar interferometry. Landslides 2010, 7, 291–301.
65. Tapete, D.; Casagli, N.; Luzi, G.; Fanti, R.; Gigli, G.; Leva, D. Integrating radar and laser-based remote sensing techniques for monitoring structural deformation of archaeological monuments. J. Archaeol. Sci 2013, 40, 176–189.
66. Heritage, G.L.; Large, A.R.G. Laser Scanning for the Environmental Sciences; John Wiley & Sons: Chichester, UK, 2009; p. 302.
67. Vosselman, G.; Maas, H.G. Airborne and Terrestrial Laser Scanning; Taylor and Francis Group: Boca Raton, FL, USA, 2010; p. 320.
68. Eisenbeiss, H.; Sauerbier, M. Investigation of UAV systems and flight modes for photogrammetric applications. Photogramm. Rec 2011, 26, 400–421.
69. Tapete, D.; Fanti1, R.; Cecchi, R.; Petrangeli, P.; Casagli, N. Satellite radar interferometry for monitoring and early-stage warning of structural instability in archaeological sites. J. Geophys. Eng 2012, 9, S10–S25.

CHAPTER 15

Rainfall and Landslide Correlation Analysis and Prediction of Future Rainfall Base on Climate Change

Moung-Jin Lee*

[1] Korea Adaptation Center for Climate Change(KACCC), Korea Environment Institute(KEI), Sejong-si, Korea

ABSTRACT

The aim of this study is to analyze the quantitative relationship between the volume of rainfall and landslide occurrence in South Korea. To predict future rainfall, a future climate scenario was developed by downscaling the regional climate model (RCM) from the global climate model (GCM) based on the Intergovernmental Panel on Climate Change (IPCC) A1B scenario. In this study, for a quantitative analysis of correlation between rainfall and landslides occurrence, data on rainfall and landslides in Korea in the 2000s was analyzed using the correlation between the occurrence of landslides and rainfall volume (daily and accumulated) and the maximum hourly intensity of rainfall. Daily rainfalls exceeding 164.5 mm is categorized as high risk for landslide. A rainfall that continued for 3 days was found to affect the occurrence of landslide in Korea in the 2000s more than any other number of days during which rainfall lasted. The research area for the future climate change scenarios (A1B) covers the entire area of South Korea. Annual average rainfall had increased by 271.23 mm during 1971–2100. The development of downscaling method using GIS and verification with observed data could reduce the uncertainty of future climate change projection.

Keywords: correlation analysis, rainfall, landslide, climate scenario, statistical downscale, verification

1. INTRODUCTION

A landslide occurs when part of a slope suddenly collapses because of rapid changes in nature such as a torrential downpour, a typhoon, or an earthquake. In South Korea, the rainy season, extending from June to September, coincides with the period when landslides occur. Casualties and property damage have been on the increase, and recently, there is frequent localized heavy rain of several hundred millimeters.

The Fourth Report of Climate Change Assessment, published by [1], predicted that climate change would persist for several more centuries due to the greenhouse gases that have already been discharged, even though they could be mitigated by efforts such as a reduction in greenhouse gas emissions. The abnormal climate and localized torrential rain caused by climate change may lead to higher landslide rates. An analysis of the causes of landslides, with a multilateral, holistic view, is needed, given this expected increase in landslides due to climate change. Damage from landslides can be minimized by assessing landslide vulnerability [2].

In order to prevent and reduce landslide casualties and property damage caused by climate change, it is necessary to develop scientific methods for landslide analysis related to future climate change. It is important to predict landslide occurrence caused by future changes in rainfall through an understanding of the correlation between past and future rainfall and landslides and to develop methodologies to quantitatively predict changes in rainfall due to climate change. In addition, necessary is advancement in methods to identify landslide locations and a methodology to analyze and verify the relationship between landslides and climate change [3].

Landslide occurrence factors can be divided into internal and external factors. Internal factors include natural ones such as geological structure, topography, soil quality, and forest, while external factors include natural ones, such as rainfall, erosion of rivers and shores, and earthquakes, as well as artificial ones such as cut-embankment, logging, estate development, and quarrying. Landslides readily occur when a slope with internally adverse factors is subject to adverse external factors. In South Korea, landslides are intensified between June and August with localized torrential rain, a period when landslides occur frequently due to concentrated torrential rainfall. Reference [4] researched types and frequency of landslides by intensified torrential rain, centered on the urban areas of Avigliano. Reference [5] studied the frequency of landslides in Ethiopia by analyzing the relation between rainfall and topography, including landscape and forest distribution. Reference [6] analyzed landslide susceptibility based on the distribution of land moisture.

As mentioned above, studies on climate change and landslides' cause of occurrence have been individually conducted. Analysis of landslide occurrence due to future climate change, by relating the two, is still in its early stages, and awaiting global recognition. This study is one of the first attempts to directly correlate analysis of rainfall and landslide occurrence. Future landslide frequency would increase given the consistent effects of climate changes and with studies into the analysis of prediction of future rainfall based on climate scenario, such as the IPCC A1B scenario.

In this study, relations between landslide and rainfall, domestically from 1991 to 2010, are analyzed, and by linking these data to future climate changes, rainfall pattern is analyzed.

2. METHODOLOGY AND CONTENTS

2.1. Analysis of Relation Between Landslide and Rainfall in The 2000s

A database for analyzing the correlation between landslides and rainfall in South Korea in the 1990s and 2000s was assembled. The period of collection is 19 years, from July 1991 to December 2010, the data being for landslides nationwide in South Korea. The references for locations and dates of landslides included newspaper articles, national and local media broadcasts, and reports by the Korea Institute of Geoscience & Mineral Resources (KIGAM) and the Korea Institute of Construction Technology (KICT). There is a great deal of data for landslides in the 2000s from various sources, but data for landslides in the 1990s, sourced from relevant academic articles and national media, are limited. This is why this section focuses on the relation between landslides and rainfall in the 2000s, data from the 1990s being included only for reference.

Locations and dates of landslides were extracted from the collected data. A climate database was assembled using observational data from the Automatic Weather Station (AWS) for the area closest to each landslide over the preceding 5 days. Centered on the date of the landslide, the daily rainfall data from 5 days before to 5 days after the landslide was organized. The total organized data comprised 186 landslide locations and daily rainfall. The 186 landslide spots were not averaged out to be the spot of the landslide but indicate points where landslides occurred *en masse*. The analysis of the relation between landslides and rainfall is conducted by two means. First, rainfall data was collected for the same day of the year as each landslide in order to analyze the correlation between rainfall on the day of landslides and that in the 2000s. Second, daily rainfall was added to each following day in order to analyze the relation between landslides and cumulative rainfall. Additionally, in order to identify the effect of daily rain intensity on landslide occurrence, cumulative rainfall for 1, 3, and 5 days were analyzed.

2.2. Prediction of Future Rainfall Change Based on Climate Scenarios

Abnormal weather and concentrated torrential rainfall are becoming more frequent around the world, with concentrated torrential rainfall predicted to occur consistently with climate change [7]. The prediction of future rainfall changes, reflected in climate change, can be divided into predictions using rainfall probability and predictions using future climate change scenarios. A prediction using rainfall probability is for calculating the frequency of occurrence of the same rainfall by inputting past climate data—a probable statistic method based on past data. It can be used with past extreme climate events such as concentrated torrential rainfall and typhoons and can predict to a certain temperature degree. However, it requires processing of data from the same period in adjacent observation spots with the same method in order to do the interpolation because the calculated results are depicted as points representing the observation spots.

Predictions using future climate change scenarios are conducted after the development of the regional climate model (RCM) by individual research centers of the relevant nation from the global climate model (GCM) presented by the Data Distribution Center (DDC) of the IPCC. The GCM is to predict the future by using the atmospheric general circulation model (AGCM) and humanistic, economic models, predicting temperature increases and changes in rainfall by the prediction of CO_2 discharge. The GCM has several problems: first, future predictions are dependent on the results of the prediction model; second, its spatial resolution is between 200 and 400 km and it is not appropriate for nationwide studies such as the ones covering South Korea; and third, it cannot predict extreme climate events such as concentrated torrential rainfall and typhoons. In order to avoid these problems of the GCM, a study to develop the RCM with its spatial resolution at the regional scale has been conducted. The RCM is to be built on the basis of the GCM by connection with the GCM and humanistic, economic status at regional levels. RCM can be improved to a spatial resolution of 20–25 km, but it is still dependent on the results of the prediction model of the future climate change, and it is hard to predict changes in extreme climate events. This study, in order to improve its spatial resolution, improved the existing spatial interpolation by applying the temperature and rainfall lapse rate to conduct the specification of the future climate change scenario at scales covering South Korea.

2.3. Spatial Statistical Downscaling Of Climate Change Scenario

As with climate change scenarios, the spatial data used in climate change studies should be connected continuously in space. However, the climate data observed in the past and those for predicting the future consist of representative values in the form of points, and for areas without such representative values, the values should be inferred by spatial statistic interpolation. Interpolation indicates a method by which continuous spatial distribution data is built from the inference of values of nonobserved and adjacent spots from the values of observed spots.

Spatial interpolation has various methods of different characteristics. This study selected Co-Kriging, a geostatistics technique in which eigenvalues for the relevant spots are predicted by the linear combination of already-known adjacent values. In this technique, the adjacent, actually measured values are linearly combined for interpolation, and values are estimated by using such statistical methods. Co-Kriging estimates values by the statistical analysis of many adjacent measured values, indicating that it reflects correlative intensity among the adjacent measured values as well as the distance to actually measured values. Co-Kriging is advantageous for identifying overall trends. The applied Co-Kriging equation is as shown in Eq. (1):

$$Z^*(u) - m(u) = \sum_{\alpha}^{n(u)} \lambda_a [Z(u_a) - m(u_a)] \tag{1}$$

where *u*, *ua* are the locations estimated and locations of known data, *Z*(*u*), *m* (*u*) are the estimation from adjacent data used, *m*(*u*), *m*(*ua*) are the estimate d values of *Z*(*u*) and *Z*(*ua*), and *λa*(*u*) is the Co-Kriging weighted value (weight).

As mentioned, Co-Kriging is used in this study because it is advantageous for identifying trends across a wide area. To calculate estimated values on nonobserved spots, the effects of observed values of the relevant observed spot are more reflected as the relationship of the adjacent observed spots and the linearity becomes closer. In other words, the types of linearity between nonobserved spots and adjacent observed spots are inferred and reflected as inversely proportional to distance and observed values.

Among the topographical factors, altitude has the greatest effect on changes in rainfall and temperature, as climate elements. Temperature, of the climate factors, is particularly affected by altitude, decreasing as altitude increases under the troposphere [8].

When an interpolation of the general geographic information system is applied, accurate estimation is difficult because of the severe skewness of spatial dependence by other factors. In this study, temperature and rainfall serve as variables, and altitude is a factor damaging spatial dependence. In order to overcome this problem, a form of Co-Kriging with which altitude data can be directly considered is selected over general Co-Kriging. Co-Kriging is a method used to interpolate data in the process of spatial estimation. Temperature and rainfall data has a linear correlation to altitude data [8], and thus altitude data can present additional information to reduce inferred measured values in estimating values for temperature and rainfall in nonobserved spots.

When the data used in all the study areas, including digital elevation data, is used with Co-Kriging, it is known to cause uncertainty in getting a Co-Kriging system weight matrix because the correlativity of adjacent altitude data is larger than that of sample values of temperature or rainfall [9]. In addition, altitude data, existing at or adjacent to estimated locations, may conceal the effects of distant altitude data. In order to reduce such errors, this study used collocated Co-Kriging in which additional data can be used globally in the existing Co-Kriging, limited only by additional data values at estimated locations in Eq. (2):

$$Z_{CK}^{*}(u)=\sum_{a=1}^{n(u)}\lambda_{a}^{ck}(u)z(u_{a})+\lambda^{ck}(u)\left[y(u)-m_{Y}+m_{z}\right] \tag{2}$$

where $y(u)$: is the altitude value at location u that is not sampled and $my + mz$ is the average value of altitude data and data of temperature or rainfall.

For the altitude data in this study, a digital map (1:25,000) is used as primary data, and the spatial resolution is transformed to digital elevation model (DEM) data (30 m). The locations of the nation's 75 weather stations are plotted on a map, and the land section of the land cover map (1:25,000) is used to build the data for the coastline of South Korea. Because the observed values from the 75 weather observation spots between 1971 and 2000 differ in altitude above sea level and in the heights above surface level of thermometers and rain gauges, a modification equation for the temperature lapse rate was presented by [10]. The temperature lapse rate was based on altitude (Eqs. (3) and (4)). As for rainfall data, the rainfall lapse rate is considered on the basis of altitude [11] (Eqs. (5) and (6)). Eq. (5) shows that the rainfall value should be increased by 74% per 1 km for October-April (Cold Season). For the period May-September (Warm Season), Eq. (6) reflects a 46% reduced rainfall value:

$$|\tau| = 0.00688 + 0.0015\cos 0.0172(i-60)\text{Average Temperature Lapse Rate} \tag{3}$$

where
|τ| is the absolute value of air temperature lapse rate based on annual dates, and *i* is the annual date (1/1 day = 1, 12/31 day = 365):

$$\mathrm{T} = T_i + \text{Elevation}(m) \times |\tau| \tag{4}$$

where *T* : is the temperature corrected by the air temperature lapse rate, an d *Ti*: is the daily temperature (air-temperature lapse rate corrected before the temperature):

$$R = R_i \times (1.74)^{\frac{E}{1000}} \tag{5}$$

$$R = R_i \times (0.46)^{\frac{E}{1000}} \tag{6}$$

where *R* represents rainfall and *E* represents elevation (m).

3. RESULT

3.1. Relation Between Landslide and Rainfall of The Day

Figure 1 shows representative types of landslides and rainfall. The *X*-axis averages periods of rain and the *Y*-axis averages periods of occurrence per year as frequent rainfall. Rainfall in the diagram is calculated from the AWS's rainfall data within the areas of landslides.

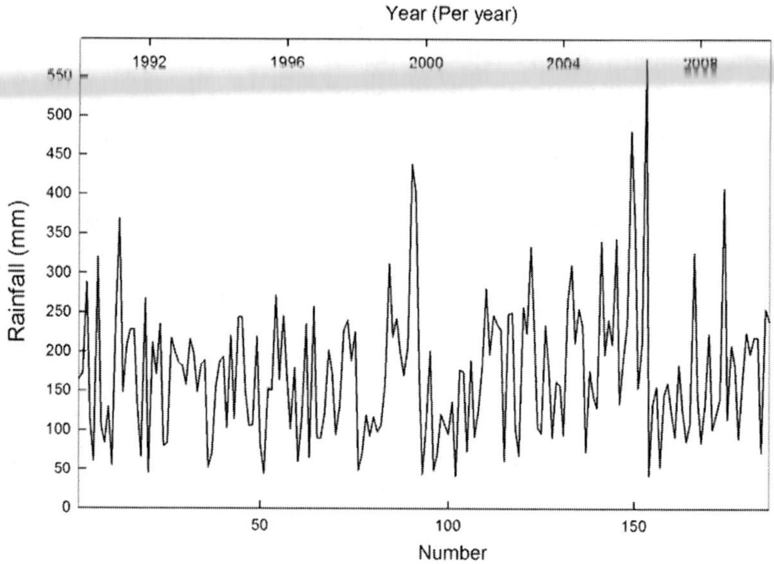

FIGURE 1. 1-day maximum rainfall of the study area in the 2000s.

Regarding the characteristics of rainfall on days related to landslides in the 2000s, landslide dates are clustered around July 14 and 16, July 22 and 23, August 25 and 26, and September 13 and 16. Examining average rainfall at the time of landslide occurrence, landslides seem to usually occur when rainfall is around 170 mm. Landslides occurring when rainfall was less than 100 mm are considered to result from temporary construction or high cumulative rainfall.

3.2. Relation Between Landslide and Cumulative Rainfall

A comparison of cumulative rainfall before landslides with rainfall on the day of landslides was conducted in order to analyze the relation between rainfall on the day of a landslide and cumulative rainfall as a possible cause of landslides. Cumulative rainfall before landslides (1, 3, and 5 days) and rainfall on the day of landslides are compared per year. The horizontal (X) axis indicates rainfall on the day of the landslide, while the vertical (Y) axis indicates the amount of accumulated rainfall over 1, 3, and 5 days prior to the landslide. **Figures 1–4** show that landslides are caused by rainfall when a region over the 45° central border line is spotted, while they are considered to be caused by cumulative rainfall when a region under the line

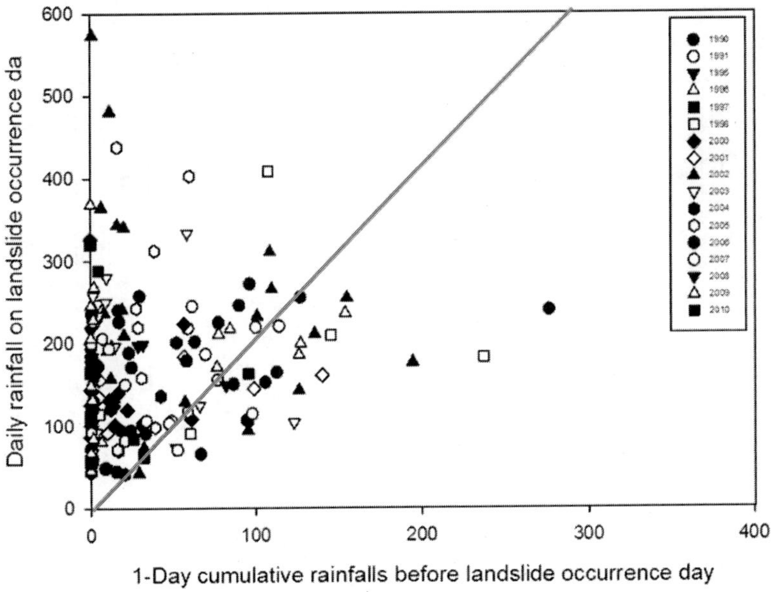

FIGURE 2. 1-day cumulative rainfall before landslide occurrence day.

is spotted. In an analysis of rainfall on the day of a landslide and the cumulative rainfall of 1 day before the landside for the 2000s, landslide spots over and under the 45° central border line are clearly separated. There are 162 spots over and 24 spots under the central border lines (**Figure 2** and **Table 1**). **Figure 2** and **Table 1** include rainfall on the day of the landslide and cumulative rainfall for 3 days before the landslide, where the spots are distributed over and under the 45° central border line, indicating that landslides are equally caused (spots over the border: 86; spots under the border: 80) by rainfall on the day of the landslide. The cumulative rainfall data for 3 days prior is shown in **Figure 3** and **Table 1**. Rainfall on the day of the landslide and cumulative rainfall for 5 days prior are shown in **Figure 4** and **Table 1**. Landslides are spotted more on the area under rather than over the 45° central border line (spots over the border: 56; spots under the border: 130), indicating that landslides are caused by 5-day cumulative rainfall rather than by rainfall on the day of occurrence. Note that 3-day cumulative rainfalls, therefore, may be more indicative of whether landslides are caused by rainfall on the day of occurrence or by cumulative rainfall. Many landslides in the 2000s are closely related to 3-day cumulative rainfall, and thus landslides in the 2000s may be affected more by cumulative rainfall than by rainfall on the day of occurrence. As a result, 1-day rainfall and 3-day cumulative rainfall, rather than cumulative rainfall on other days, had a higher correlation with landslide occurrence.

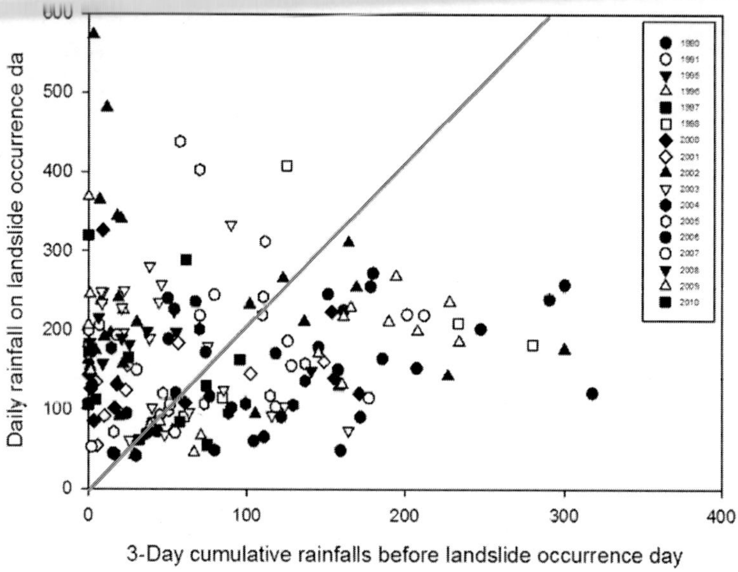

FIGURE 3. 3-day cumulative rainfall before landslide occurrence day.

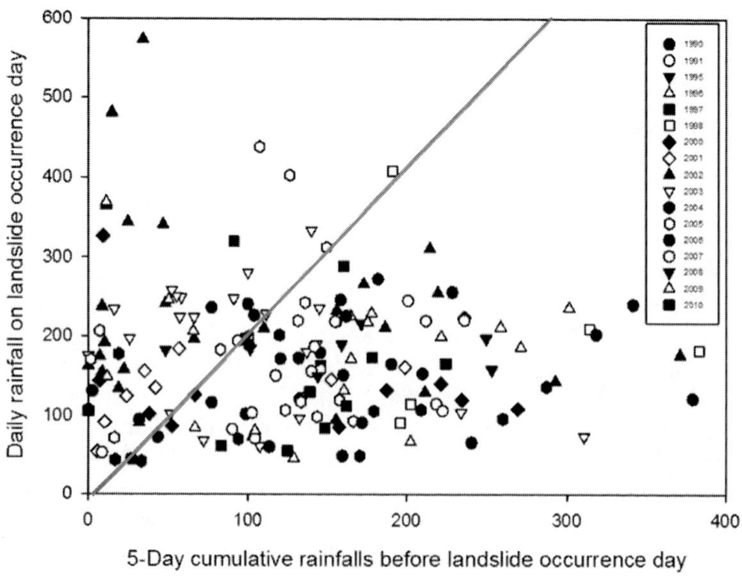

FIGURE 4. 5-day cumulative rainfall before landslide occurrence day.

TABLE 1. Summary of landslide spot counts over and under the 45° central border line (number of spots).

Classification	Over	Under
Daily rainfall and 1-day cumulative rainfalls before failure data (**Figure 2**)	162	24
Daily rainfall and 3-day cumulative rainfalls before failure data (**Figure 3**)	86	80
Daily rainfall and 5-day cumulative rainfalls before failure data (**Figure 4**)	56	130

3.3. Scenario For Future Climate Changes In South Korea

The period for the study results built from spatial statistical downscaling of the KMA-RCM is 1971–2100, and average temperature and rainfall per month are produced. Temperature and rainfall are selected from the climate change models because they are the two factors that can express future changes in climate fragmentally and representatively, with higher practical utility (**Figure 5**).

Rainfall was analyzed by accumulation of the monthly average, decreasing from 947.38 mm in Year 1971 to 886.02 mm in Year 2000, a drop of 61.35 mm. Results of future climate change data processing showed that the rainfall estimate increased from 1002.12 mm in Year 2001 to 1218.60 mm in Year 2100, or by 216.48 mm. The annual average rainfall is showed by 271.23 mm from Year 1971 to Year 2100 (**Figure 6**). **Figure 6**, where the *X*-axis is the year and the *Y*-axis is average rainfall, respectively, shows a linear increase.

Regarding the annual cumulative rainfall for this 130-year period, 1839.36 mm (2003, 2043, and 2083 year) is the highest, and 847.37 mm (2018, 2058, and 2098 year) is the lowest, a difference of 991.99 mm, while 1817.03 mm (Year 1991) is the second highest, and 853.5 mm (Year 1998) is the second lowest. When years with the same rainfall distribution were organized, 71 are separated. In the distribution maps, built from past climate data, single rainfall is shown as 1817.03 mm and 1701.14 mm. When downscaling based on the KMA-RCM is conducted, the changes in cumulative rainfall as a future change in climate show the repetition of the same distribution with an interval of 40 years, with an overall increase in rainfall [12].

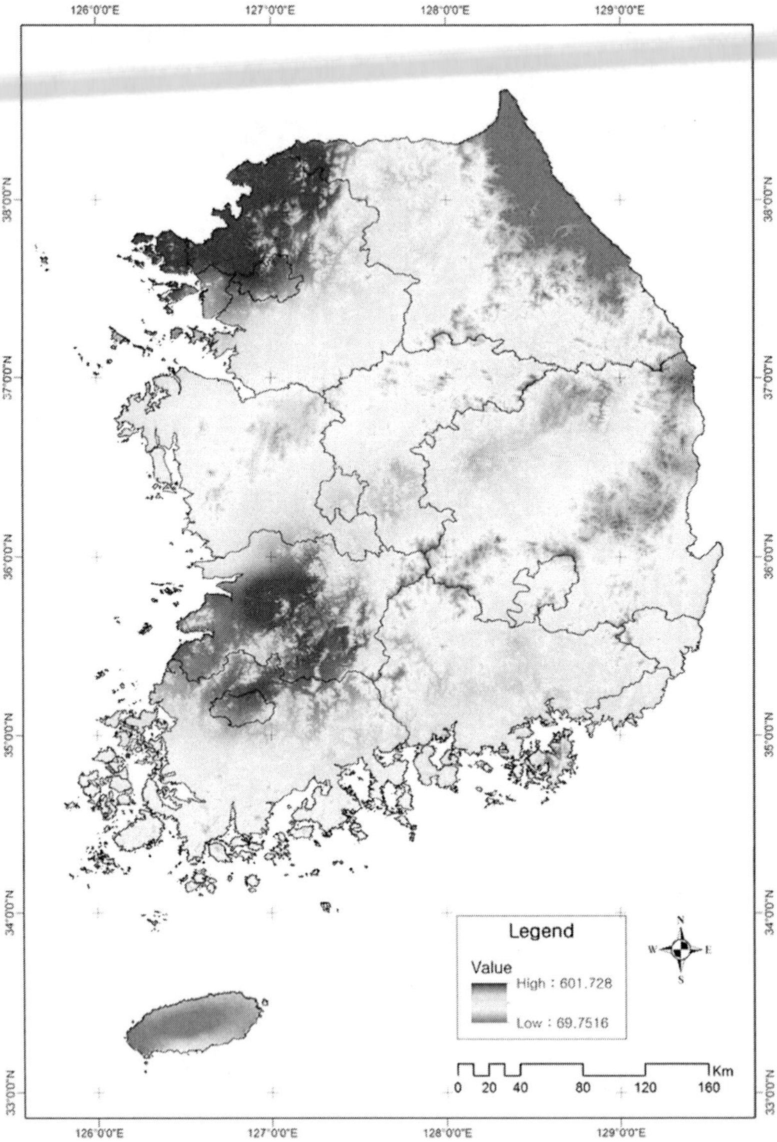

FIGURE 5. The average rainfall in July 2071 (based on KMA-RCM).

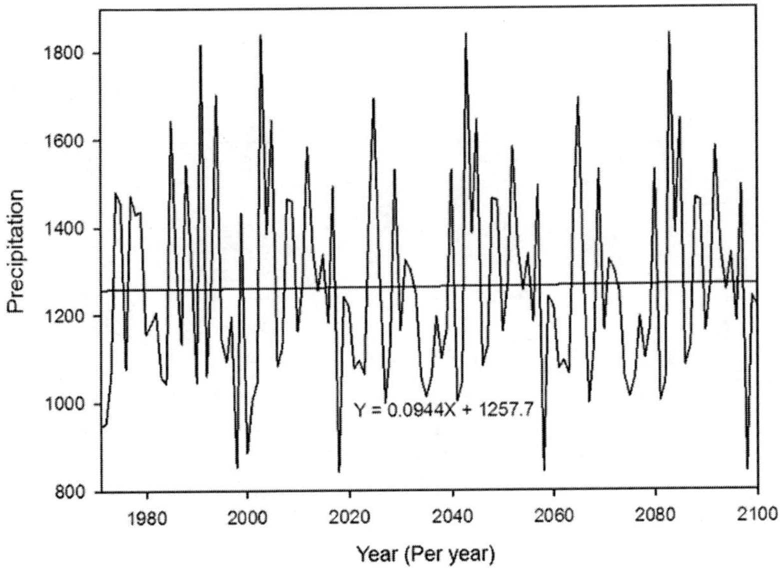

FIGURE 6. 1971–2100 annual rainfall change in South Korea.

3.4. 1971–2100 Annual Rainfall Change In South Korea: Verification Of Climate Change Scenario Result

In order to reduce uncertainty occurred in the process of spatial statistical downscaling of future climate change scenarios, the analysis of the correlation between this research's results and actual measurements was carried out. Weather data after Year 2000 was selected as the reference year and 75 AWS, which were closest to 75 ground observation points, and data from January 2001 to August 2010 existed was selected and the analysis of correlation between temperature and rainfall data and data obtained through future climate change scenario prediction from 2001 to 2010 was carried out. The data was analyzed and the coefficient of correlation between future climate change scenario prediction data and actual measurements from AWS was over 0.98 on average in the case of temperature and 0.56 in the case of rainfall. This was lower than that of the temperature, but in consideration of uncertainty in rainfall prediction on the climate change scenario, it was a high correlation.

TABLE 2. Result of correlation between statistical downscaling of future climate change scenario and AWS in Years 2001–2010 (rainfall).

Area	Correlation coefficient	Area	Correlation coefficient	Area	Correlation coefficient
Gangneung	0.41	Busan	0.63	Jangheung	0.49
Ganghwa	0.70	Sancheong	0.53	Jeongu	0.65
Geochang	0.52	Seosan	0.50	Jeongeup	0.65
Goheung	0.49	Seoul	0.64	Jecheon	0.64
Gwangju	0.46	Sokcho	0.45	Jinju	0.48
Gumi	0.62	Suwon	0.62	Cheonan	0.65
Gunsan	0.61	Suncheon	0.61	Cheorwon	0.68
Geumsan	0.60	Andong	0.63	Chupungnyeong	0.67
Namwon	0.57	Yangpyeong	0.62	Chuncheon	0.67
Namhae	0.40	Yeongdeok	0.45	Chungju	0.58
Daegwallyeong	0.59	Yeongwol	0.60	Taebaek	0.57
Daegu	0.49	Yeongju	0.64	Tongyeong	0.39
Daejeon	0.73	Yeongcheon	0.50	Pohang	0.45
Masan	0.45	Ulsan	0.46	Hapcheon	0.49
Mungyeong	0.70	Uljin	0.35	Haenam	0.49
Baengnyeongdo	0.50	Uiseong	0.60	Hongcheon	0.60
Boyeong	0.59	Icheon	0.57	Heuksando	0.50
Boeun	0.76	Incheon	0.64	Yeosu	0.42
Bonghwa	0.66	Imsil	0.66		
Busan	0.29	Jangsu	0.65		

Note that 75 AWS where rainfall data existed from January 2001 to August 2010, 75 observation points were extracted and summarized in **Table 2.** In the case of correlation on rainfall summarized in **Table 2**, most weather stations showed a significantly high correlation, but in some cases of correlation on rainfall shown in **Table 2**, the range of rainfall variability was significantly high depending on the weather station.

In the analysis of correlation on the amount of rainfall, Ganghwa showed a correlation of 0.70, which was significantly high, but Busan showed a correlation of 0.29, which was lower than that of other areas. It was analyzed that generally the correlation between future climate change scenario prediction data and actual measurements was significantly high, but in the case of rainfall, the variability for each year and season was high. South Korea shows abundant rainfall in summer, and the standard deviation of rainfall in summer varies more than two times between regions, and the rainfall variability for each region is also significantly high.

In addition, the range of rainfall variability near the coast is higher than that in the inland area in South Korea, and the east coast has the higher rainfall variability than the west coast [13]. In the result of correlation between the rainfall and future climate change scenario prediction data, the bottom 10%, which had the lowest correlation, showed a correlation of less than 0.40, and the corresponding area included Busan, Uljin, Tongyeong, and Namhae. In order to analyze the correlation of relevant area more precisely, the analysis of correlation for each season was carried out additionally. An area with a low correlation was selected because a seasonal change in an area with a low correlation was more closely related with rainfall change [13]. In the analysis result, it was analyzed that the correlation between three regions except for Uljin was lower during summer and winter, with a high range of variance in the meteorological factors (**Table 3**).

TABLE 3. Result of seasonal correlation coefficient between statistical downscaling of future climate change scenario and AWS in Years 2001–2010 (rainfall).

Area	Correlation coefficient	Correlation coefficient (summer and winter)	Correlation coefficient (spring and fall)
Busan	0.29	0.24	0.45
Uljin	0.35	0.30	0.16
Tongyeong	0.39	0.37	0.50
Namhae	0.40	0.39	0.43

4. DISCUSSION AND CONCLUSION

In this study, the pattern of rainfall generating a landslide in the past was analyzed, and the threshold of landslide occurrence by rainfall was also analyzed. Based on the analysis result, the possibility of landslide occurrence in the future was analyzed by analyzing the rainfall of future climate change scenarios.

When the relation between rainfall during the 2000s and landslides in South Korea is quantitatively analyzed, 1-day rainfall and 3-day cumulative rainfall had higher correlations with landslides. Based on this, the landslide occurrence threshold in the study area is defined to be 202 mm for 1-day rainfalls and 449 mm for 3-day cumulative rainfalls. The results of this analysis of rainfall probability show the ratio of the occurrence threshold consistently increases as the target year increases in the study area, the same tendency is seen in the future climate change scenario. Conclusively, the study area has seen increasing rainfall as time passes, and damage, such as that from the 2006 landslide, may increase gradually in the future.

As the target year increases, the accuracy increases for the 202–449 mm thresholds of rainfall probability and the future climate change scenario. In all of the methods applied, the accuracy is higher for the 449 mm threshold than for 202 mm of rainfall probability. A 3-day cumulative rainfall affects landslide occurrence more than a 1-day rainfall in the relation between rain and landslides.

In addition, a rainfall change from Years 1971 to 2100 was analyzed through the analysis of the climate change scenario based on KMA-RCM in this study. The main result of this study can be summarized as follows.

First, the downscaling technique using rainfall lapse-rate technique was developed by using Co-Kriging among geographic information spatial-interpolation techniques. As a research result, the average rainfall between 1971 and 2100 was drawn. The result showed that the average rainfall increased by 271.23 mm.

Second, the analysis of correlation between the average rainfall between 2001 and 2010 and actual measurements from 75 AWSs from regional-scale climate change scenario was carried out. As a result of analyzing the correlation between the future climate change scenario prediction data and actual measurements during the same period, it was concluded that the correlation on the rainfall was 0.56.

A study to downscale KMA-RCM with a spatial resolution of 27 km into the climate change scenario with a spatial resolution of 1 km, which could express the local level climate change, was carried out. In order to reduce the uncertainty in the climate change scenario occurred during this process, the correlation between actual measurements after the reference year entered on the future climate change scenario and research results were analyzed. The result verified that the significance of downscaling of results through KMA-RCM is high, and the average climate pattern (30 years in the past) and the weather pattern from Years 2001 to 2010 were similar. In the case of rainfall, temperature, humidity, and local characteristics (e.g., topography and characteristics of ground surface, etc.) have interacted in the rainfall process complexly, showing a nonlinear relationship [13], and the result of this study was also affected in the same way.

The results of this analysis of correlation between rainfall and landslide show the cumulative rainfall consistently increases in South Korea, the same tendency is also seen in the future climate change scenario.

A result of correlation between rainfall and landslide and the future rainfall change also shows future rain events for quantitative analysis of climate change in South Korea. Changes in rainfall in South Korea are shown to be larger.

The occurrence of landslides is directly caused by intensive rainfall. If there is a change in rainfall, it will lead to a change in the occurrence of landslides.

However, this study is unable to reflect extreme conditions according to climate change, and, even though, the correlation between the result of this study and actual measurements was significantly high, it is necessary to improve the methodology to future climate change scenario prediction continuously. It is necessary to supplement methodologies regarding extreme rainfall and extreme climate events in order to reduce the uncertainty in the future climate change scenario in future studies.

5. ACKNOWLEDGEMENTS

This research was supported by Korea Environment Institute funded by the Basic Science Research Program through the National Research Foundation of Korea (NRF) funded by the Ministry of Education (NRF-502014R1A1A1002704).

REFERENCES

1. IPCC. Climate Change 2007—IPCC 4th Assessment Report, Working Group 1—The Physical Science basics. IPCC; 2007. p. 996
2. Creutin J D, Delrieu G, Lebel T. Rain measurement by raingage-radar combination: a geostatistical approach. Journal of Atmospheric and Oceanic Technologies. 1998. 5: 102–115. DOI:
3. Daly C, Helmer E H, Maya Quinones. Mapping the climate of Puerto Rico, Vieques and Culebra. International Journal of Climatology. 2003. 23: 1359–1381.
4. Borge M. Accuracy of radar rainfall estimates for streamflow simulation. Journal of Hydrology. 2002. 267: 26–39.
5. Temesgen B, Mohammed M U, Korme T. Natural hazard assessment using GIS and remote sensing methods, with particular reference to the landslides in the Wondogenet Area, Ethiopia. Physics and Chemistry of the Earth, Part C: Solar, Terrestrial & Planetary Science. 2001. 26: 665–675.
6. Regmi N R, Giardino J R, Vitek J D. Modeling susceptibility to landslides using the weight of evidence approach: Western Colorado, USA. Geomorphology. 2010. 115: 172–187.
7. Daly C. Guidelines for assessing the suitability of spatial climate data sets. International Journal of Climatology. 2006. 26: 707–721.
8. Benestad R E. Empirically downscaled multimodel ensemble temperature and precipitation scenarios for Norway. Journal of Climate. 2002. 15: 3008–3027.

9. Hewitson B C, Crane R G. Consensus between GCM climate change projections with empirical downscaling: precipitation downscaling over South Africa. International Journal of Climatology. 2006. 26: 1315–1337.

10. Comriea A C, Broylcsb B. Variability and spatial modeling of fine-scale precipitation data for the Sonoran Desert of south-west Arizona. Journal of Arid Environments. 2002. 50(4): 573–592.

11. Smith C D, The relationship between monthly precipitation and elevation in the Alberta Foothills during the Foothills orographic precipitation experiment. Cold Region Atmospheric and Hydrologic Studies; Chapter 10. The Mackenzie GEWEX Experience. Springer Berlin Heidelberg, 2008. pp. 167–185.

12. IPCC TGICA. General Guidelines on the use of Scenario Data for Climate Impact and Adaptation Assessment. Version 2. IPCC; 2007. pp. 120–122

13. Lee M J, Park I, Won J S, Lee S. Landslide hazard mapping considering rainfall probability in Inje, Korea. Geomatics, Natural Hazards and Risk. 2016. 7(1): 424–446

CHAPTER 16

Impact of Event Landslides on Road Networks: A Statistical Analysis of Two Italian Case Studies

Marco Donnini I Elisabetta Napolitano I Paola Salvati I Francesca Ardizzone I Francesco Bucci I Federica Fiorucci I Michele Santangelo I Mauro Cardinali I Fausto Guzzetti

Marco Donnini , Elisabetta Napolitano , Paola Salvati , Francesca Ardizzone , Francesco Bucci , Federica Fiorucci , Michele Santangelo ,Mauro Cardinali ,Fausto Guzzetti

ABSTRACT

Despite abundant information on landslides, and on landslide hazard and risk, in Italy, little is known on the direct impact of event landslides on road networks and on the related economic costs. We investigated the physical and economic damage caused by two rainfall-induced landslide events in Central and Southern Italy, to obtain road restoration cost statistics. Using a GIS-based method, we exploited road maps and landslide event inventory maps to compute different metrics that quantify the impact of the landslide events on the natural landscape and on the road networks, by road type. The maps were used with cost data obtained from multiple sources, including local authorities, and specific legislation, to evaluate statistically the unit cost per metre of damaged road and the unit cost per square metre of damaging landslide, separately for main and secondary roads. The obtained unit costs showed large variations which we attribute to the different road types in the two study areas and to the different abundance of landslides. Our work confirms the long-standing conundrum of obtaining accurate landslide damage data and outlines the need for reliable, standardized methods to evaluate landslide damage and associated restoration costs that regional and local administrations can use rapidly in the aftermath of a landslide event. We conclude recommending that common standardized procedures to collect landslide cost data following each landslide event are established, in Italy and elsewhere. This will allow for more accurate and reliable evaluations of the economic costs of landslide events.

Keywords: Landslide Landslide event Landslide damage Road network Cost estimation Economic impact Italy

INTRODUCTION

In Italy, geohydrological hazards, including landslides, cause serious economic damage and represent a severe threat to the population (Salvati et al. 2014). In this country, rich information exists on landslides, their physical characteristics and consequences, while little is known on their economic impact.

The damage caused by landslides (and other natural hazards) in Italy has been and is currently being fixed primarily using public resources, often following specific legislation. As a result, the few available landslide cost estimates are given in reports and documents produced by local, regional and national government administrations and authorities. Lack of a rigorous strategy and of a single and accurate method to collect damage and cost information, combined with a complex network of agencies involved in reconstruction activities and cost reimbursement, make the collection of reliable landslide cost data challenging. In addition, the long periods required to complete restoration activities following a landslide event result in supplementary costs that are seldom accounted, making it even more difficult to obtain comprehensive landslide cost estimates. Lack of a single method to collect damage costs also hampers the efforts of regional and local administrations that immediately after a landslide event have to obtain reasonably accurate damage estimates to request the declaration of the state of emergency in their territory and emergency funding to reimburse the incurred and the foreseen restoration costs.

Here, we investigate the economic impact of landslides triggered by two rainfall events in Central and in Southern Italy. We focus on the damage caused by landslides to the road network, and we use the case studies to test a general methodology to assess statistically the direct economic impact of landslides on the road networks. In the first case, we analyse the impact of landslides caused by an intense rainfall event that hit the coastal area of the Messina province, Sicily, Southern Italy, on 1 October 2009. In the second case, we study the impact of landslides occurred as a result of prolonged rainfall in two mountainous municipalities, Acquasanta Terme and Roccafluvione, in the Marche region, Central Italy, in November and December 2013. For both events, we first evaluate the physical impact of the event landslides on the road networks. Next, we quantify the cost per unit length of damaged road and per unit area of damaging landslide. To evaluate the landslide impact on the road network, we use accurate event landslide inventory maps prepared shortly after the events, road network maps and cost data obtained from multiple sources available from local authorities and national administration offices. The proposed methodology allowed us to estimate the average cost per unit length of damaged road and per unit area of damaging landslide, after intense and prolonged rainfall events that triggered more than a thousand landslides. The obtained average costs can be used by local administrations and authorities for a preliminary assessment of the costs of the damage caused by landslides.

BACKGROUND

Different classifications of costs produced by natural hazards exist in the literature. The classifications are based on various types of information, including the type of the event, the category of the costs and the extent of the geographical area. Meyer et al. (2013), in a recent review of methods to assess the costs of natural hazards, classified the costs as direct, business interruption, indirect, intangible and risk mitigation costs. Cost assessments may be estimated after an event (*ex post*) or considering potential future events (*ex ante*). *Ex post* assessments are the most common type of evaluations in the literature and aim at quantifying incurred economic damage to inform local, regional or national governments, stakeholders and business and to provide quantitative information useful to determine compensations or to decide on the possible levels of economic support (Karunasena and Rameezdeen 2010; McCarty and Smith 2005; Sächsische Staatskanzlei 2003). *Ex ante* cost assessments are less frequent in the literature and are usually conducted to support governmental and administrative decision-making and to decide among risk mitigation options and adaptation strategies. Hilker et al. (2009), using the Swiss flood and landslide damage database that includes information on direct economic impact, estimated the total cost of natural geohydrological events in Switzerland, from 1972 to 2007. Boonyanuphap (2013), working in northern Thailand, used interviews with stakeholders involved in land rehabilitation, to execute a cost benefit analysis. The analysis was performed comparing the net welfare gain of different land rehabilitation measures after landslide events.

Exploiting a landslide inventory map, a susceptibility model and information obtained through interviews, Vranken et al. (2013) proposed a methodology to estimate the direct and indirect damage caused by landslides at the regional scale in the Flanders, Belgium. Estimates were calculated differently for private properties and public infrastructures. For private properties, losses to land, buildings and production were considered, whereas for public infrastructures, the costs of prevention measures and restoration actions were estimated. Heam et al. (2008) used a landslide inventory map to estimate damage restoration costs along the national road network in the People's Democratic Republic of Lao. The authors considered direct and indirect costs, including the cost due to road disruptions and their consequent temporary interruption, and the related environmental and social costs. Costs to improve slope stability in the design and construction phases were compared to potential benefits for ordinary and emergency maintenance. Klose et al. (2015) investigated direct costs of landslides along transportation routes in Lower Saxony, Germany, between 1980 and 2010, and proposed a methodology for an *ex post* assessment based on the compilation, modelling and extrapolation of the landslide losses at different spatial scales over time. Using a landslide database, a regional landslide susceptibility model and cost data, the average annual cost per kilometre of highway was ascertained, for two case studies.

In Italy, abundant information is available on landslide events and their consequences. The National Research Council's Aree Vulnerate Italiane (AVI—areas affected by landslides or floods in Italy) project, an archive inventory of historical landslide (and flood) events in Italy (Guzzetti et al. 1994), lists information on more than 32,000 landslides from 1918 to 2001 (Guzzetti and Tonelli 2004). Information on the type and magnitude of damage for a subset of the AVI inventory is available in SICI, the Information System on Hydrological and Geomorphological Catastrophes in Italy (http://sici.irpi.cnr.it/). The national inventory of landslides in Italy (Inventario dei Fenomeni Franosi in Italia (IFFI), Trigila et al. 2010, 2015) has mapped 528,903 landslides, covering a total area of 22,176 km^2 (7.3% of national territory). For about 8% of the mapped landslides, information on the type of damage is available. Analysis of the IFFI database revealed that roads are the single category of vulnerable elements most frequently affected by damaging landslides.

The first attempt to evaluate economic costs of geohydrological (i.e. landslide and flood) events in Italy was conducted by Catenacci (1992) for the period 1944–1990. When landslides occurred together with large flood events, the author reported a single economic figure, classifying the events as "geohydrological". This made it impossible to separate the cost due to landslides from the cost due to floods. Following Catenacci (1992), and based on official amounts of public money spent after natural disasters, a report prepared jointly by the Italian National Association of Construction Contractors (Associazione Nazionale Costruttori Edili (ANCE)) and the Centre for Economic, Sociological and Market Research (Centro Ricerche Economiche, Sociologiche e di Mercato (CRESME)) (ANCE-CRESME 2012) quantified the cost of geohydrological hazards in Italy at 61.5 billion € (2011), for the 69-year period 1944–2012. This is an average cost of 0.9 billion € per year.

MATERIALS

To devise and test our new cost estimation method, we exploit two recent landslide events in two study areas in NE Sicily, Southern Italy, and in the Marche region, Central Italy (Fig. 1). The two study areas differ for their spatial extent, terrain elevation and morphology, geology, prevalent landslide types and distribution and abundance of the vulnerable elements (i.e. the roads). The two case studies were chosen because of the availability of landslide event inventory maps, the road network maps and the cost data. The latter in Italy are typically available in different ways: (1) as cost estimations produced by local authorities and (2) as money allocated by law. We used the first for the 2013 Marche case study and the second for NE Sicily event. For both the areas, official damage road (ODR) lists were also provided by official documents. All the information that were collected giving costs (estimated by local authorities or allocated by law) related to landslide damage were associated to a unique road and organized in a database.

NE Sicily case study

On 1 October 2009, a very intense rainfall event hit the Messina province, NE Sicily, producing inundations and triggering hundreds of landslides that caused 31 deaths, seven missing persons and hundreds of injured people (http://polaris.irpi.cnr.it). The intense rainfall event resulted in a maximum cumulated rainfall of 225 mm in 7 h, recorded at the Santo Stefano di Briga rain gauge (Fig. 2c) (Regione Siciliana, Centro Funzionale Regionale 2009). The rainfall caused inundations and a large number of soil slides and debris flows (Ardizzone et al. 2012) in an area of approximately 60 km^2 (Mondini et al. 2011). For our study, we focus on the area where landslides were most abundant and damaging, and specifically in the Briga, Giampilieri, Divieto and Racinazzo coastal catchments, covering a total study area A_S = 22 km² (Fig. 1c, Table 1). In this area, intense rainfall triggered a total of 1480 landslides (Table 1) (Ardizzone et al. 2012).

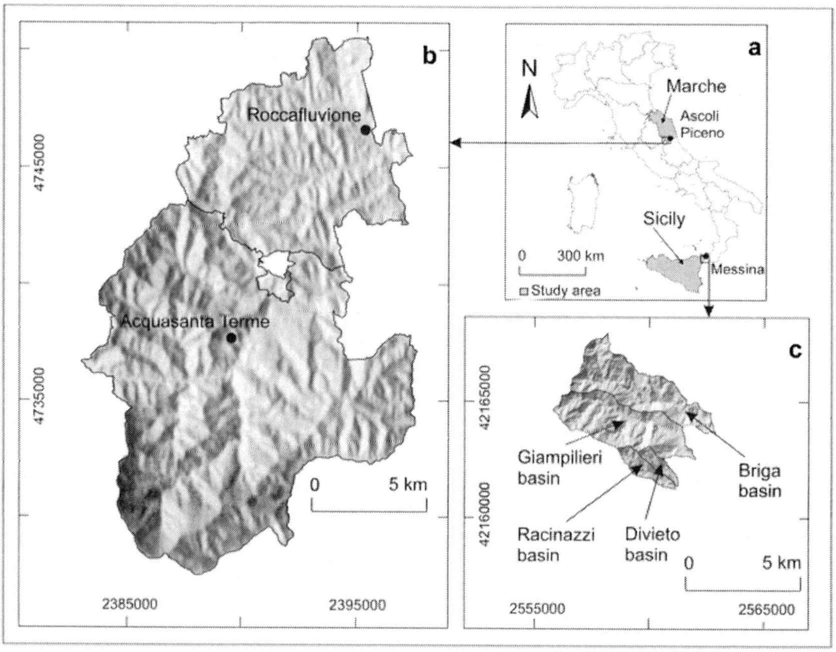

Figure 1. Location and morphology of the study areas. **a** Index map. **b** Terrain morphology in the Acquasanta Terme and the Roccaluvione municipalities, Marche, Central Italy. **c** Terrain morphology in the Briga, Giampilieri and Divieto catchments, NE Sicily, Southern Italy. Maps are in the EPSG 3004 geographical reference system

The event landslides affected roads, underpasses and the ground floors and basements of buildings in urban areas. The A18 Messina–Catania highway, the SS-114 state road 114 and the national railway connecting Messina to Catania were all disrupted by the landslides.

After the event, specific legislation was issued by the national government. First, the Prime Minister Order no. 3815/09 (*Ordinanza del Presidente del Consiglio dei Ministri*, OPCM 3815/09) of 2 October 2009 declared the "state of emergency" for the Messina Province and listed urgent civil protection works. Next, extra allocation was stated with OPCM 3865/10. The Work Plan included in the OPCM 3815/09 (updated on 12 September 2012) allocated a total amount of 193,688,223 € for risk mitigation, restoration of the hydraulic functionality of rivers and torrents, debris removal and restoration of roads, railways and public buildings (http://opcm3815.altervista.org/3815.html).

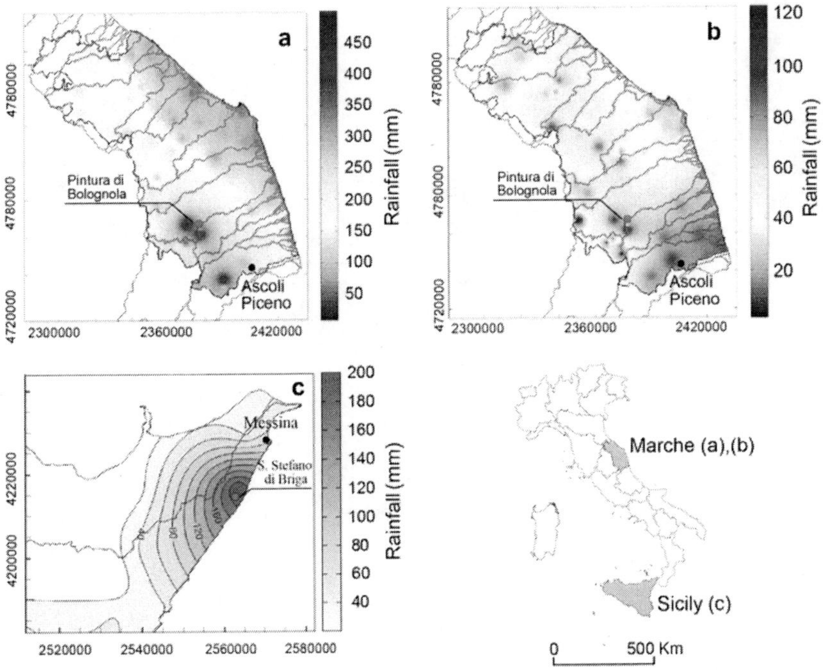

Figure 2. Distribution of the cumulated rainfall for the NE Sicily and the Marche region test cases. **a** Cumulated rainfall in the Marche region from 12:00 (UTC+1) on 10 November 2013 to 24:00 (UTC+1) on 13 November 2013 (modified after Regione Marche, Centro Funzionale per la Meteorologia, l'Idrologia e la Sismologia 2013a). **b** Cumulated rainfall for the Marche region from 24:00 (UTC+1) on 25 November 2013 to 24:00 (UTC+1) on 1 December 2013 (modified after Regione Marche, Centro Funzionale per la Meteorologia, l'Idrologia e la Sismologia 2013b). **c** Twenty-four hour cumulated rainfall of 1 October 2009 in NE Sicily (modified after Regione Siciliana, Centro Funzionale Regionale 2009). Maps are in the EPSG 3004 geographical reference system

Table 1. Descriptive statistics of event landslides for the NE Sicily and the Marche case studies

Metric	Unit	NE Sicily	Marche
A_S	km^2	22	198
N_{ES}	#	30	289
N_{SS}	#	0	424
N_{SEF}	#	0	328
N_{EF}	#	0	371
N_{DF}	#	1071	0
N_{GE}	#	375	0
N_{RF}	#	0	17
N_{CPX}	#	0	9
N_{UL}	#	3	0
N_{LT}	#	1479	1438
M_L	–	3.17	3.16
A_{Lmin}	m^2	1.9×10^0	1.6×10^0
A_{Lmax}	m^2	4.5×10^4	2.2×10^4
A_{Lavg}	m^2	1.0×10^3	4.8×10^2
A_{LT}	m^2	1.5×10^6	7.7×10^5
V_{Lmin}	m^3	8.0×10^{-2}	2.1×10^{-1}
V_{Lmax}	m^3	1.3×10^5	1.5×10^5
V_{Lavg}	m^3	1.1×10^3	1.3×10^3
V_{LT}	m^3	1.2×10^6	1.8×10^6
D_L	–	0.068	0.004
d_L	#/km^2	67.3	7.3

A_S (km^2) is the size of the study area. N_{ES} (#), N_{SS} (#), N_{SEF} (#), N_{EF} (#), N_{DF} (#), N_{GE} (#), N_{RF} (#), N_{CPX} (#) and N_{UL} (#) are, respectively, the number of earth slides, soil slides, slides-earth flows, earth flows, debris flows, gully erosions, rock falls, complex slides and undetermined landslides. N_{LT} (#) is the total number of landslides. M_L (–) is the landslide event magnitude estimated as log (N_{LT}) following Malamud et al. (2004). A_{Lmin} (m^2), A_{Lmax} (m^2), A_{Lavg} (m^2) and A_{LT} (m^2) are the minimum, maximum, average and total event landslide areas, respectively. V_{Lmin} (m^3), V_{Lmax} (m^3), V_{Lavg} (m^3) and V_{LT} (m^3) are the minimum, maximum, average and total event landslide areas, respectively. D_L (–) is the density of landslide area, calculated dividing A_{LT} (in km^2) by A_S (in km^2). d_L (#/km^2) is landslide density, calculated dividing N_{LT} (#) by A_S (in km^2)

Event landslide inventory map

For the study area, an accurate event landslide inventory map was prepared by Ardizzone et al. (2012), at 1:10,000 scale, through field surveys and the visual interpretation of pre-event and post-event, stereoscopic and pseudo-stereoscopic aerial photographs. Figure 3a, modified from Ardizzone et al. (2012), shows 1480 event landslides (Table 1): 1071 were classified as debris flows, 30 as earth slides (Cruden and Varnes 1996; Hungr et al. 2014), 375 as gully erosion (areas with superficial and channelled erosion affecting the surface deposits) and 3 landslides remained unclassified. Figure 4 portrays examples of the mapped landslides.

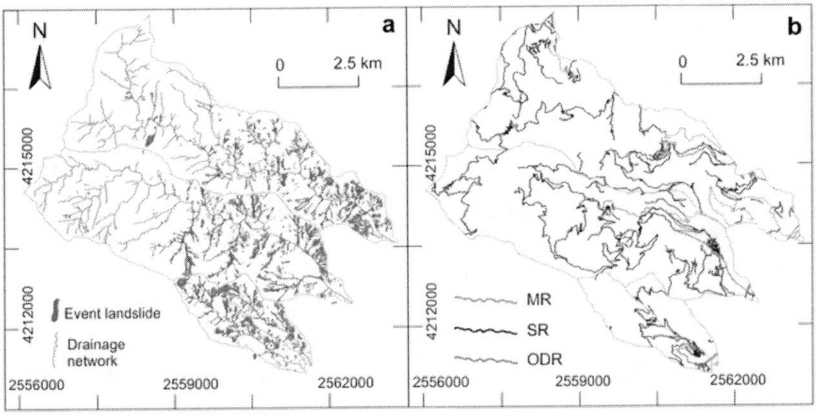

Figure 3. NE Sicily case study, encompassing the Briga, Giampilieri and Divieto catchments. **a**Landslide event inventory map, modified after Ardizzone et al. (2012). **b** Road network map. *MR* main road (*green*), *SR* secondary road (*black*), *ODR* official damage road (*red*). See text for explanation. Maps are in the EPSG 3004 geographical reference system

Road network map

We obtained a map of the road network for the study area from a commercial provider, in vector format. We classified the single roads in two classes: (i) main public roads (MRs), including highways (toll roads), state and provincial roads, and (ii) secondary public roads (SRs), encompassing municipal roads and country roads. We calculated a total road length of 128.45 km, including 23.72 km (18%) of MR and 104.73 km (82%) of SR (Table 2).

Figure 4. NE Sicily case study. Examples of rainfall-induced landslides triggered by the intesnse rainfall event. **a** Debris flow. **b** Slide

Table 2. Descriptive statistics for the road networks in NE Sicily and in Marche area

Road metric	Unit	NE Sicily	Marche
L_{RT}	km	128.45	428.14
D_{RT}	km^{-1}	5.8	2.2
L_{MR}	km	23.72	99.93
D_{MR}	km^{-1}	1.1	0.5
L_{SR}	km	104.73	328.21
D_{SR}	km^{-1}	4.8	1.7
L_{ODR}	km	12.79	271.98
D_{ODR}	km^{-1}	0.58	1.37
$ODR_{\#}$	#	5	110
$ODR_{\%}$	%	53.9	82.9

L_{RT} and D_{LT} is, respectively, the length and the density of the total road network map. L_{MR} and D_{MR} is, respectively, the length and the density of the main roads. L_{SR} and D_{SR} is, respectively, the length and the density of the secondary roads. L_{ODR} and D_{ODR} is, respectively, the length and the density of the official damaged roads. $ODR_{\#}$ is the number of official damaged roads. $ODR_{\%}$ is the percentage of official damage roads, in terms of kilometres of roads, estimated with respect to secondary roads for the Marche case study and to main roads for the NE Sicily case study

Cost data

We obtained cost data from the mentioned Work Plan included in the OPCM 3815/09, (updated on 12 September 2012). The Work Plan provides economic information for five MRs (Table 3) corresponding to the ODR. The Work Plan does not provide information on SR. In some cases, the cost data were aggregated (e.g. by including in a single item the restoration costs of a damaged road together with the restoration cost of a neighbouring damaged public building). For this reason, we manually selected only data that unambiguously referred to a single MR, but it was impossible to associate to each main road the single damage and the exact description of the restoration works. We acknowledge that this has resulted in an underestimation of the economic resources allocated for the restoration of the damaged MR in our study area. Figure 3b shows the ODR and all the other public roads (MR and SR) in the Briga, Giampilieri, Divieto and Racinazzo coastal catchments. The 12.79 km of ODR represents 53.92% of the total length of the MR in the study area (Table 2). The cost for fixing the five ODRs totalled 26,140,736 €, corresponding to an average cost per road of 5,228,147 € and per kilometre of 2,043,841 €.

Table 3. NE Sicily case study

Road	km	Cost (€)
A18		3,218,770
SS-114		9,027,000
SP32		7,199,966
SP33		2,670,000
SP35		4,025,000
Total		26,140,736

Damaged roads and associated restoration costs from the Prime Minister Order OPCM 3815/09 Work Plan, revised on 12 September 2012. Highway (Austostrada (*A*)), state road (*SS*), provincial road (*SP*)

Marche case study

In November and December 2013, prolonged and intense regional rainfall events caused abundant and widespread landslides throughout the Marche region, Central Italy, with more abundant landslides in the southern part of the region. For our experiment, we focus on two mountain municipalities, Roccafluvione and Acquasanta Terme, in the Ascoli Piceno Province, Southern Marche, covering collectively 198 km^2 (Fig. 1a, Table 1).

In the 4-day period from 10 to 13 November 2013, heavy rainfall occurred in the area, with a maximum cumulated rainfall of 499 mm recorded at Pintura di Bolognola (Fig. 2a). A few days later, between 25 November and 1 December 2013, the area was affected by heavy snowfall followed by rapid snowmelt (Fig. 2b). Finally, in the 24-h period from 24:00 (UTC+1) on 2 December to 24:00 (UTC+1) on 3 December 2013, a heavy rainfall event struck the southern part of region, with a maximum cumulated rainfall of 129 mm recorded at Pintura di Bolognola. As a result, floods and flash floods occurred along rivers and torrents, and a large number of landslides were triggered, mainly in the region's hilly and mountainous terrain. Most of the landslides affected secondary roads that connect numerous small rural settlements. The landslides caused road interruptions at multiple sites; around 100 people were isolated (Regione Marche, Centro Funzionale per la Meteorologia, l'Idrologia e la Sismologia 2013a, 2013b) and few hundreds were evacuated. The Marche regional government estimated at 398,010,736 € the total amount required to overcome the state of emergency (Dipartimento della Protezione Civile 2014).

Event landslide inventory map

In January and February 2014, we executed reconnaissance field surveys in the municipalities of Acquasanta Terme and Roccafluvione. Each landslide was mapped in the field as a polygon using Google Earth™. Single or pseudo-stereoscopic photographs were taken in the field for each landslide or group of landslides. The preliminary mapping (position, shape, size) of the individual landslides was transferred in a geographical information system (GIS), in vector format, and was checked and enhanced using the field photographs. Overall, we mapped 1438 landslides in the municipalities of Acquasanta Terme and Roccafluvione (Fig. 5a, Table 1), including 289 earth slides, 424 soil slides, 328 slide–earth flows, 371 earth flows, 17 rock falls and 9 complex landslides (Cruden and Varnes 1996; Hungr et al. 2014) (Fig. 6).

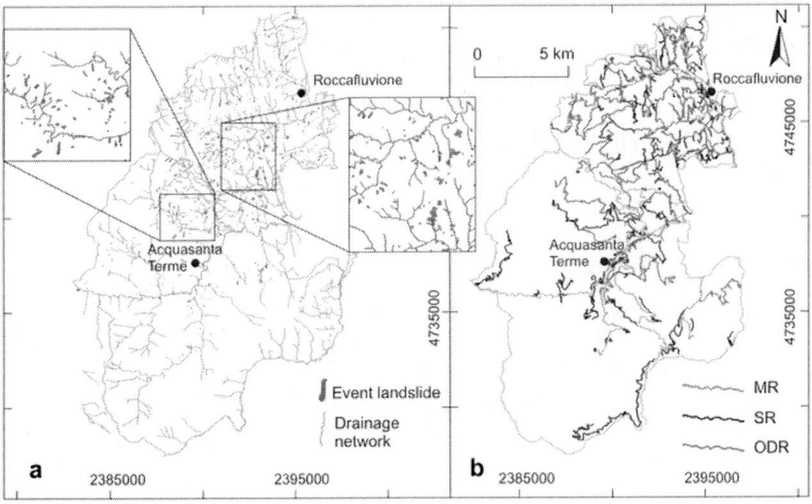

Figure 5. Marche case study, encompassing the Acquasanta Terme and Roccaluvione municipalities. **a** Landslide event inventory map, modified after Ardizzone et al. (2012). **b** Road network map. *MR* main road (*green*), *SR* secondary road (*black*), *ODR* official damaged road (*red*). See text for explanation. Maps are in the EPSG 3004 geographical reference system

During the field surveys, the damage caused by landslides found along roads was mapped and assessed. Adopting a heuristic (qualitative) approach, the different types of damage were organized in three classes (Cardinali et al. 2002; Galli and Guzzetti 2007; Reichenbach et al. 2005): (i) aesthetic (minor), where the road functionality was not compromised; (ii) functional (medium), where the road functionality was compromised; and (iii) structural (severe) where the road was severely damaged or destroyed completely. Overall, the damage was assessed as aesthetic at 140 sites, functional at 154 sites and structural at 64 sites.

Road network map

We prepared a digital map, in vector format, of the road network in the Acquasanta Terme and Roccafluvione municipalities. We started from the road information available in CAD format from regional technical maps (CTR), at 1:10,000 scale. For our scopes, the original dataset had a number of limits, including the lack of road classification attributes. To overcome these limitations, we first moved the road geographical information in a GIS environment to produce an ESRI® shapefile of the entire road network. We then updated the road information using 1-m resolution orthophotographs taken in 2012 (http://www.pcn.minambiente.it) and Bing© images taken in July 2011 and August 2012 (http://www.bing.com). Table 2 shows, for the Marche case study, a total road length of 428.14 km, including 99.93 km (23%) of MR and 328.21 km (77%) of SR.

Figure 6. Marche case study. Examples of rainfall-induced landslides triggered by the rainfall events. **a** Earth slide. **b** Soil slide. **c** Slide-earth flow. **d** Earth flow. **e** Rock fall. **f** Complex landslide

Cost data

A few days after the 2 December rainfall event, the Acquasanta Terme and Roccafluvione administrations restored partially the functionality of their SR to guarantee access to villages, hamlets and rural homes. The two municipalities compiled mandatory reports providing cost information for 110 ODRs, shown in Fig. 5b. The reports gave for each road the restoration cost without associating them information on the type of damage and on the type of restoration works. In terms of road length, the 271.98 km of ODR represents 82.87% of the total length of SR in the two municipalities (Table 2). Restoration costs for the 110 ODRs totalled 11,837,409 €, corresponding to an average cost per road of 107,613 € and an average cost per kilometre of 43,523 €. Costs estimated for restoring a single ODR ranged from 2000 € to 1,230,000 € (average 107,792 €). Figure 7b shows the distribution of the estimated costs and reveals a clear (and expected) bias towards less expensive works, with 60 works (54%) in the lowest cost class (<60,000 €). The box plot in Fig. 7a portrays the distribution of the costs, with 75% of the cost for restoring the SR below 130,000 € and 50% of the costs below 43,000 €. We attribute the relatively low cost of the single restoration works to the type of the damaged roads. The most expensive work (estimated at 1,230,000 €) consisted in the restoration of a road hit by nine different landslides.

Landslide and road statistics

A preliminary analysis of the event landslide inventory maps (Figs. 3a and 5a) allowed for general considerations on the impact of the event landslides in the study areas, in terms of size, density and number of damaging landslides.

The size and the cumulated area of the event landslides triggered by the two rainfall events (A_{LT}) were different in the two case studies (Table 1). In NE Sicily, landslides ranged in size 1.9×10^0 m^2 $< A_L < 4.5 \times 10^4$ m^2 (average = 1.0×10^3 m^2), for a total landslide area $A_{LT} = 1.5 \times 10^6$ m^2, and in the Marche case study, landslides were 1.6×10^0 m^2 $< A_L < 2.2 \times 10^4$ m^2 (average = 4.8×10^2 m^2), with $A_{LT} = 7.7 \times 10^5$ m^2. In terms of total landslide area A_{LT}, the event in NE Sicily had twice the impact of the Marche event.

Determining the volume of a landslide is known to be a challenging task that requires information on the surface and subsurface geometry of the slope failure, an information difficult and expensive to collect. The volume estimation of a large population of landslides (hundreds to several thousand failures) can be estimated only by applying empirical

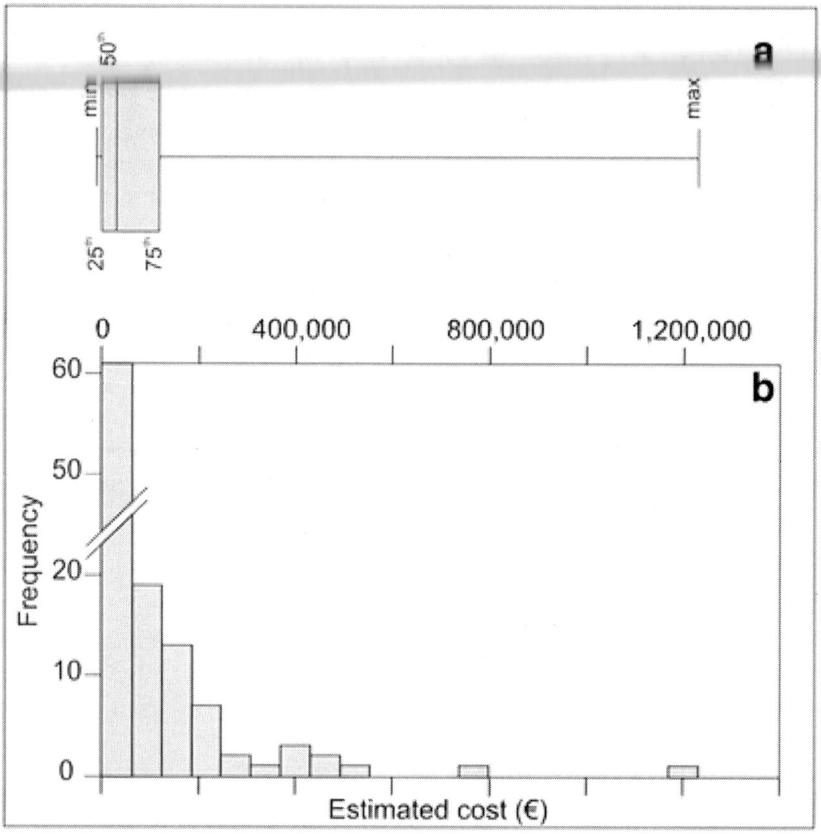

Figure 7. Marche case study. Cost estimates for restoring official damage roads (ODRs) in the Acquasanta Terme and Roccaluvione municipalities. **a** Box plot. **b** Frequency histogram. Plots prepared using PAST software (Hammer et al. 2001)

relationships that link the surface area to the volume of the individual landslides (Guzzetti et al. 2009). Table 1 shows landslide volumes (V_L) that were estimated applying the empirical power law equation proposed (i) by Guzzetti et al. (2009) for landslides of the slide type and (ii) by Innes et al. (1983) for landslide of the flow type (debris flow, earth flow). For the other failure types, we did not estimate the volume of the single failures because no empirical relationship linking V_L to A_L was available to us. Table 1 shows that, in NE Sicily, landslide volume was in the range 8.0×10^{-2} m^3 < V_L < 1.3×10^5 m^3 (average = 1.1×10^3 m^3), corresponding to a total landslide volume $V_{LT} = 1.2 \times 10^6$ m^3, whereas in the Marche area landslide volume ranged 2.1×10^{-1} m^3 < V_L < 1.5×10^5 m^3 (average = $1.3 \times$

10^3 m^3), for a total landslide volume $V_{LT} = 1.8 \times 10^6$ m^3. We note here that an accurate comparison of the total landslide volume V_{LT}mobilized by the two events was not possible, because—as explained before—the volume of some of the failures was not estimated.

The density of the event landslide area D_L, derived dividing A_{LT} by the size of the study area A_S, was 0.069 in NE Sicily and 0.004 in the Marche area (Table 1). This means that D_L, an alternative measure of the impact of the landslide events on the landscape, in the Marche areas was only 5.6% of the corresponding density in Sicily and that the event in NE Sicily had a much higher impact on the landscape than the event in the Marche study area.

Significant differences in the two study areas (Table 2) emerged from the comparison of the statistics of the road networks (Figs. 3b and 5b). In NE Sicily, the total length of the roads was $L_{RT} = 128.45$ km, corresponding to a road density $D_{RT} = 5.8$ km/km^2, whereas in the Marche area, the total length of the roads was $L_{RT} = 428.14$ km, a density $D_{RT} = 2.2$ km/km^2, 37% of the road density in NE Sicily. Since roads were classified as MR and SR, we calculated the road density separately for the two road categories. The density of the MR was $D_{MR} = 1.1$ km/km^2 in NE Sicily and $D_{MR} = 0.5$ km/km^2 in the Marche area, corresponding to 46.3% of the density of MR in NE Sicily. The density of the SR was $D_{SR} = 4.8$ km/km^2 in NE Sicily and $D_{SR} = 1.7$ km/km^2 in the Marche area, corresponding to 35% of the density of MR in NE Sicily.

METHOD

Since it is difficult for each landslide affecting roads to obtain information on the type of damage and cost, we devised a general (i.e. not site or area specific) method to assess the impact of landslides on the road network and to evaluate the related direct unit costs. The method exploits a GIS approach, which is exemplified in Fig. 8. The aim of the method is to quantify (i) the geographical impact of landslides on the road network, (ii) the unit cost per metre of damaged road (€/m) and (iii) the unit cost per square metre of damaging landslide (€/m^2). The approach exploits the geometric intersection of a road network map, where the roads are shown by lines, with an event landslide inventory map for the same area, where landslides are shown by polygons. To account for the width of the roads, we considered a fixed 12-m buffer zone centred on the road centreline, represented by a line in the road network maps (Fig. 3b for NE Sicily and Fig. 5b for Marche).

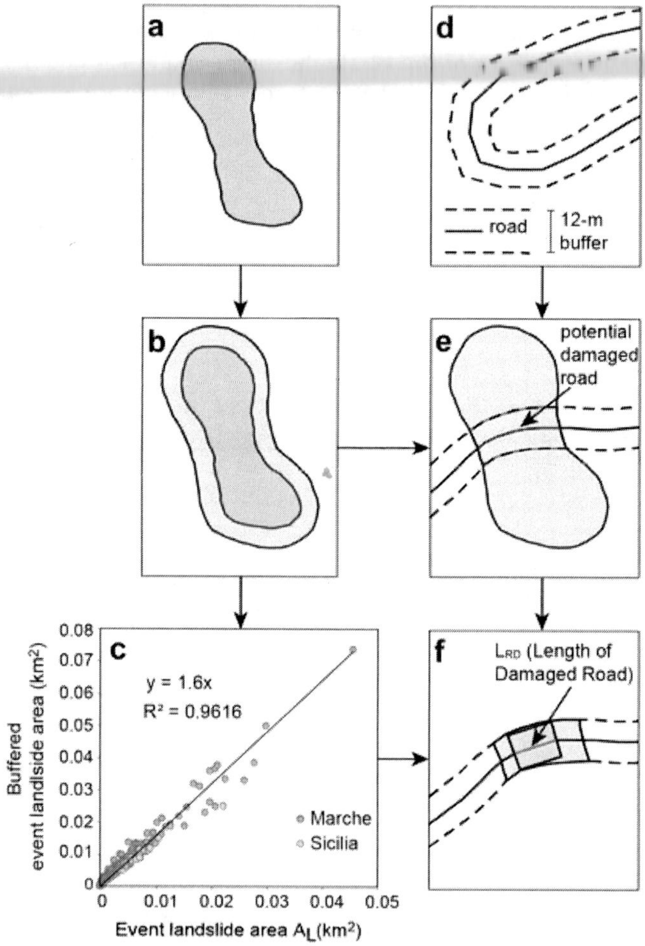

Figure 8. Schematic representation of the GIS approach used to ascertain quantify the impact of event landslides on the road network. **a** Polygon represents an event landslide. **b** Five metre buffer drawn around the event landslide to account for mapping uncertainty. **c**Relationship between the area of the event landslides (shown in **a**) and the area of the buffered event landslides (shown in **b**). *Blue dots* are event landslides for the Marche case study, and *orange dots* are event landslides for the NE Sicily case study. **d***Continuous line* represents a road (centreline), and *dashed lines* show boundaries of a 12-m buffer drawn around the road centreline. **e** Intersection between the 5-m buffered event landslide (shown in **b**) and the 12-m buffered road (shown in **d**). The *yellow area*shows the landslide–road intersection area. **f** The area of the *green rectangle* is obtained modifying the area of the *yellow polygon* (shown in **e**) using the scaling relationship shown in **c**. *Red line* is L_{RD}, the length of road damaged by the event landslide (shown in **a**), obtained dividing the area of the green rectangle by the 12-m constant road width. See text for explanation

To account for possible mapping errors during the landslide reconnaissance field survey (Santangelo et al. 2010), and in the visual interpretation of the aerial photographs (Santangelo et al. 2015), we considered a 5-m buffer drawn around the landslides shown in the NE Sicily (Fig. 3a) and in the Marche (Fig. 5a) event landslide inventory maps. We selected these two buffer sizes based on field observations. Application of the 5-m buffer resulted in an increase in the size (area) of the single landslides (Fig. 8b). The relationship between the original landslide area and the buffered landslide area was estimated using a dependency between the two areas obtained though linear regression (least square method) (Fig. 8c). For the purpose, we intersected, in the same geographical area, the buffered road network map with the buffered landslide event inventory map (Fig. 8e).

The geometric intersection allowed us to outline the sections of the roads that were (potentially) damaged by landslides (the yellow areas in Fig. 8e). This area depends on the size of the two buffers that we applied to the landslides (5 m) and to the roads (12 m). Therefore, a direct computation of the total length of roads intersected (i.e. potentially damaged, blue line in Fig. 8e) by landslides would have resulted in an overestimation of the actual (real) length (length of damaged road (L_{RD})), biasing the subsequent analyses. To avoid the bias, we scaled the intersection polygons based on the linear regression between the original landslide area and the buffered landslide area. Specifically, we applied a 0.625 scaling factor, which corresponds to the inverse of the angular coefficient of the regression line shown in Fig. 8c.

To calculate L_{DR}, we considered the downscaled intersection polygon to be equivalent to a 12-m-large rectangle aligned along the road (green rectangle in Fig. 8f), with the L_{DR} representing the base of the rectangle (red line in Fig. 8f). We excluded from the analysis the intersection polygons with a downscaled area of less than 12 m^2, which corresponds to less than 1 m of damaged road. For each intersection polygon, we then linked the corresponding landslide area in the GIS database.

The procedure was applied to (i) all the roads in the road maps, to evaluate the geographical impact of landslides on the entire road networks, and (ii) the subset of the ODR (roads with associated restoration costs), to ascertain the unit cost per metre of damaged road (€/m) and per square metre of damaging landslide (€/m^2).

The intersection between the event landslide inventory maps and the ODR highlighted that the former did not intersect all of the latter. This occurred because the cost data associated to the roads did not differentiate between landslides and other types of rainfall-induced damaging events, including, e.g. fallen trees, damaged drainage systems, debris or mud along the roads.

For this reason, some of the damaged roads were not associated to any specific landslide.

The estimation of the unit cost per metre of damaged road (€/m) and per square metre of damaging landslide (€/m²) was performed applying the following procedure. First, we calculated the total road cost *C*, as the sum of the cost to restore each single road,

$$C = \sum_{i=1}^{i=n} c_i, \ [€] \quad (1)$$

where *c* is the cost to restore road *i* and *n* is the total number of damaged roads. We then divided *C* by the number of roads in the ODR subset N_{ODR}, to obtain the average road cost C−C−, in euro.

Finally, we computed the average unit cost per 1 m of damaged road C−RC−R, as

$$\overline{C}_R = \frac{\overline{C} \times N_{ODR}}{L_{RDT}}, [€/m] \quad (2)$$

where N_{ODR} is the number of ODR intersected by the event landslides in the inventory and L_{RDT} is the total length of roads damaged by the event landslides (in m).

The average unit cost per square metre of damaging event landslide area C−LC−L, was estimated as

$$\overline{C}_L = \frac{\overline{C} \times N_{ODR}}{A_{L\cap RT}}, [€/m^2] \quad (3)$$

where N_{ODR} is the number of ODR intersected by event landslides and $A_{L\cap RT}$ is the total area of the event landslides that intersected the damaged roads (in m²).

RESULTS

To quantify the impact of the event landslides on the road network, and the related restoration costs, we applied in the Marche and NE Sicily case studies our proposed method (and the GIS approach that implements the method) shown in "Method" section, using (i) the event landslide inventory maps (Figs. 3a and 5a), (ii) the available road network maps (Figs. 3b and 5b) and (iii) the official cost data available to us (Table 3 for NE Sicily). The event landslide inventory maps were intersected with (i) all the roads in the road network maps (total road network map (TRN)) and with (ii) roads for which restoration costs are known (ODRs).

The GIS intersections of the event landslide inventory maps with the ODR highlighted that in Marche case study, about 60% of the ODR were intersected by more than one landslide type, and in NE Sicily case study, all the ODRs were intersected by two landslide types.

To verify if a relationship exists between the type of landslide and the type of damage, we compared, for Marche case study, (i) the types of mapped landsides and (ii) the type of qualitative damage estimations (aesthetical, functional and structural) made during the field surveys. The results (Table 4) shows that more than 50% of the landslides caused aesthetical and functional damage and that no significant differences exist among the landslide type, in terms of damage to the road network.

Table 4. Marche case study

	Aesthetical [# (%)]	Functional [# (%)]	Structural [# (%)]	Tot [#]
Complex slide	2 (22%)	6 (67%)	1 (11%)	9
Earth flow	29 (48%)	26 (43%)	6 (10%)	61
Rock fall	2 (33%)	4 (67%)	0 (0%)	6
Earth slide	26 (30%)	37 (43%)	23 (27%)	86
Slide earth flow	18 (22%)	34 (42%)	29 (36%)	81
Soil slide	63 (55%)	47 (41%)	5 (4%)	115
Total	140 (39%)	154,843%)	64 (18%)	358

Comparison between the type of mapped landside and the type of qualitative damage estimation (aesthetical, functional and structural) made during the field surveys

Table 5 shows the results of the GIS intersections of the event landslide inventory maps with (i) the TRN and (ii) the ODR. In NE Sicily, a total of 520 landslides intersected the road network ($N_{L∩R}$), including $N_{L∩R} = 95$ (18.3%) intersecting MR and $N_{L∩R} = 425$ (81.7%) intersecting SR. In the Marche area, $N_{L∩R} = 681$ landslides intersected roads, including $N_{L∩R} = 167$ (24.5%) affecting MR and $N_{L∩R} = 514$ (75.5%) affecting SR. Interestingly, the relative proportions of damaged (intersected) roads in the two road categories (MR, SR) are similar in the two study areas. The size of the event landslides that have intersected the road networks ($A_{L∩R}$) in the two study areas is listed in Table 6. The table shows that the size of landslides in NE Sicily was one order of magnitude larger than in Marche area. We used the method exemplified in Fig. 8 to obtain measures of each damaged road segment intersected by an event landslide L_{RD}, in metres, and we reported the results for the two case studies in Table 6. The total length of the damaged roads (L_{RDT}) in NE Sicily was twice than in the Marche area. Compared to L_{RT}, the percentages of damaged roads were 13.3% (MR) and 8.58% (SR) in NE Sicily and 1.37% (MR) and 1.53% (SR) in the Marche area. We conclude that the event landslides in NE Sicily impacted the MR more than twice and the SR 1.8 times than in the Marche area.

Table 5. Results of the GIS intersections between the event landslide inventory map and the two road layers, for the two case studies

	Road layer	$N_{L\cap R}$ #	$N_{L\cap R}$ %	$N_{R\cap L}$ #	N_{ODR} %	L_{ODR} %	Road type
NE Sicily	TRN	95	18.3%	Na			MR
		425	81.7%	Na			SR
		520		Na			Total
	ODR	86		4	80%	96.57%	MR
Marche	TRN	167	24.5%	Na			MR
		514	75.5%	Na			SR
		681		Na			Total
	ODR	392		71	64.5%	79.9%	SR

TRN is the complete road network map, and ODR is the official damaged road. $N_{L\cap R}$ (#) is the total number of intersected landslides. $N_{L\cap R}$ (%) is the percentage of intersected landslides in the MR and in the SR. $N_{R\cap L}$ (#) is the number of road intersected by landslides, applicable only to ODR because in the TRN, the number of roads is not available. N_{ODR} (%) is the percentage of the number of ODR intersected by the event landslides. L_{ODR} (%) is the percantage of the total length of ODR intersected by the event landslides. MR is the main roads, and SR is secondary roads

Table 6. For the two case studies, statistics of the size of the landslides that intersected the roads ($A_{L\cap R}$) and of the damaged road length (L_{RD}) estimated through the GIS analysis for the main roads (MRs), the secondary roads (SRs) and the official damaged roads (ODRs)

		NE Sicily			Marche		
Metric	Unit	MR	SR	ODR	MR	SR	ODR
				$A_{L\cap R}$			
Min	m^2	1.3×10^1	3.5×10^0	1.3×10^1	1.5×10^1	2.9×10^0	3.1×10^0
Max	m^2	1.2×10^4	4.5×10^4	2.8×10^4	3.8×10^3	1.0×10^4	1.0×10^4
Mean	m^2	1.2×10^3	1.8×10^3	2.3×10^3	3.2×10^2	3.6×10^2	3.9×10^2
Total	m^2	1.1×10^5	7.7×10^5	1.9×10^5	1.4×10^4	1.7×10^5	1.6×10^5
				L_{RD}			
Min	m	1.0×10^0	1.1×10^0	1.0×10^0	1.0×10^0	1.0×10^0	1.0×10^0
Max	m	2.9×10^2	5.9×10^2	1.2×10^2	2.3×10^1	7.3×10^1	4.6×10^1
Mean	m	2.1×10^1	2.1×10^1	1.6×10^1	5.8×10^0	7.8×10^0	7.7×10^0
Total	m	3.1×10^3	9.0×10^3	1.3×10^3	1.4×10^3	5.0×10^3	3.9×10^3

A goal of our work was to assess the restoration unit cost for the different road categories: MR and SR. To obtain this result, we applied the method described in "Method" section using the ODRs, i.e. the subset of the roads in each study area for which (i) information exists that they were damaged and (ii) official estimates of the restoration costs are available. The descriptive statistics of the road networks (Table 2) revealed that the ODRs were 5 MRs in NE Sicily and 110 SRs in the Marche area.

Table 2 also shows that the total length of the ODR was L_{ODR} = 12.8 km in NE Sicily (53.92% of all MR in the area) and L_{ODR} = 272 km (82.87% of all SR in the area) in the Marche area. Differences exist also in the density of the official damaged road D_{ODR}, calculated as L_{ODR}/A_S. D_{ODR} is 0.58 in NE

Sicily and 1.37 in the Marche area. The two figures are difficult to compare, but they confirm a different impact of event landslides on the road networks in the two study areas.

Analysis of the intersection between the event landslides and the ODR, shown in Table 5, reveals that in NE Sicily, four of the five ODRs (80%) were intersected (anywhere) by event landslides shown in the inventory map (Fig. 3b), representing 96.6% of the total length of the ODR, L_{ODR}. We checked the missing intersection along the SS-114 and found that it was located outside our study area, which is smaller than the entire area affected by the 1 October 2009 rainfall event (Mondini et al. 2011). In the Marche area, 71 of the 110 ODRs (64.5%) were intersected (anywhere) by event landslides shown in the inventory map (Fig. 5b). The proportion represents 79.9% of L_{ODR}.

In NE Sicily, event landslides affected 53.92% of the MR, whereas in the Marche area, the percentage was an outstanding 82.87% for SR (Table 2). This is despite the fact that the landslide event in the Marche area was significantly less severe than in NE Sicily. The landslide density was $D_L = 67.27$ landslide per square kilometre in NE Sicily, significantly higher than $D_L = 7.36$ for the Marche area. We hypothesize that the figures measure a higher vulnerability of the SR, compared to the MR.

For the ODR, Table 6 summarizes statistics (a) for the size of landslides intersecting roads ($A_{L\cap R}$) and (b) for the length of the damaged roads (L_{RD}) estimated through the GIS analysis, for the two case studies. In NE Sicily, the size of landslides intersecting ODR was on average = 2.3×10^3 m, for a total landslide area intersecting ODR $A_{L\cap RT} = 1.9 \times 10^5$ m^2. In the Marche area, the size of landslides intersecting ODR was on average = 3.9×10^2 m^2 and $A_{L\cap RT} = 1.6 \times 10^5$ m^2. In terms of average $A_{L\cap R}$, the size of the landslides in NE Sicily was one order of magnitude larger than in the Marche area. Focusing on the damaged road length L_{DR}estimated through the GIS analysis, inspection of Table 6 reveals that in NE Sicily, the landslide–road intersections was on average = 1.6×10^1 m, for a total landslide–road intersection length $L_{RDT} = 1.3 \times 10^3$ m. In the Marche area, L_{DR} was on average = 7.7×10^0 m and $L_{DRT} = 3.9 \times 10^3$ m. In conclusion, the L_{RDT} for ODR the in the Marche area was three times the corresponding L_{RDT} in NE Sicily.

Finally, we estimated the average unit cost per metre of damaged road (using Eq. (2)) and per square metre of damaging landslide area (using Eq. (3)). The total road cost C was 26,140,736 € in NE Sicily and 11,837,409 € in the Marche area. The number of ODR intersected by landslides I was 4 in NE Sicily and 71 in the Marche area ($N_{L\cap R}$ in Table 5). The total length of damaged roads L_{RDT} was 1.3×10^3 m in NE Sicily and 3.9×10^3 m in the

Marche area (Table 6), and the total area of event landslides that have intersected a road A_{LORT} was 1.9×10^5 m^2 in NE Sicily and 1.6×10^5 m^2 in the Marche area (Table 6). Using these figures, we calculate C–RC–R = 16,287 €/m and C–LC–L = 110 €/m^2 for NE Sicily and C–RC–R = 1958 €/m and C–LC–L = 49 €/m^2 for the Marche area.

DISCUSSION

We tested our method for the quantitative assessment of the geographical impact of event landslides on a road network and for the evaluation of the related economic costs, in two different study areas in NE Sicily and in the Marche region, Italy (Fig. 1). The two case studies differ significantly: The Marche study area (A_S = 198 km^2) is nine times larger than the study area in NE Sicily (A_S = 22 km^2) but received a slightly lower number of event landslides (N_{LT} = 1438) than those occurred in NE Sicily (N_{LT} = 1480) (Table 1). Thus, despite the magnitude of the two landslide events, measured by the logarithm of the total number of event landslides, $M_L = \log_{10}(N_{LT})$ (Malamud et al. 2004) is about the same for the two landslide events (M_L = 3.17 in NE Sicily and M_L = 3.16 in the Marche area), the landslide density $d_L = N_{LT} / A_S$ was significantly larger in NE Sicily (d_L = 67.3) than in the Marche area (d_L = 7.3). The density of the landslide areas $D_L = A_{LT} / A_S$ was also significantly larger in NE Sicily (D_L = 0.068) than in the Marche area (D_L = 0.004) (Table 1). We maintain that the different metrics measure differently the impact of the landslide events on the natural landscape in the two study areas, with the impact significantly larger in NE Sicily than in the Marche area. We sustained that our method is applicable to a large range of landslide events that have caused different impacts on the landscape and the road networks.

The applicability of our method depends on the availability of the necessary data, i.e. (i) an accurate event landslide inventory map, (ii) a map of the roads in the area affected by the landslide event and (iii) as accurate as possible cost data for the landslide damage. Heam et al. (2008), Klose et al. (2015) and Vranken et al. (2013) have shown that geomorphological landslide inventories are valuable sources of information to evaluate annual landslide restoration costs. Our study extends these findings to event landslide inventories. Where an event landside inventory map is available, together with a map of the road network and reliable damage information and cost data, the geographical and the economic impact of the landslide event can be estimated effectively and with relatively little effort in a GIS.

Maps of the road (or transportation) networks are readily available from topographic base maps in digital format or from other digital cartographic sources (e.g. the global OpenStreetMap project, http://openstreetmap.org). For event inventory maps, the situation is different. Despite their usefulness, and the many techniques and methods available to prepare them, event landslide maps are not prepared systematically after landslide-triggering events (Guzzetti et al. 2012). This limits greatly the possibility to perform *ex post* analyses of the impact of event landslides on transportation networks. We estimate that the cost for the production of event landslide inventories comparable to those available for our two study areas is in the range between 40,000 € and 50,000 €, for an area of 100–200 km^2. The cost for the production of the Marche event landslide inventory was about 0.4% of the cost estimated for repairing the landslide damage. We hypothesize that the proportion is similar, or remains very low, in other geographical areas and in different physiographic and societal environments. We conclude that the very low cost necessary for the production of an event landslide inventory, compared to the landslide repair costs, justifies the systematic production of event inventory maps immediately after an event (Guzzetti et al. 2012).

Our work also confirmed known problems related to the collection of reliable information on road damage and repair/restoration cost data. Klose et al. (2015), working in Germany, found that landslide restoration costs were very difficult to obtain and, where available, their accuracy and reliability was difficult to evaluate. In Italy, and in many other countries, a standard procedure for the collection of landslide damage and related cost data does not exist. Hallegatte and Przyluski (2010) sustained that after a disaster, media, insurance companies and international institutions publish numerous assessments of the cost of the disaster based on different methodologies and approaches, and consequently, they often reach different results. Similarly for geohydrological hazards, different sources often provide lumped sums without separating costs due to landslides and other damaging events and costs for restoration actions from those for mitigation activities. Lack of standards contributes negatively to the paucity of damage information and cost data.

In the Marche case study, the cost data refer to the restoration of 110 damaged SRs. Our GIS analysis has identified 71 geographical intersections between event landslides and SR (64.5%) with 31 damaged locations along SR not identified by the GIS analysis. In NE Sicily, the cost data accounted for the total amount of money allocated to repair five MRs, including a highway viaduct, while our GIS analysis revealed four geographical intersections, missing one damaged site. We explain the differences (mismatches) with (i) mistakes in the attribution of the causes of the damage in the municipality reports, including, e.g. damage caused by fallen trees attributed to a landslide, and (ii) landslide mapping errors during the reconnaissance field survey (Santangelo et al. 2010). Despite the inconsistencies, which we consider acceptable for the scope of the analysis, we maintain that the very large percentage of roads intersected by the event landslides (L_{RD}) measures the severity of the impact caused by the event landslides on the road networks, in the two study areas.

The GIS-based approach allowed estimating average unit costs per metre of damaged road (C–RC–R) and per square metre of damaging landslide (C–LC–L). Results were different in the two study areas. In the Marche area, where SR were considered, we found C–LC–L = 1958 €/m and C–RC–R = 49 €/m^2. In NE Sicily, where MR were considered, we found C–LC–L = 16,287 €/m and C–RC–R = 110 €/m^2. We explain the large range in the costs with the type of the roads: MR, including a highway in NE Sicily, and SR in the Marche area, and MR is certainly more expensive to construct, repair and restore than SR.

We can compare our event-based cost estimates with similar estimates available in the literature. Heam et al. (2008) used information on the national road network in the People's Democratic Republic (PDR) of Lao and road maintenance costs for the period 2004–2007 to estimate the annual average landslide expenditure per kilometre, which was in the range US$1000–1500. Klose et al. (2015) estimated at US$52,000 the average cost per kilometre of highway at risk of landslides in the Lower Saxon Uplands, NW Germany, in the period 1980–2010. Our unit cost estimates are quite different, and we attribute the difference to multiple factors, including the facts that (i) our estimates are for single events, whereas estimates for the PDR of Lao and for the Lower Saxon Uplands cover periods of time during which multiple landslide events may have occurred; (ii) we used restoration costs after intense (in the Marche area) vs. very intense (in NE Sicily) rainfall events, that produced a different impact on the landscape and the road networks (Tables 5 and 6); (iii) different types of roads are present in the different areas, with different engineering characteristics and maintenance levels; and (iv) in the considered areas, the social and economic environments differ, conditioning the remediation and maintenance policies and costs.

CONCLUSIONS

We proposed a method to estimate the physical and economic impact of populations of event landslides on road networks. To test the method, we used two recent rainfall-induced landslide events in NE Sicily and in Marche region. The two case studies differ for their geographical extent, the geologic and morphologic settings, the landslide types (Table 1) and the type and density of the roads (Table 2). The method produced reasonable results in both study areas. We conclude that the method is general (i.e. not site or area specific) and suited to evaluate the impact of landslides damaging roads in different geographical and physiographic settings. Our two case studies were of rainfall-induced landslides. The method can be applied to evaluate the impact of populations of event landslides caused by other triggers, including, e.g. rain-on-snow events, rapid snowmelt and earthquakes, where preliminary estimates of landslide costs have to be obtained, e.g. to request the declaration of a state of emergency and to obtain fundings to reimburse incurred and foreseen restoration costs.

The applicability of the method depends on the availability and the accuracy of the necessary information, and particularly the availability of an accurate event landslide inventory map and of reliable restoration cost data. Despite their undisputed usefulness, event landslide inventory maps are not prepared systematically after landslide-triggering events (Guzzetti et al. 2012), and landslide restoration cost data are difficult to obtain (Klose et al. 2015), and their reliability and accuracy are difficult to evaluate.

We computed different metrics to quantify the impact of the two studied landslide events on the natural landscape (including the total number of event landslides N_{LT}, the landslide event magnitude M_L (Malamud et al. 2004), the total area of the event landslides A_{LT}, the landslide density d_L and the density of landslide area D_L, Table 1) and on the road networks (including the size of the landslides that intersected the roads $A_{L \cap R}$ and the size of the damaged road length L_{RD}, Table 6) and found that the different metrics provided complementary measures of the impacts. We conclude that different, complementary metrics should always be used to quantify the impact of a population of event landslides on a landscape, a road network or any other type or class of vulnerable elements.

Our work confirmed the difficulty in collecting accurate cost data for the direct damage caused by landslides and in using the cost data for the evaluation of the economic impact of the event landslides. Systematic availability of landslide cost data will allow for accurate and reliable *ex ante* evaluations of the economic costs of landslide events. The unit cost estimates that we statistically derived can be applied to quantify the impact of populations of event landslides when preliminary estimates of landslide costs have to be obtained, e.g. to request the declaration of a state of emergency and to obtain fundings to reimburse incurred and foreseen restoration costs.

ACKNOWLEDGEMENTS

The work was conducted in the framework of the FP7-LAMPRE project and the national project "Valutazione Economica dei Disastri Naturali" (Economic Evaluation of Natural Disasters) funded by Fondazione Assicurazioni Generali. F. Bucci and M. Santangelo were supported by a grant of the Italian National Department of Civil Protection. M. Donnini was supported by a grant of the Fondazione Assicurazioni Generali. F. Fiorucci was supported by a grant of the Umbria Region under contract POR-FESR 861.2012, and E. Napolitano was supported by a grant of the Ministero dell'Istruzione, dell'Università e della Ricerca-Progetti di Rilevante Interesse Nazionale (MIUR-PRIN) project. We are grateful to Massimiliano Alvioli and Ivan Marchesini (CNR IRPI) for their work in the field and to Stefano Lo Parco (Acquasanta Terme Municipality), Carlo Trobbiani (Roccafluvione Municipality) and Marilena Romanelli (private consultant) for their help in collecting cost data.

REFERENCES

1. ANCE-CRESME (2012) Lo stato del territorio italiano. Insediamento e rischio sismico ed idrogeologico. Roma, 284 pp, (in Italian)
2. Ardizzone F, Basile G, Cardinali M, Casagli N, Del Conte S, Del Ventisette C, Fiorucci F, Garfagnoli F, Gifli G, Guzzetti F, Iovine G, Mondini AC, Moretti S, Panebianco M, Raspini F, Reichenbach P, Rossi M, Tanteri L, Terranova O (2012) Landslide inventory map for the Briga and the Giampilieri catchments, NE Sicily, Italy. Journal of Maps 8(2):176–180
3. Boonyanuphap J (2013) Cost-benefit analysis of vetiver system-based rehabilitation measures for landslide-damaged mountainous agricultural lands in the lower northern Thailand. Nat Hazards 69:599–629.
4. Cardinali M, Carrara A, Guzzetti F, Reichenbach, P (2002) Landslide hazard map for the Upper Tiber River basin. Pubblication CNR GNDCI n. 2634, Scale 1:1,200,000, (in English & Italian)
5. Catenacci V (1992) Memorie descrittive della Carta Geologica d'Italia Vol. 47/1992, (in Italian)
6. Cruden DM, Varnes DJ (1996) Landslide types and processes. In: Turner AR, Schuster RL (eds) Landslides: Investigation and Mitigation, Sp. Rep. 247. Transportation Research Board, National Research Council, National Academy Press, Washington, pp 36–72
7. Galli M, Guzzetti F (2007) Landslide vulnerability criteria: a case study from Umbria, Central Italy. Environ Manag 40:649–664.
8. Guzzetti F, Tonelli G (2004) Information system on hydrological and geomorphological catastrophes in Italy (SICI): a tool for managing landslide and flood hazards. Nat Hazards Earth Sys 4:213–232
9. Guzzetti F, Cardinali M, Reichenbach P (1994) The AVI project: a bibliographical and archive inventory of landslides and floods in Italy. Environ Manag 18:623–633
10. Guzzetti F, Ardizzone F, Cardinali M, Rossi M, Valigi D (2009) Landslide volumes and landslide mobilization rates in Umbria, Central Italy. Earth Planet Sc Lett 279:222–229

11. Guzzetti F, Mondini AC, Cardinali M, Fiorucci F, Santangelo M, Chang K-T (2012) Landslide inventory maps: new tools for an old problem. Earth Science Reviews 11246–66
12. Hallegatte S, Przyluski V (2010) The economics of natural disasters: concepts and methods. World Bank Policy Research Working Paper No. 5507. Available at SSRN: https://ssrn.com/abstract=1732386
13. Hammer Ø, Harper DAT, Ryan PD (2001) PAST: paleontological statistics software package for education and data analysis. Palaeontol Electron 4(1):9
14. Hearn GJ, Hunt T, Aubert J, Howell JH (2008) Landslide impacts on the road network of Lao PDR and the feasibility of implementing a slope management programme. International Conference on Management of Landslide Hazard in the Asia-Pacific Region. Sendai, Japan
15. Hilker N, Badoux A, Hegg C (2009) The Swiss flood and landslide damage database 1972–2007. Nat Hazards Earth Sys 9:913–925
16. Hungr O, Leroueil S, Picarelli L (2014) The Varnes classification of landlslide types, an update. Landslides 11:167–194
17. Innes JL (1983) Lichenometric dating of debris-flow deposits in the Scottish Highlands. Earth Surf Proc Land 8:579–588
18. Karunasena G, Rameezdeen R (2010) Post-disaster housing reconstruction: comparative study of donorvs owner-driven approaches. International Journal of Disaster Resilience in the Built Environment 1(2):173–191
19. Klose M, Damm B, Terhorstet B (2015) Landslide cost modeling for transportation infrastructures: a methodological approach. Landslides 12:321–334.
20. Malamud BD, Turcotte DL, Guzzetti F, Reichenbach P (2004) Landslide inventories and their statistical properties. Earth Surf Process Landforms 19:687–711
21. McCarty C, Smith SK (2005) Florida's 2004 hurricane season: local effects. Florida Focus, University of Florida, Florida
22. Meyer V, Becker N, Markantonis V, Schwarze R, van den Bergh JCJM, Bouwer LM, Bubeck P, Ciavola P, Genovese E, Green C, Hallegatte S, Kreibich H, Lequeux Q, Logar I, Papyrakis E, Pfurtscheller C, Poussin J, Przyluski V, Thieken AH, Viavattene C (2013) Review article: assessing the costs of natural hazards—state of the art and knowledge gaps. Nat Hazards Earth Sys 13:1351–1373
23. Mondini AC, Guzzetti F, Reichenbach P, Rossi M, Cardinali M, Ardizzone F (2011) Semi-automatic recognition and mapping of rainfall induced shallow landslides using satellite optical images. Remote Sens Environ 115:1743–1757.
24. Reichenbach P, Galli M, Cardinali M, Guzzetti F, Ardizzone F (2005) Geomorphological mapping to assess landslide risk: concepts, methods and applications in the Umbria region of central Italy. Landslide hazard and risk, Chichester, Wiley, pp 429–468
25. Sächsische Staatskanzlei (2003) Schadensausgleich und Wiederaufbau im Freistaat Sachsen, Dresden, Germany, (in German)
26. Salvati P, Bianchi C, Fiorucci F, Giostrella P, Marchesini I, Guzzetti F (2014) Perception of flood and landslide risk in Italy: a preliminary analysis. Nat Hazard Earth Sys. 14:2589–2603.
27. Santangelo M, Cardinali M, Rossi M, Mondini AC, Guzzetti F (2010) Remote landslide mapping using a laser rangefinder binocular and GPS. Natural Hazard Earth Sys 10:2539–2546.
28. Santangelo M, Marchesini I, Bucci F, Cardinali M, Fiorucci F, Guzzetti F (2015) An approach to reduce mapping errors in the production of landslide inventory maps. Nat Hazard Earth Sys 15:2111–2126.
29. Trigila A, Iadanza C, Spizzichino D (2010) Quality assessment of the Italian Landslide Inventory using GIS processing. Landslides 7:455–470
30. Trigila A., Iadanza C., Bussettini M., Lastoria B., Barbano A. (2015) Dissesto idrogeologico in Italia: pericolosità e indicatori di rischio. Rapporto 2015. ISPRA, Rapporti 233/2015

31. Vranken L, Van Turnhout P, Van Den Eeckhaut M, Vandekerckhove L, Poesen J (2013) Economic valuation of landslide damage in hilly regions: a case study from Flanders, Belgium. Sci Total Environ 447:323–336
32. Government acts
33. Dipartimento di Protezione Civile (2014) Audizione del Capo del Dipartimento Prefetto Franco Gabrielli inerente la situazione nazionale delle emergenze connesse al dissesto idrogeologico a seguito delle recenti avversità atmosferiche. Presidenza del Consiglio dei Ministri e Senato della Repubblica, XIII Commissione, mercoledì 5 marzo 2014, (in Italian)
34. Regione Marche, Centro Funzionale per la Meteorologia, l'Idrologia e la Sismologia (2013a) Rapporto di evento 10–13 novembre 2013, Regione Marche, 38 pp, (in Italian)
35. Regione Marche, Centro Funzionale per la Meteorologia, l'Idrologia e la Sismologia (2013b) Rapporto di evento 25–27 novembre 2013 e dicembre 2013, Regione Marche, 60 pp, (in Italian)
36. Regione Siciliana, Centro Funzionale Regionale (2009) Rapporto sugli eventi meteo che hanno colpito la Provincia di Messina il giorno 1 ottobre 2009, Dipartimento della Protezione Civile, Regione Siciliana, 60 pp, (in Italian)

CHAPTER 17

Slope Stability Assessment of the Sarcheshmeh Landslide, Northeast Iran, Investigated Using InSAR and GPS Observations

Mehrdad Akbarimehr [1,*], Mahdi Motagh [2] and Mahmud Haghshenas-Haghighi [1]

[1]*Department of Surveying and Geomatics Engineering, University of Tehran, Tehran 14395-515, Iran*

[2]*Helmholtz Centre Potsdam GFZ German Research Centre for Geosciences, D-14473 Potsdam, Germany*

ABSTRACT

The detection and monitoring of mass movement of susceptible slopes plays a key role in mitigating hazards and potential damage associated with creeping slopes and landslides. In this paper, we use observations from both Interferometric Synthetic Aperture Radar (InSAR) and Global Positioning System (GPS) to assess the slope stability of the Sarcheshmeh ancient landslide in the North Khorasan province of northeast Iran. InSAR observations were obtained by the time-series analysis of Envisat SAR images covering 2004–2006, whereas repeated GPS observations were conducted by campaign measurements during 2010–2012. Surface displacement maps of the Sarcheshmeh landslide obtained from InSAR and GPS are both indicative of slope stability. Hydrogeological analysis suggests that the multi-year drought and lower than average precipitation levels over the last decade might have contributed to the current dormancy of the Sarcheshmeh landslide.

Keywords: landslide; slope stability; displacement; InSAR; GPS

1. INTRODUCTION

Landslides are one of the most destructive natural hazards that threaten human lives and influence the socioeconomic condition of many countries. A landslide is basically identified as slope failure that is caused by gravitational force as the primary factor [1]. Other triggering factors, such as rainfall, earthquakes, floods or human activity may also influence the occurrence of this geologic event. Because of its specific geologic, morphologic, climatic and tectonic settings, Iran is one of the most landslide-prone areas in the world. By the year 2000, approximately 2600 landslides, which were responsible for 162 deaths, 176 fully destroyed houses and 170 damaged roads, were detected and mapped by experts from various government agencies in the country [2]. Figure 1 shows a landslide distribution map of Iran. As shown in Figure 1, most of the landslides are concentrated on the rim of the Alborz and Zagros mountains.

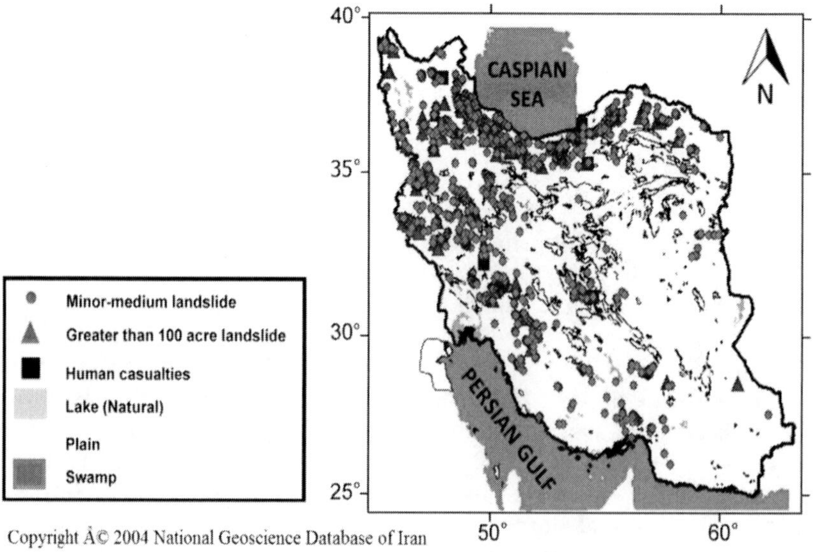

Figure 1. Landslide distribution map of Iran [3].

Detection and monitoring of landslides is necessary to decrease and mitigate associated losses. However, this is challenging in slow-moving landslides because of their complicated mechanisms [4]. Measurement of the surface displacement is also important to predict models of slope failure and to make risk assessments. Remote satellite techniques and *in situ* field observations are common methods to investigate landslides and monitor slope instability [5–11]. In the present study, we use Interferometric Synthetic Aperture Radar (InSAR) and Global Positioning System (GPS) observations to investigate the slope stability of the Sarcheshmeh landslide, northeast Iran.

The Sarcheshmeh landslide is a northwesterly-directed slope in the province of North Khorasan in northeastern Iran (Figure 2). A village called Sarcheshmeh with approximately 400 inhabitants is at the foot of this slope, which is 1.6 km long in the east-west direction and 600 m long in the north-south direction, with an approximate area of 100 acres.

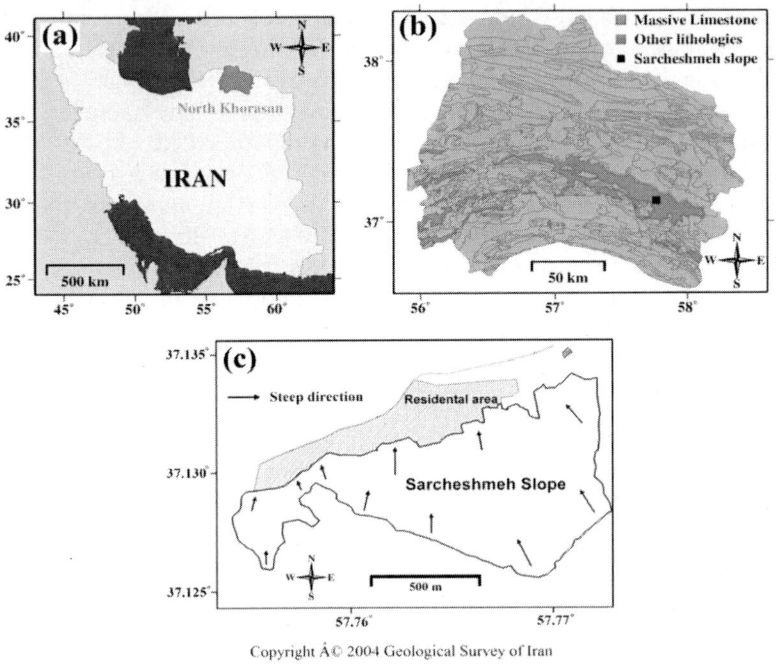

Figure 2. Geology map of the study area: (**a**) Provincial map showing the location of the North Khorasan province (the green area); (**b**) surface geological map of the North Khorasan province with the location of the Sarcheshmeh slope marked by the black rectangle [14]; (**c**) outline of the Sarcheshmeh slope; the black arrows show the main direction of potential motion.

The area has an annual precipitation of approximately 300 mm, which occurs during rainy periods from September to April [12], resulting in a relatively high level of groundwater resulting in springs on the slope (Figure 3b). The materials involved in sliding are mixtures of soil and rocks in alternating formations of thin-bedded to massive silt clay and limestone; the latter constitutes the majority of the surface geology in northeast Iran (Figure 2b). The general morphology of the study area is dominated by a mountainous landscape, which includes V-shaped valleys. Preliminary geomorphological analysis suggests that the Sarcheshmeh slope was part of a greater ancient landslide with a depth of approximately 15 m that sustained more than 10 m downward movement [13]. Numerous scarps and bluffs with height differences ranging from 2 to 30 m are indicative of the ancient activity of this landslide (Figure 3c). All the current information about the Sarcheshmeh landslide is derived from morphological and surface geology analysis. The current rates of motion are not well known and the creeping mechanism is poorly understood. Because of the placement of the residential area, which is just downstream of the Sarcheshmeh landslide, monitoring is essential to assess the progressive deformation of this potentially dangerous slope and to avoid or mitigate any damage caused by sudden slope failure. Because of the complicated nature of the landslide phenomenon, information about the previous activity of the slope could help us better recognize effective components that are involved in probable slope creeping. Thus, we can benefit from InSAR techniques and extensive archives of SAR data that could assist us in studying and analyzing past mechanisms of slope motion. Additionally, to assess the current rate of deformation on the Sarcheshmeh landslide, we benefitted from the use of GPS geodetic measurements with point-wise reliable results.

2. DATA AND METHODOLOGY

The dataset used in this paper is divided into two categories: SAR satellite images and GPS observations. To evaluate and analyze the past mechanisms of Sarcheshmeh surface deformation, we used the SAR interferometry (InSAR technique). The current rate of deformation was investigated using GPS periodic measurements.

Figure 3. (**a**) A panorama photo from the north toward the study slope, which is approximately indicated by the red line; (**b**) a close-up view from the south and upper part toward the north and down the slope. Cluttered topography is obviously depicted in the photo. Additionally, two natural springs and a deep track, which was created by water erosion, are shown on this slope; (**c**) a close-up view from the west to a scarp on the slope, which is formed as a result of former activity from the landslide.

Space-borne Interferometric Synthetic Aperture Radar (InSAR) has developed rapidly over the past 20 years and has proven to be a valuable tool for topographic mapping and surface deformation measurements [15,16]. Because of its dense spatial coverage and competitive accuracy, InSAR has now become one of the most preferred geodetic methods to study surface deformation associated with slow-moving landslides [17–20]. Advanced InSAR methods such as Persistent Scatterer (PS) [4,21–23] and Small Baseline Subset (SBAS) [24–26], which can overcome artifacts arising from atmospheric noise and temporal and spatial baseline decorrelation, enable us to benefit from displacement time-series analysis using a stack of SAR data. Both techniques work better in urban areas because of the high density of man-made objects, which increase the likelihood of finding coherent pixels in a stack of SAR data over time. Because of the non-urban nature of the landslide area in Sarcheshmeh (it is covered partially with grass and other permanent vegetation and used for farming), we used the SBAS technique for surface displacement time-series analysis. The SBAS method uses all possible interferograms with a perpendicular and temporal baseline below a certain threshold to minimize decorrelation effect and topographic errors. The SBAS interferograms are then inverted after phase unwrapping for deformation time series retrieval.

GPS is a high precision positioning technique, which can be used to provide point-wise, semi-continuous maps of crustal deformation that is associated with geophysical phenomena using a variety of static or kinematic techniques [27]. Due to operating in all weather conditions and independent of direct visibility between stations and 3D precise positioning of specified points, GPS positioning is a beneficial tool in landslide monitoring [28–30]. This technique also has limitations, such as choosing the appropriate location for stations so that they have good sky views and are less affected by multipaths [27]. The choice of the observation mode and processing techniques depends on several practical considerations, such as the accessibility, cost, number of points and required precision [27].

Both GPS and InSAR are complementary to one other in that GPS provides detailed ground-truth vector information of displacement for specific localities compared to the extensive spatial coverage provided by InSAR line-of-sight (LOS) observations [31–33]. A combination of different ground and space-based geodetic monitoring systems provides promising tools to assess and evaluate geological hazards [31,32].

2.1. InSAR Analysis

For this study we used ASAR images acquired over the study area by the Envisat satellite from the European Space Agency (ESA). Sixteen descending images (Figure 4) in I2 mode (incidence angle 23°) were obtained for Sarcheshmeh spanning the years 2004–2006 (Table 1). The data were processed using the time-series approach of Small Baseline Subset (SBAS) in a two-step process: We initially produced a network of small baseline differential interferograms and subsequently did the time-series analysis to derive a one-dimensional Line-of-Sight (LOS) time-series and displacement map of surface deformation on the slope.

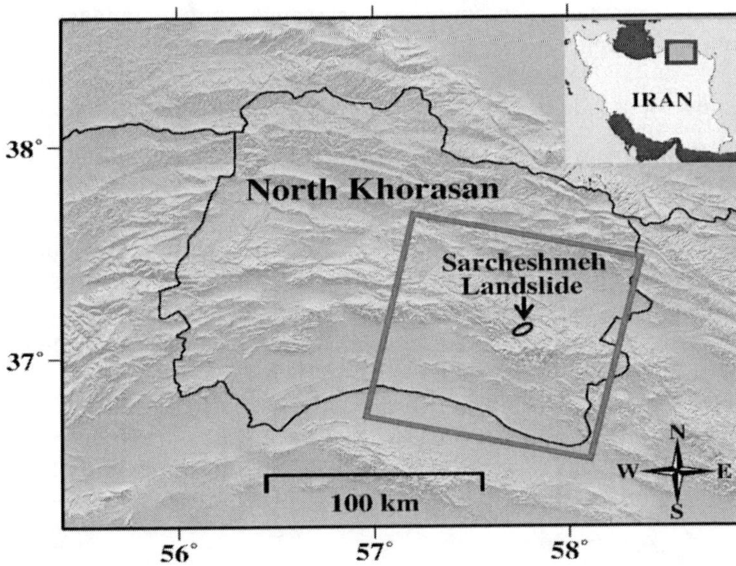

Figure 4. The study area in northeastern Iran (upper inset); the red box delineates the area covered by the Envisat frame in this study. Approximate location of the Sarcheshmeh landslide is specified by a black ellipse.

Table 1. List of used Envisat SAR images, date of acquisition and other specifications.

No	Acquisition Date	Mission	Sensor	Orbit	Track	Pass
1	2004.03.23	Envisat	ASAR	10781	206	Descending
2	2004.06.01	Envisat	ASAR	11783	206	Descending
3	2004.09.14	Envisat	ASAR	11783	206	Descending
4	2004.10.19	Envisat	ASAR	13787	206	Descending
5	2004.11.23	Envisat	ASAR	14288	206	Descending
6	2004.12.28	Envisat	ASAR	14789	206	Descending
7	2005.02.01	Envisat	ASAR	15290	206	Descending
8	2005.03.08	Envisat	ASAR	15791	206	Descending
9	2005.04.12	Envisat	ASAR	16292	206	Descending
10	2005.05.17	Envisat	ASAR	16793	206	Descending
11	2005.06.21	Envisat	ASAR	17294	206	Descending
12	2005.07.26	Envisat	ASAR	17795	206	Descending
13	2005.08.30	Envisat	ASAR	18296	206	Descending
14	2005.10.04	Envisat	ASAR	18296	206	Descending
15	2005.11.08	Envisat	ASAR	19298	206	Descending
16	2006.02.21	Envisat	ASAR	20801	206	Descending

The interferometric processing was performed using a standard repeat-pass method implemented in DORIS software, developed by the Delft Institute for Earth-Oriented Space Research (DEOS), Delft University of Technology, which is freely available to the public [34].

The topography-related phase correction and geocoding of the interferograms in the interferometric processing were estimated using a three arcsec Digital Elevation Model (DEM) from the Shuttle Radar Topography Mission [35]. The DEOS precise orbital information of Envisat that was estimated from the DORIS instruments provided by ESA was used during processing to eliminate orbital phase distortion [36]. We primarily formed 41 interferograms for Sarcheshmeh and then used 34 with a maximum temporal baseline of 311 days and a perpendicular baseline of 494 m for the time-series analysis. The time-series analysis was conducted using the Small Baseline (SBAS) approach that is implemented in StaMPS software [21,22,37]. Figure 5 presents the final network of interferograms used in the SBAS processing.

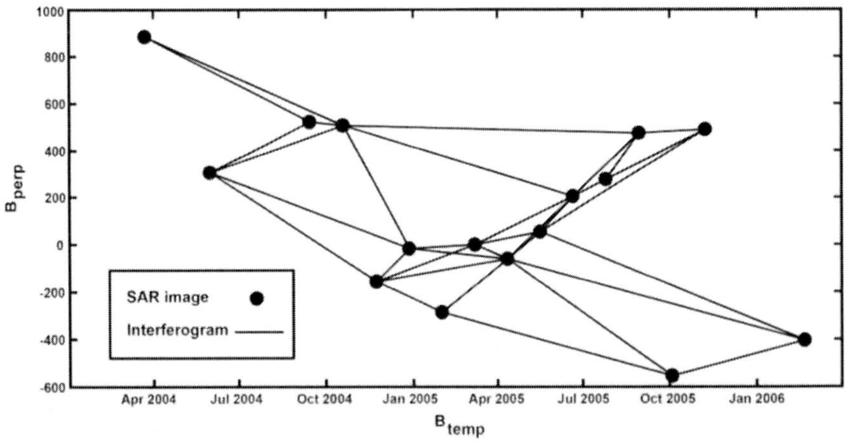

Figure 5. The Small Baseline (SBAS) network of selected interferograms used for the time-series analysis; horizontal and vertical axes represent the temporal and spatial baselines of selected pairs, respectively.

2.2. GPS Analysis

To evaluate current surface deformation, we used GPS and created a geodetic network of ~20 stations, including reference and monitoring points in and around the landslide area. Constructed stations are pillars made of reinforced concrete, anchored to a depth of 1 m. The top of the pillars was 0.5 m above the ground. The reference network consists of three pillars constructed on stable localities from the geological perspective, such as bedrock around and outside the landslide area (marked as BM1, BM2, and BM3 in Figure 6). The monitoring stations were distributed near the cracks and sliding masses inside the landslide area (MP1–MP17 in Figure 6). The required accuracy for monitoring slow movement phenomena such as creeping landslides should be on the order of centimeters to millimeters [29]. GPS relative positioning in static mode can achieve millimeter accuracy after considering measuring system error sources [28]. In this study the GPS field observations were conducted in static mode with four Trimble dual frequency GPS receivers. Four campaigns of GPS field measurements were performed in October 2010, May and August 2011 and April 2012 to derive the contemporary rate of surface deformation on the Sarcheshmeh slope. Figure 6 shows the established network of control and monitoring stations on a 3D topographic map of the area.

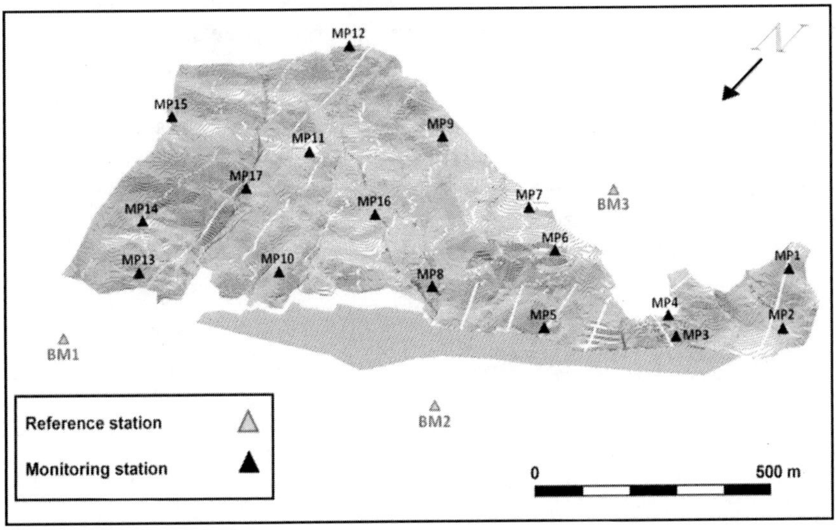

Figure 6. The spatial distribution of Global Positioning System (GPS) stations used in this study, overlaid on a 3D view of the topography map with 1 m contour interval.

GPS observations were performed in static mode, in which monitoring points were positioned by the differential method with respect to reference points. Each monitoring point was observed from at least two reference stations. Additional observations allowed us to control the position of monitoring stations to detect any probable errors. The average session duration for baseline determination between reference points was approximately 7 h, whereas the session duration for the baselines between the monitoring and reference points was 40 min. The collected raw data in each epoch of observation after converting to standard RINEX format were post-processed to achieve geodetic baselines between stations from space baselines between stations and satellites using precise ephemeris [38]. Figure 7 presents the final network of baselines used for positioning computation using GPS. To obtain compatibility of coordinates with the Iranian National Geodetic Network (INGN), we connected our network to a nearby permanent station from this network. A least-squares optimization method was then applied to adjust the network to determine the unique precise coordinates of stations during each campaign. It should be noted that reference network adjustment computation (baselines between reference stations BM1–BM3) was performed first in each epoch to ensure the stability of reference stations between two sequential campaigns. The displacements of the monitoring stations were computed from the difference between the coordinates of monitoring points determined in each epoch. The standard error analysis using a 95% confidence interval was applied to calculate the relative error ellipse and other statistical parameters that were associated with displacements using a variance-covariance matrix of observed baselines. It is worth nothing that computed formal errors by GPS software are usually over-optimistic [39]. Therefore a scale factor was computed and applied to the formal GPS errors based on the average of the weighted root-mean-square residual to the linear fit, computed for all stations [40–42].

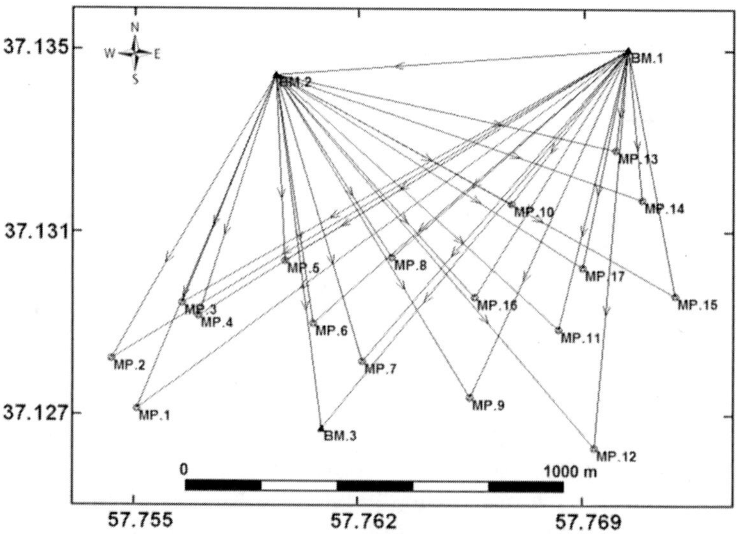

Figure 7. Schematic view of GPS network observations: black lines represent intersectional considering baselines between stations.

3. RESULTS

3.1. InSAR Results

Figure 8 illustrates several differential interferograms with good coherency in geographical coordinates after removing the orbital and topography effects. Black dotted ellipses specify the Sarcheshmeh slope on the interferograms. As observed in Figure 8, short and long-wavelength disturbances that may have arisen from tropospheric artifacts are easily recognizable in the interferograms, in turn making the estimation of slow creep signal using differential interferograms challenging. If there had been any major rapidly completed sliding process or sudden surface collapse during the observation period in Sarcheshmeh, it would have created either large-scale phase distortion or phase decorrelation in individual short-term interferograms, a pattern which was not observed either in the analyzed interferograms or during the field check-ups. However, because of atmospheric artifacts, it is difficult to evaluate the exact amount of long-term creep rate on the entire slope using only classical repeat-pass interferometry. Therefore, we focused on time-series analysis to precisely determine the rate of displacement during 2004–2006.

Using time-series analysis enables us to benefit from a tangible reduction in atmospheric artifact effects by averaging generated interferograms and forming a continuous surface deformation map. Figure 9a,b shows the mean velocity map of the Sarcheshmeh landslide in oblique geometry from the front view toward the slope and orthogonal geometry, respectively. The InSAR time-series analysis of the Envisat data detects more than 600 pixels that remain coherent during the time-interval covered by the Envisat data and can therefore be used for the displacement analysis. The mean deformation rate of the slope with respect to the reference area selected on the residential area downslope is estimated at 1.2 ± 4 mm/year.

To better illustrate the time evolution of surface deformation, we selected 11 points (triangular points marked in Figure 9b) and plotted their time-series (Figure 10). These points are selected close to the GPS stations to enable comparison with the current rates of deformation at these localities. For the sake of brevity, we only demonstrate the time-series plots at TS8, TS9, TS10 and TS17 in Figure 10 and list the complete results in Table 2. As shown in Figure 10, we did not observe a significant downward trend that is indicative of slope motion in the time-series plots. All of the selected points show a dominant flat trend punctuated in time by atmospheric turbulence. Time-series analysis reveals that the displacement of the slope is lower than the error threshold of the InSAR time-series. The average standard deviation of the mean deformation velocity for points presented in Table 2 is approximately 4 mm/year. Slight fluctuations of the time-series diagram in Figure 10 are also unavoidable because of tropospheric residual errors.

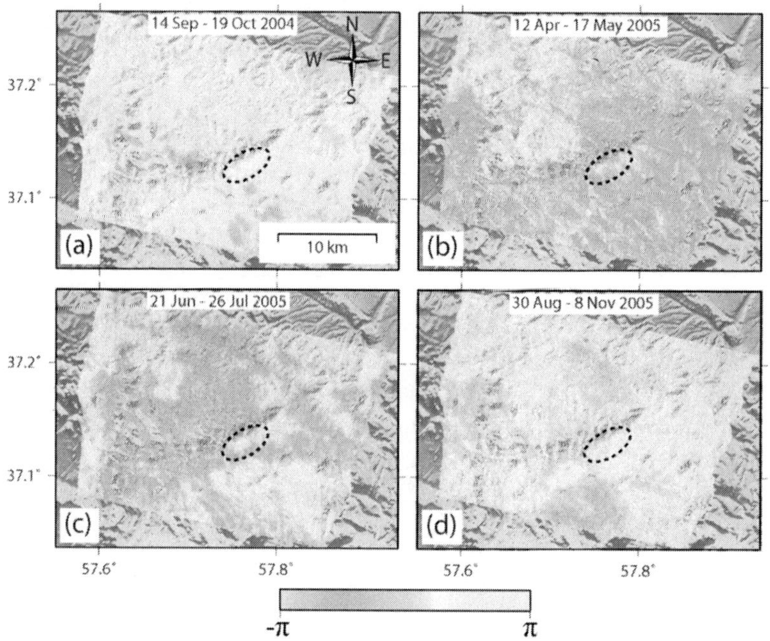

Figure 8. An example of four geocoded differential interferograms from the cropped part of SAR data after removing orbital and topography effects; black dotted ellipses indicate the location of Sarcheshmeh slope in the interferograms. One full color cycle, representing a difference in phase of $-\pi$ to π, corresponds to 2.8 cm of LOS range change.

3.2. GPS Results

Because GPS supplies a higher degree of horizontal accuracy than vertical accuracy [27], displacement results from GPS measurements are just computed and investigated for a 2D horizontal network to assess the current rate of deformation on the Sarcheshmeh landslide. Table 3 shows the numerical results for the horizontal displacement of the reference stations after four epochs of campaign measurements. Among the three reference stations, BM1 and BM2 did not show statistically significant motion during four epochs of observations and the recorded

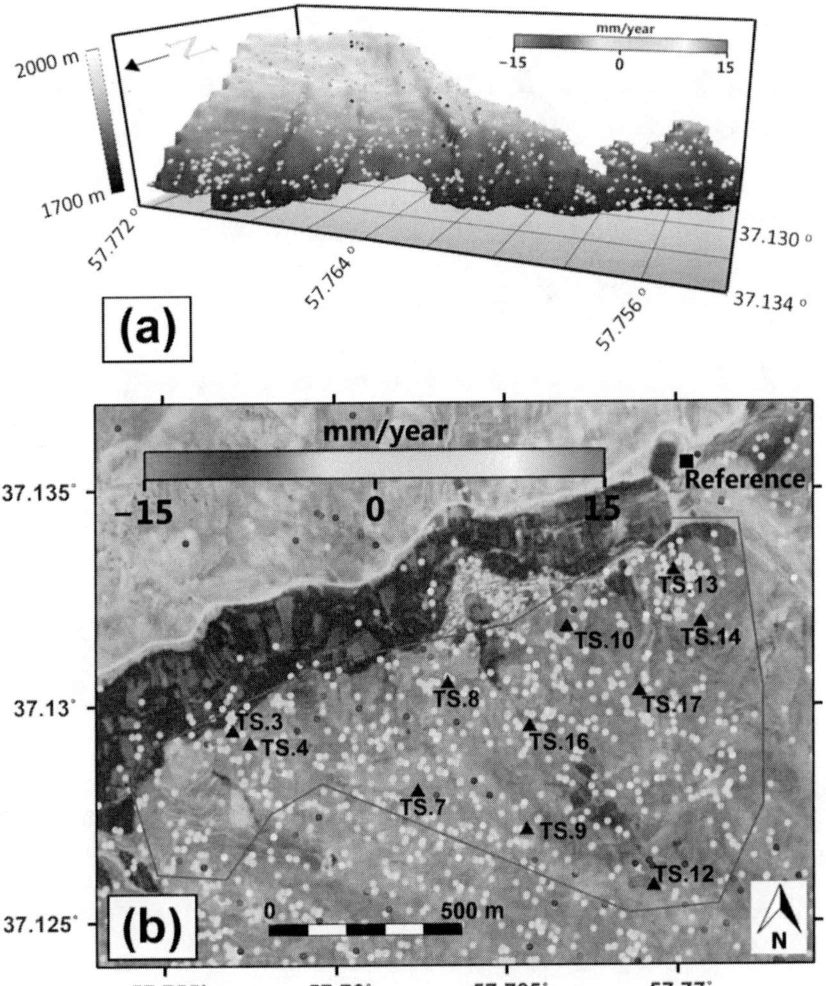

Figure 9. (**a**) Mean velocity map of the Sarcheshmeh slope in oblique geometry derived from InSAR time-series analysis; (**b**) a close up view of the mean velocity map in geographical coordinates; selected points for the time-series analysis are indicated by bold triangular points and a reference point is indicated by a bold rectangular point.

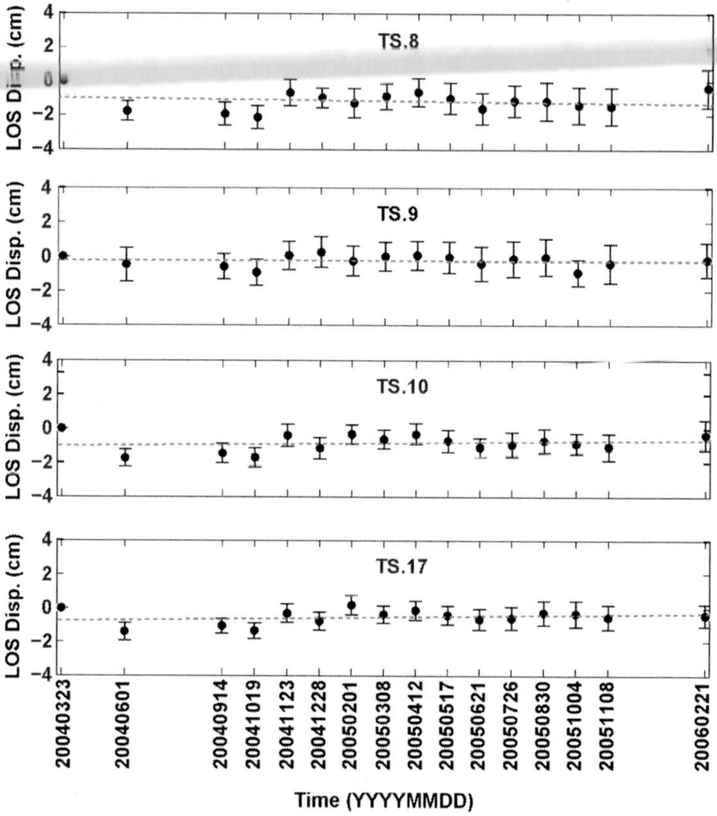

Figure 10. Four samples of InSAR time-series diagrams at TS8, TS9, TS10 and TS17; for the location of the points, please see Figure 9b. The solid circles on the diagrams represent the displacement values and the dashed red line is a one-dimensional polynomial trend fitted to displacement values. The displacement values and error bars have been computed for a window size of 100 m centered on each pixel.

displacements between the sequential campaigns were lower than the computed standard errors. This indicates that these reference stations are on stable ground and that the difference between corresponding coordinate components in sequential epochs results from stochastic error originating from various GPS environmental and instrumental sources of errors [27]. However, for the reference stations BM3, we observed significant horizontal motion of 7 mm between epochs one and two. Therefore, we excluded this station from our reference network for displacement analysis of the monitoring stations.

Table 2. Statistical results that were extracted from InSAR time-series analysis, which was applied for selected points in Figure 9b (displacements are in the line-of-sight (LOS) direction).

Point Name	Disp. Velocity (cm/year)	Standard Deviation (cm/year)
TS.3	0.1	±0.4
TS.4	0.1	±0.4
TS.7	0.2	±0.4
TS.8	0.2	±0.5
TS.9	0.0	±0.4
TS.10	0.2	±0.4
TS.12	−0.4	±0.5
TS.13	0.3	±0.4
TS.14	0.2	±0.4
TS.16	0.2	±0.4
TS.17	0.2	±0.3

Table 3. GPS horizontal displacement from the reference network during the 16 month time interval between the first and the fourth campaign with 95% confidence interval.

Point Name	Epoch 1 to 2 (5 Months)		Epoch 2 to 3 (3 Months)		Epoch 3 to 4 (8 Months)	
	E-W (mm)	N-S (mm)	E-W (mm)	N-S (mm)	E-W (mm)	N-S (mm)
BM1	0 ± 1	0 ± 1	0 ± 5	0 ± 7	0 ± 5	0 ± 8
BM2	−2 ± 2	0 ± 2	1 ± 5	0 ± 7	0 ± 5	−1 ± 8
BM3	−6 ± 1	−4 ± 2	−3 ± 7	2 ± 9	4 ± 7	−5 ± 9

Performing four epochs of repeated GPS observations on a monitoring network also did not reveal any significant deformation on the Sarcheshmeh slope from October 2010 to April 2012 (Table 4). Figure 11 is a schematic display of an average horizontal displacement field map resulting from the first and the fourth campaign on the Sarcheshmeh slope. GPS horizontal displacement vectors for most of the stations do not show significant displacement at 95% confidence intervals. According to results in Table 4, all GPS stations were stable during the observed time interval. The only exceptions are MP.13 and MP.14, which showed significant horizontal motions of up to 1 cm and 14 cm, respectively, beyond the bounds of the estimated error ellipses (Figure 11 and Table 4). However, field check-ups performed following this observation indicated that the motion at these stations was not related to sliding of the entire slope but was locally triggered by human interference and water infiltration from nearby farming.

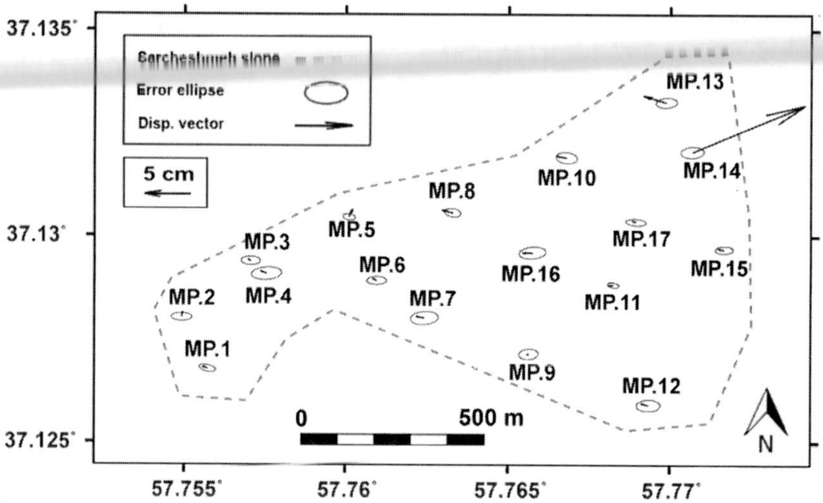

Figure 11. Schematic display of an average horizontal displacement field on the Sarcheshmeh slope spanning the first and the fourth GPS campaign with 95% confidence ellipses.

Table 4. GPS horizontal displacement for monitoring stations with 95% confidence interval.

Point Name	Epoch 1 to 2 (5 Months)		Epoch 2 to 3 (3 Months)		Epoch 3 to 4 (8 Months)	
	E-W (mm)	N-S (mm)	E-W(mm)	N-S (mm)	E-W (mm)	N-S (mm)
MP1	−4 ± 4	−2 ± 6	−1 ± 4	3 ± 6	1 ± 4	2 ± 6
MP2	−1 ± 4	6 ± 4	−3 ± 6	−4 ± 6	4 ± 4	6 ± 6
MP3	−4 ± 4	−2 ± 4	−2 ± 12	−3 ± 14	4 ± 12	7 ± 14
MP4	−4 ± 4	0 ± 4	2 ± 6	1 ± 6	−3 ± 6	3 ± 8
MP5	4 ± 4	12 ± 4	−3 ± 4	−4 ± 6	2 ± 4	4 ± 6
MP6	−3 ± 2	3 ± 4	0 ± 2	0 ± 4	0 ± 4	1 ± 6
MP7	−4 ± 2	4 ± 4	0 ± 4	−4 ± 6	−3 ± 4	2 ± 8
MP8	−6 ± 6	4 ± 8	−5 ± 4	−1 ± 6	2 ± 4	1 ± 6
MP9	−1 ± 2	−3 ± 4	−2 ± 4	−4 ± 10	2 ± 4	7 ± 10
MP10	−7 ± 6	−1 ± 6	2 ± 4	3 ± 4	−3 ± 4	1 ± 4
MP11	−3 ± 4	1 ± 4	−2 ± 4	−1 ± 4	1 ± 4	−1 ± 4
MP12	−2 ± 4	2 ± 4	0 ± 4	−3 ± 6	−3 ± 4	4 ± 6
MP13	−9 ± 2	6 ± 4	−6 ± 2	−1 ± 4	−5 ± 2	6 ± 4
MP14	−4 ± 2	1 ± 4	130 ± 2	57 ± 4	25 ± 2	17 ± 4
MP15	−2 ± 2	−2 ± 4	−1 ± 4	2 ± 8	−2 ± 4	2 ± 8
MP16	−5 ± 2	1 ± 6	2 ± 4	2 ± 6	−6 ± 4	−3 ± 4
MP17	−1 ± 4	0 ± 4	−4 ± 4	−1 ± 6	2 ± 4	4 ± 6

Figure 12 shows the displacement time series of some GPS stations (MP8, MP9, MP10 and MP17) corresponding to InSAR displacement time series (TS8, TS9, TS10 and TS17) illustrated in Figure 10. As clearly shown in Figure 12, the displacement values and plotted error bars are indicative of the station's stability.

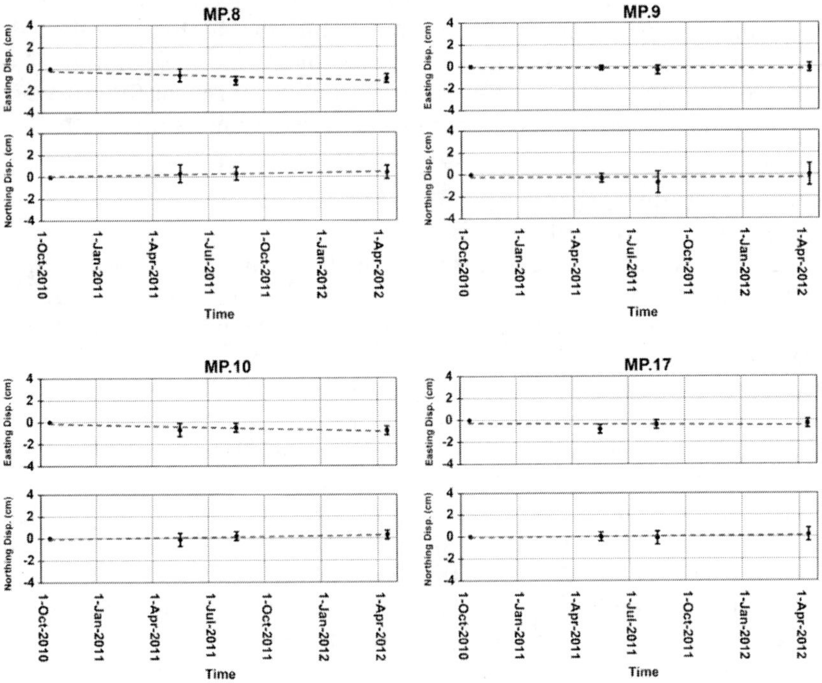

Figure 12. Four samples of GPS results with 95% error bars; for the location of points please see Figure 11. The solid circles represent displacement values, and the dashed blue line is a one-dimensional polynomial trend fitted to the displacement value.

4. DISCUSSION

In this study, InSAR was applied to Envisat data to assess the history of surface deformation on the Sarcheshmeh landslide, and repeated GPS campaign measurements were performed to assess the rate of surface deformation in the area. Although both methods were implemented in different time intervals, both indicate current stability of the Sarcheshmeh slope.

Standard errors analysis in GPS measurement does not show significant displacement on the landslide at 95% confidence interval. As shown in Figure 12, the cumulative displacement for most of the GPS stations on the landslide during the 2010–2012 time interval falls within the bounds of the estimated error ellipse; the only exceptions were MP 13 and MP 11 which were affected by localized deformation. The large uncertainty of the estimated displacements may arise from the limited number of campaign-mode measurements that we used in this study. The uncertainties of site displacements decrease if more repeat GPS surveys are performed to span longer times or we use continuous GPS instead of campaign-mode measurements [43].

For the InSAR investigation, the main limitations of C-band repeat-pass interferometry to reliably evaluate the long-term creep rate over the Sarcheshmeh landslide are atmospheric artifacts in interferograms with short temporal baselines and temporal decorrelation in interferograms with long time intervals (~1 year); the area is mountainous, partially covered with grass and other permanent vegetation and is used for farming, all of which contribute to the disturbance of the InSAR signal. However, SBAS time-series analysis using Envisat data enables identification of more than 600 coherent pixels that illustrate slope stability in the years 2004–2006 within the error threshold of InSAR observations. The mean deformation rate of the slope with respect to the reference area selected on the residential area downslope is estimated at 1.2 ± 4 mm/year. However, as the main direction of the landslide is towards the NW, any coherent motion related to landslide would have appeared as negative LOS in descending InSAR results. The positive value for the mean deformation and its relatively large uncertainty is an indication of the large impact of residual tropospheric effects on InSAR time-series analysis that compromises the precision of the derived deformation signal using a limited number of SAR images used in this study [44].

The Sarcheshmeh slope has been characterized as an ancient creeping hillside. Cluttered topography and the existence of scarps, trenches and deep grooves (Figure 3) are all geologically indicative of the creeping history of this slide. This area is located on a vast, massive limestone (Figure 2); thus, creeping can be facilitated by the groundwater level and high precipitation rate among other factors. Surface water can loosen dry soil and make sliding easier on the slippery bedrock. Water penetration also leads to slope overloading and increased slope weight. Therefore, gravity can interact more strongly in downward creeping, especially on weakly consolidated masses. The current stability of the Sarcheshmeh landslide might result from a combined reduction in the ground water levels and precipitation rate. Sarcheshmeh is in an area in Iran with a relatively high level of groundwater and high rate of seasonal precipitation. Multiple springs on and around the slope are indicative of relatively high levels of groundwater.

However because of regional population growth, the expansion of agricultural activity and, subsequently, the irregular use of underground water resources, this area has been subject to tangible reduction in groundwater level over the last few decades. Figure 13a shows a watershed basin segmentation map of North Khorasan. Different colors in Figure 13b are indicative of the groundwater level status of the segments. As shown in this figure, the Esfarayen basin is in a critical forbidden area from the groundwater level aspect because of irregular water extraction. The Sarcheshmeh slope, marked by a green dot in Figure 13a, is in the Esfarayen basin; from 1993 to 2010, the average level of ground water in the basin to which the Sarcheshmeh slope belongs has declined more than 13 m (Figure 13c). Additionally, the precipitation rate has also decreased by more than 30% during 2004–2007 (Figure 14). All of these factors might have contributed to the increased resistance of the slope to downward motion.

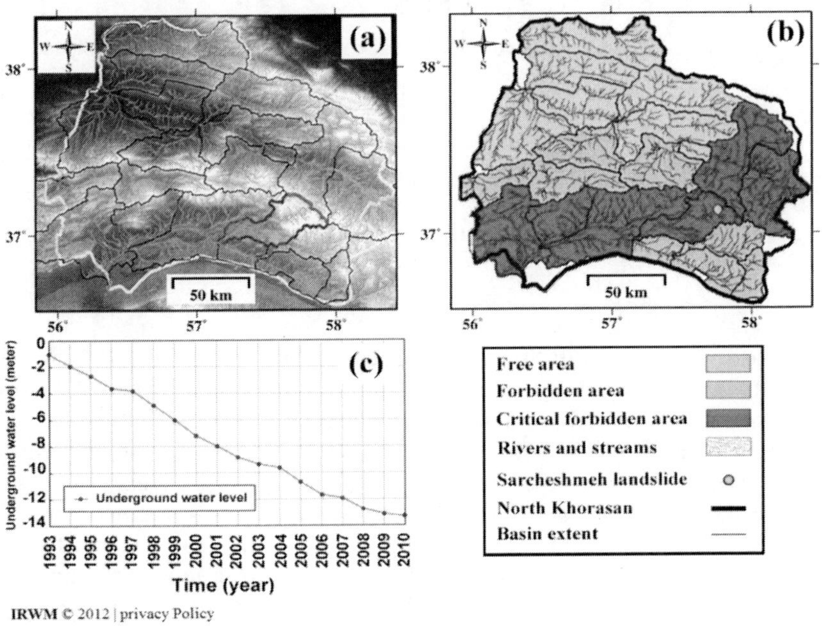

Figure 13. (**a**) Watershed basin segmentation map of the North Khorasan province. The red polygon delineates the watershed basin of Esfarayen to which the Sarcheshmeh slope belongs; (**b**) groundwater level status of each basin in the North Khorasan province indicated using different colors; (**c**) Groundwater level diagram of the Esfarayen basin which shows significant reduction from 1993 to 2010 [45].

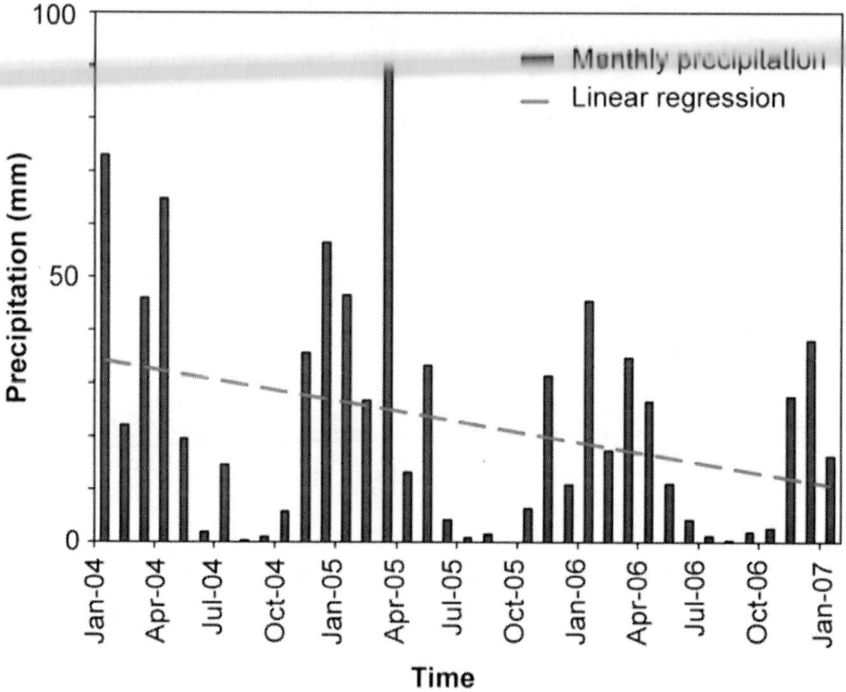

Figure 14. Accumulated monthly precipitation in Sarcheshmeh during 2004–2007 [12].

Despite the recent stability of the Sarcheshmeh slope, our GPS results indicate that human-induced farming activity might trigger sudden motion on this slope. As mentioned in the GPS results section, at two stations, namely MP.13 and MP.14, we recorded significant horizontal movement during our GPS campaigns. Field check-ups revealed that the motion on these stations was due to the development of a new farmland nearby. This farmland was irrigated manually from groundwater resources, facilitating the penetration of water to creep-prone masses. Therefore, saturated weak consolidated soil became heavy and slippery and started to move downward because of gravity. If similar practices become widespread, this accidental event and its consequences, which result from human interference, can be a critical warning regarding the Sarcheshmeh slope. Current geodetic investigation shows that this ancient landslide is primarily dormant in the current period. However, even in the absence of major triggering factors such as earthquakes and heavy rainfall, the landslide can be reactivated by other factors, such as water infiltration from farming. Therefore, preventive measures and subsequent proceedings are still necessary to mitigate and reduce probable losses and causalities. The development of a surface and subsurface drainage system to control and conduct runoff outside the slope, construction of a new residential area in a safe area far from the landslide zone and establishment of an early warning system to provide information concerning critical moments are some executable plans to mitigate any damages.

5. CONCLUSION

In this study, time-series analysis of Envisat InSAR data and repeated GPS observations using campaign measurements were employed to assess the slope stability of the Sarcheshmeh ancient landslide in Northeast Iran. No previous studies have examined the past and most recent movement of this landslide, and as a result its creeping mechanism was poorly understood. We were able to perform InSAR time-series analysis using Envisat data and detected more than 600 coherent pixels on the landslide, which showed no statistically significant LOS motion in the years 2004–2006. Horizontal campaign GPS data collected between 2010 and 2012 at 17 stations did not reveal any significant site motions within the estimated error ellipse of GPS results either. We interpret these results as indicating that the Sarcheshmeh landslide is currently inactive. Hydrological analysis suggests that the current dormancy of the landslide might be due to the absence of hydrological triggering factors in the region over the last decade.

Although the InSAR and GPS results in this paper are promising, we find that the uncertainty and precision of the results hamper our ability to detect and analyze the very slow creep signal in this region (<2–3 mm/year). Part of this uncertainty arises from not enough sampling of the slow creep signal in time in both Envisat data with 35-day repeat interval and campaign-mode GPS measurements. For future research, higher temporal and spatial resolution monitoring using observations from new SAR data such as TerraSAR-X (with a resolution of 1–3 m and a repeat time of 11 days) and placing continuously recording GPS sites in the landslide area are recommended. This would allow for more detailed sampling of deformation signal in time and space, in turn mitigating better the atmospheric and noise effects for evaluating slow landslide creep, and improvement of our capability to analyze the magnitude and seasonal dependence of a potential landslide motion in response to triggering conditions.

ACKNOWLEDGMENTS

This work was supported by Initiative and Networking Fund of the Helmholtz Association in the frame of Helmholtz Alliance "Remote Sensing and Earth System Dynamics" and the German Federal Ministry of Research and Technology (BMBF) within the framework of PROGRESS (Potsdam Research Cluster for Georisk Analysis, Environmental Change and Sustainability). It is a pleasure to thank those people who made this research possible, especially those who participated in the GPS field observations. We also thank the Forest, Range and Watershed Management (FRWM) organization of Iran for their support during this research project. Special thanks go to Bahman Akbari for initiating the project and to Mrs. Mohammadi and Fereydoun Masoumpour who graciously helped with the preparation of our field work. SAR data were provided by the European Space Agency (ESA) via proposal AOALO3740. Copyrights of Figures 1, 2b and 13c belong to National Geoscience Database of Iran, Geological Survey of Iran and Iran Water Resources Management Company, respectively. We are grateful to three anonymous reviewers and guest editor, Marco Scaioni, whose comments greatly improved the quality of the manuscript.

CONFLICT OF INTEREST

The authors declare no conflict of interest.

REFERENCES

1. Sidle, R.C.; Ochiai, H. Landslides: Processes, Prediction, and Land Use; American Geophysical Union: Washington, DC, USA, 2006; Volume 18, p. 312.
2. Landslide in Iran, Available online: http://landslide.ir (accessed on 20 December 2012).
3. National Geoscience Database of Iran, Available online: http://www.ngdir.ir/landslide/LandSlideInfo.asp (accessed on 20 December 2012).
4. Greif, V.; Vlcko, J. Monitoring of post-failure landslide deformation by the PS-InSAR technique at Lubietova in Central Slovakia. Environ. Earth Sci 2012, 66, 1585–1595.
5. Malet, J.-P.; Maquaire, O.; Calais, E. The use of Global Positioning System techniques for the continuous monitoring of landslides: Application to the Super-Sauze earthflow (Alpes-de-Haute-Provence, France). Geomorphology 2002, 43, 33–54.
6. Brückl, E.; Brunner, F.; Kraus, K. Kinematics of a deep-seated landslide derived from photogrammetric, GPS and geophysical data. Eng. Geol 2006, 88, 149–159.
7. Delacourt, C.; Allemand, P.; Berthier, E.; Raucoules, D.; Casson, B.; Grandjean, P.; Pambrun, C.; Varel, E. Remote-sensing techniques for analysing landslide kinematics: A review. Bull. Soc. Geol. Fr 2007, 178, 89–100.

8. Strozzi, T.; Ambrosi, C.; Raetzo, H. Interpretation of aerial photographs and satellite SAR interferometry for the inventory of landslides. Remote Sens 2013, 5, 2554–2570.
9. Kimura, H.; Yamaguchi, Y. Detection of landslide areas using satellite radar interferometry. Photogramm. Eng. Remote Sensing 2000, 66, 337–344.
10. Lacroix, P.; Zavala, B.; Berthier, E.; Audin, L. Supervised method of landslide inventory using panchromatic SPOT5 images and application to the earthquake-triggered landslides of Pisco (Peru, 2007, Mw8.0). Remote Sens 2013, 5, 2590–2616.
11. Othman, A.A.; Gloaguen, R. River courses affected by landslides and implications for hazard assessment: A high resolution remote sensing case study in NE Iraq–W Iran. Remote Sens 2013, 5, 1024–1044.
12. National Aeronautics and Space Administration. Rainfall Analysis Tools, Available online: http://disc2.nascom.nasa.gov/Giovanni/tovas/rain.GPCP.2.shtml (accessed on 11 January 2013).
13. Water Science, Available online: http://waterscience.blogfa.com (accessed on 20 December 2012).
14. Geological Survey of Iran. States Information, Available online: http://gsi.ir /States/Lang_en/StateId_51/Action_LastUpdate/index.html (accessed on 25 June 2011).
15. Hanssen, R.F. Radar Interferometry: Data Interpretation and Error Analysis; Kluwer Academic Publishers: New York, NY, USA, 2001.
16. Massonnet, D.; Feigl, K.L. Radar interferometry and its application to changes in the Earth's surface. Rev. Geophys1998, 36, 441–500.
17. Strozzi, T.; Farina, P.; Corsini, A.; Ambrosi, C.; Thüring, M.; Zilger, J.; Wiesmann, A.; Wegmüller, U.; Werner, C. Survey and monitoring of landslide displacements by means of L-band satellite SAR interferometry. Landslides 2005, 2, 193–201.
18. Tantianuparp, P.; Shi, X.; Zhang, L.; Balz, T.; Liao, M. Characterization of landslide deformations in three gorges area using multiple InSAR data stacks. Remote Sens 2013, 5, 2704–2719.
19. Motagh, M.; Wetzel, H.-U.; Roessner, S.; Kaufmann, H. A TerraSAR-X InSAR study of landslides in southern Kyrgyzstan, Central Asia. Remote Sens. Lett 2013, 4, 657–666.
20. Rott, H.; Scheuchl, B.; Siegel, A.; Grasemann, B. Monitoring very slow slope movements by means of SAR interferometry: A case study from a mass waste above a reservoir in the Ötztal Alps, Austria. Geophys. Res. Lett 1999, 26, 1629–1632.
21. Hooper, A.; Zebker, H.; Segall, P.; Kampes, B. A new method for measuring deformation on volcanoes and other natural terrains using InSAR persistent scatterers. Geophys. Res. Lett. 2004, 31.
22. Hooper, A.; Segall, P.; Zebker, H. Persistent scatterer interferometric synthetic aperture radar for crustal deformation analysis, with application to Volcán Alcedo, Galápagos. J. Geophys. Res. 2007, 112.
23. Tofani, V.; Raspini, F.; Catani, F.; Casagli, N. Persistent Scatterer Interferometry (PSI) technique for landslide characterization and monitoring. Remote Sens 2013, 5, 1045–1065.
24. Berardino, P.; Fornaro, G.; Lanari, R.; Sansosti, E. A new algorithm for surface deformation monitoring based on small baseline differential SAR interferograms. IEEE Trans. Geosci. Remote Sens 2002, 40, 2375–2383.
25. Lanari, R.; Casu, F.; Manzo, M.; Zeni, G.; Berardino, P.; Manunta, M.; Pepe, A. An overview of the small baseline subset algorithm: A DInSAR technique for surface deformation analysis. Pure Appl. Geophys 2007, 164, 637–661.
26. Akbari, V.; Motagh, M. Improved ground subsidence monitoring using small baseline SAR interferograms and a weighted least squares inversion algorithm. IEEE Geosci. Remote Sens. Lett 2012, 9, 437–441.
27. Hofmann-Wellenhof, B.; Lichtenegger, H.; Collins, J. Global Positioning System. Theory and Practice; Springer: Wien, Austria, 1993; p. 347.

28. Zhou, P.; Zhou, B.; Guo, J.; Li, D.; Ding, Z.; Feng, Y. A demonstrative GPS-aided automatic landslide monitoring system in sichuan province. J. Glob. Position. Syst 2005, 4, 184–191.
29. Gili, J.A.; Corominas, J.; Rius, J. Using Global Positioning System techniques in landslide monitoring. Eng. Geol 2000, 55, 167–192.
30. Squarzoni, C.; Delacourt, C.; Allemand, P. Differential single-frequency GPS monitoring of the La Valette landslide (French Alps). Eng. Geol 2005, 79, 215–229.
31. Motagh, M.; Djamour, Y.; Walter, T.R.; Wetzel, H.U.; Zschau, J.; Arabi, S. Land subsidence in Mashhad Valley, northeast Iran: Results from InSAR, levelling and GPS. Geophys. J. Int 2006, 168, 518–526.
32. Yin, Y.; Zheng, W.; Liu, Y.; Zhang, J.; Li, X. Integration of GPS with InSAR to monitoring of the Jiaju landslide in Sichuan, China. Landslides 2010, 7, 359–365.
33. Strozzi, T.; Delaloye, R.; Kääb, A.; Ambrosi, C.; Perruchoud, E.; Wegmüller, U. Combined observations of rock mass movements using satellite SAR interferometry, differential GPS, airborne digital photogrammetry, and airborne photography interpretation. J. Geophys. Res. 2010, 115.
34. Kampes, B.; Hanssen, R.; Perski, Z. Radar Interferometry with Public Domain Tools. Proceedings of FRINGE, ESA ESRIN, Frascati, Italy, 1–5 December 2003; Available online: http://earth.esa.int/fringe03/proceedings/papers/22_kampes.pdf (accessed on 25 May 2013).
35. Farr, T.G.; Kobrick, M. The shuttle radar topography mission. Rev. Geophys 2007, 45.
36. Scharroo, R.; Visser, P. Precise orbit determination and gravity field improvement for the ERS satellites. J. Geophys. Res 1998, 103, 8113–8127.
37. Hooper, A. A multi-temporal InSAR method incorporating both persistent scatterer and small baseline approaches. Geophys. Res. Lett. 2008, 35.
38. National Geodetic Survey. GPS Orbit Data, Available online: http://www.ngs.noaa.gov/orbits/orbit_data.shtml(accessed on 30 November 2012).
39. Wesche, C.; Eisen, O.; Oerter, H.; Schulte, D.; Steinhage, D. Surface topography and ice flow in the vicinity of the EDML deep-drilling site, Antarctica. J. Glaciol 2007, 53, 442–448.
40. Lowry, A.R.; Hamburger, M.W.; Meertens, C.M.; Ramos, E.G. GPS monitoring of crustal deformation at Taal Volcano, Philippines. J. Volcanol. Geotherm. Res 2001, 105, 35–47.
41. Hessami, K.; Nilforoushan, F.; Talbot, C.J. Active deformation within the Zagros mountains deduced from GPS measurements. J. Geol. Soc 2006, 163, 143–148.
42. McClusky, S.; Balassanian, S.; Barka, A.; Demir, C.; Ergintav, S.; Georgiev, I.; Veis, G. Global Positioning System constraints on plate kinematics and dynamics in the eastern Mediterranean and Caucasus. J. Geophys. Res.-Solid Earth2000, 105, 5695–5719.
43. Langbein, J. Noise in GPS displacement measurements from Southern California and Southern Nevada. J. Geophys. Res.-Solid Earth 2008, 113.
44. Casu, F.; Manzo, M.; Lanari, R. A quantitative assessment of the SBAS algorithm performance for surface deformation retrieval from DInSAR data. Remote Sens. Environ 2006, 102, 195–210.
45. Iran Water Resources Management Company. North Khorasan Province, Available online: http://wnkh.ir/showpage.aspx?pgid=8&ids=29&? (accessed on 11 January 2013).

Index